Agriculture, Environment and Development

Antonio Augusto Rossotto Ioris ·
Bernardo Mançano Fernandes
Editors

Agriculture, Environment and Development

International Perspectives on Water, Land and Politics

Second Edition

palgrave
macmillan

Editors
Antonio Augusto Rossotto Ioris
School of Geography and Planning
Cardiff, UK

Bernardo Mançano Fernandes
São Paulo State University
São Paulo, Brazil

ISBN 978-3-031-10266-0 ISBN 978-3-031-10264-6 (eBook)
https://doi.org/10.1007/978-3-031-10264-6

This Palgrave Macmillan imprint is published by the registered company Springer Nature Switzerland AG
The registered company address is: Gewerbestrasse 11, 6330 Cham, Switzerland

To the members of the Landless Workers Movement (MST), La Via Campesina and the many indigenous organisations around the world, who represent all those who know and practice true food production and life giving agriculture.

Contents

Contributors

Abbas Máriam Observatório do Meio Rural (OMR), Maputo, Mozambique

Antunes Wolney Felippe Junior State University of Campinas (Unicamp), Campinas, Brazil

Borras Saturnino M. Jr. International Institute of Social Studies (ISS), Den Haag, The Netherlands

Borsatto Ricardo S. Federal University of São Carlos (UFSCar), São Carlos, Brazil

Bowness Evan Food and Agriculture Institute, University of the Fraser Valley, Abbotsford, BC, Canada

Bruna Natacha Observatório do Meio Rural (OMR), Maputo, Mozambique

Cazella Antonio Ademir Federal University of Santa Catarina (Ufsc), Florianópolis, Brazil

Correia Kevin Campos Federal University of ABC (UFABC), Santo André, Brazil

da Costa Oliveira Neto Adolfo Federal University of Pará (UFPA), Belém, Brazil

de Camargo Macedo André State University of Campinas (Unicamp), Campinas, Brazil

Dulci Luiza Perseu Abramo Foundation, São Paulo, Brazil

Fernandes Bernardo Mançano São Paulo State University (UNESP), São Paulo, Brazil

Frederico Samuel São Paulo State University (UNESP), São Paulo, Brazil

Ioris Antonio Augusto Rossotto Cardiff University, Cardiff, UK

James Dana Centre for Sustainable Food Systems, University of British Columbia, Vancouver, BC, Canada

Jara Cristian Emanuel Universidad Nacional de Santiago del Estero, Santiago del Estero, Argentina

Lang Tim Centre for Food Policy, City University of London, London, UK

Magnanti Natal João Centro Vianei de Educação Popular, Lages, Brazil

Mann Alana Sydney Environment Institute, University of Sydney, Sydney, NSW, Australia

Marsden Terry Cardiff University, Cardiff, UK

Meek David University of Oregon, Eugene, OR, USA

Millstone Erik Science Policy Research Unit, University of Sussex, Sussex, UK

Origuéla Camila Ferracini São Paulo State University (Unesp), São Paulo, Brazil

Paz Raúl Gustavo Universidad Nacional de Santiago del Estero, Santiago del Estero, Argentina

Pereira Lorena Izá São Paulo State University (UNESP), São Paulo, Brazil

Santos Milton University of Sao Paulo, Sao Paulo, Brazil

Silva Marina Pereira Federal University of ABC (UFABC), Santo André, Brazil

Souza-Esquerdo Vanilde Ferreira State University of Campinas (Unicamp), Campinas, Brazil

Tarlau Rebecca The Pennsylvania State University, State College, PA, USA

Welch Clifford Andrew Federal University of São Paulo (Unifesp), São Paulo, Brazil

Wittman Hannah Centre for Sustainable Food Systems, University of British Columbia, Vancouver, BC, Canada

Zimerman Artur Federal University of ABC (UFABC), Santo André, Brazil

List of Figures

List of Tables

1

Agriculture, Environment and Development: International Perspectives and a Critical Agenda of Investigation

Antonio Augusto Rossotto Ioris

Agri-food production and agrarian development constitute some of the most challenging areas of public policymaking, ecological conservation and socio-environmental justice nowadays. It is not by chance that George Monbiot affirmed, in a *The Guardian* article on 19 May 2022, that 'the banks collapsed in 2008—and our food system is about to do the same'. Major deficiencies in food systems have become unmistakeably apparent, including conflicting values, uncertain supply chains, power imbalances and low institutional capacity and unequal rights and entitlements. One could say, with no exaggeration, that the problems of agri-food vividly capture the maelstrom of present-day globalised economy and increasingly alienated societies (considering not only political, marketplace and cultural alienation, but also alienation in the strict politico-economic sense of a fundamental separation from the control

A. A. R. Ioris (✉)
Cardiff University, Cardiff, UK
e-mail: IorisA@cardiff.ac.uk

A. A. R. Ioris and B. Mançano Fernandes (eds.), *Agriculture, Environment and Development*, https://doi.org/10.1007/978-3-031-10264-6_1

of production and consumption). In the words of Raj Patel, the food system is a battlefield, though few realise quite how many casualties there have been. Despite the growing relevance of those questions, the most common outcome is a sort of unspoken impasse: the agri-food problems are acknowledged, remains high on the agenda, but it is hard to see where responses and meaningful changes would come from. Most approaches still reflect the influences of a productivist (and heavily carbonised) paradigm, which is basically a narrow attempt to produce more and make a certain type of food too easily available on market shelves. Production, distribution and consumption have been left in placeless foodscape that is disconnected from genuine social demands and the ecological basis of communities and locations. However, contrary to the ideology of a placeless system, agri-food processes unfold at different scales with profound historical and geographical significance, as they involve territorialised transformations that interconnect disputes and collaborations between social groups across different localities, countries and regions.

If we think carefully about supermarket shelves and commercial advertisements, the nutritional and cultural dimensions of agriculture and food often seem to be overlooked today, as the sector is increasingly dominated by industrialised goods, standardised diets and intercontinental transactions. A significant proportion of the food consumed today, particularly in the Global South, comes from obscure and often unreliable sources, which are more influenced by market pressures and shareholder expectations than by health requirements, farmer demands or nutritional and environmental concerns. As Friedmann (2005: 124) points out, 'agriculture and food have all along invisibly underpinned relations of property and power in the world system'. Agriculture, similar to many other activities in the contemporary world, has been substantially transformed by the application of intensification technologies, the globalisation of agri-food markets and the financialisation of production and consumption. As denounced by Vandana Shiva (2016: ix), we 'are facing a deep and growing crisis rooted in how we produce, process, and distribute our food... An inefficient, wasteful, and nonsustainable model of food production is pushing the planet, its ecosystems, and its diverse species to the brink of destruction'. Shiva further adds that 'food, whose primary purpose is to provide nourishment and health, is today the single biggest health problem in the world: nearly one billion people suffer from hunger and malnutrition, two billion suffer from diseases like obesity and diabetes, and countless other suffer from diseases, including cancer, caused by the poisons in our food'. The image of an intense, globalised and money-making agriculture is commonly captured in language of stock markets, commodity trade and hyper-processed food.

The uncomfortable transformation of food into money and chemistry, instead of life and nutrition, is directly associated with hyper-modern farming and the insertion into the cross-scale features of capitalist agriculture along the line of the so-called 'long Green Revolution' (Patel 2013). This route consolidated the current duality of corporate agribusiness and subsistence farming in the country, a situation that was reflected in duplicity of public policies adopted by the national government (with much greater emphasis on agribusiness exports). In the frontiers of global agribusiness expansion, as in the case of Amazon since the 1960s, export-oriented monocultural farming inevitably depended on the supply of food and resources from small-scale family agricultural units and, despite its high-tech equipment, on permanent or seasonal labourers with multiple ethnic and identitary references (Ioris 2020a). Rapidly spreading agribusiness production of soybean and sugarcane entailed a combined process of socio-spatial transformation and socio-ecological regulation required to bring nature and society to the realm of market transactions. The conversion of nature and society into the sphere of exploitation and capital accumulation was based on the prior disruption of socionatural relationships and the gradual configuration of a new socionatural order. Production for the market gradually underpins the social and individual life in the agribusiness frontier. However, as already consolidated in the politico-economic centres, the market is not simply the space for the exchange of goods, services and money, but the comprehensive translation of everything into a world in which everything has price and can be traded by those who are apparently equal, but in effect highly asymmetric (especially in terms of the labour market). Instead of organising production tasks to supply food needs and nutritional demands, the division of labour ends up mirroring equalising market priorities and, at the same time, following the structurally uneven balance of politico-economic power.

Nonetheless, the most common reaction to mounting food insecurity and market variability is the intensification and productivist paradigm. Basically, the argument goes, if there is a lack of food, the answer is to produce more and make food more easily available in time and space (while issues of distribution and socio-spatial inequalities are much less important). Productivism betrays a strong influence of Western positivism and its search for the universal, superior 'truth' to be brought to the rest of humankind. The narrative is that because of steady population growth and less significant productivity gains, there is a growing threat to the ability of entire countries and regions to feed themselves, but with rising prices of basic food commodities the response was to produce more. This is evidently the principal rationality adopted by diplomacy, national governments and the corporate business sector. It directly transforms food insecurity—without asking questions about past legacies, injustices and responsibilities—into a source of political gain (in the form of promises made by politicians), the legitimisation of government strategies (as a response to social demands) and profits (as investments and production platforms will need to be funded and goods sold in the market). An important facet of the productivist response to food insecurity is the claim that the attention should be focused on technology, machinery, more land to be cultivated and more inputs to be transformed into higher outputs (especially in terms of caloric output). In that way, agriculture's cultural and ecological dimensions are disregarded. Agriculture is, in practice, equated to biochemistry, engineering and business administration, as much as food is reduced to a long table of nutrients rather than an important element of family and community relations. Those excuses pave the way to technological fixes that, in the end, represent a form of neo-Malthusianism by the backdoor (Ioris 2020b).

The aggressive modernisation and industrialisation of agri-food can be described as a disturbing movement from food-as-nutrition and agriculture-as-social-integration to a situation in which agri-food operations are carried out primarily to circulate and accumulate capital (Ioris 2017). In other words, the role of agriculture as a source of nutrition and livelihood is being increasingly supplanted by the imperatives of money and profit, which is happening in the wider context of a globalised market-based society in which everything is prey to commodification. This is the first main contention of this article: notwithstanding many other biophysical, cultural and circumstantial problems, the basic disconnection between food and nutrition, together with the firmly established nexus between food and money-making, are the twin causes of the widespread agri-food crisis faced by governments and society. The ideological and practical reduction of food to the realm of commodities, exploitation and profit also represents a decisive barrier to the resolution of nutritional and environmental problems. In synthetic terms, this has constituted a gradual shift, particularly since World War II, from *agriculture-cum-food* to *agriculture-cum-agribusiness*. The origin of the term agribusiness is historically attributed to Davis and Goldberg (1957), who defined agribusiness (based on previous references made by other authors to agro-industry and commercial agriculture) as multiple operations involving the manufacture and distribution of farm supplies and the storage, processing and distribution of agricultural commodities. Agribusiness became ever since associated with a large agro-industrial complex under concentrated forms of corporate ownership and management. In some countries, agribusiness increasingly encompasses large- and small-scale commercial agricultural production, as well as agro-industry and associated services.

Seen from a critical perspective, agribusiness appears to be more than just a commercial agriculture regime practiced in high-tech farms, but rather constitutes a particular approach to the management of rural properties, the mobilisation of resources and the financing and commercialisation of production. The fluid meaning of agribusiness, which is exceptionally broad and contains many activities directly or indirectly associated with agriculture, helps to reinforce the sense that the sector plays an increasingly important role in most national economies and global commercial exchanges. Related to the difficulty to accurately define agribusiness, sustainability and sustainable development are other highly contested concepts that encapsulate a spectrum of viewpoints, interests and experiences. Progress towards higher levels of agricultural sustainability and better environmental governance are directly impacted by the controversial nature of sustainable development and by the acceptability of sustainability notions. In that context, the ambivalent connotation of agribusiness is not accidental, but the slippery boundaries of the sector make it convenient to be used by politicians and economic groups to justify a particular course of action. For instance, the powerful agrifood sector of the United States claims to be, at least rhetorically, engaged with the sustainability agenda, but the practice demonstrates that this commitment is superficial and aimed to 'green' their established business strategies. The agribusiness sector is conventionally associated with capital-intensive farming and the extensive integration of agricultural production and food chains. Those processes are often described as clear evidence of technological efficiency, economic success and consumer benefit arising from large-scale production, extended logistic systems and international trade transactions.

A very concrete, striking sense of the contrast between agribusiness and traditional family agriculture can be grasped by standing near a fence that divides an indigenous reservation from the surrounding private farms, an exercise relatively easy to undertake in many parts of North and South America. On the one side, artificial pastures and fields with homogenous, often transgenic, crops cultivated on soils prepared with the use of heavy machinery, fossil fuels and high doses of agrochemicals (fertilisers, herbicides, pesticides, etc.). On the other side of the fence, a much more complex ecosystem with shrubs, grasses and trees (the actual composition depending on external pressure and past uses of the land) and a diversity of animals and microorganisms that is several orders of magnitude greater. The landscape managed by the indigenous communities also holds contrasting knowledges, experiences and practices cherished by people with distinctive features and languages, although 'at face value' their socio-economic condition is routinely comparable with the destitute material situation of poor peasants and urban homeless (Ioris 2021). Contemporary agribusiness farms are spaces where a great amount of capital circulates (at the expense of bank loans and foreign inputs) but very little is left behind, apart from socio-ecological impacts. Most technical operations were developed and tested in agronomic research centres located thousands of kilometres away by scientists who never heard the name of the region where the agribusiness farms are located. Agribusiness is fundamentally predicated upon the estrangement and subsequent indifference for the unjust differences created in the course of agrarian development. As observed by Manuel and Posluns (2019: 188), 'it is not necessary to travel all the way to the Artic to see the traditional colonial system operating with the full assistance of modern technology'.

Large agribusiness farms, enthusiastically celebrated by politicians, economists, urban elites and the dominant mass media, are cenotaphs of indifference for the socio-ecological differences that ultimately make them work and that were viciously appropriated to satisfy exogenous demands. It is easy to confirm that paradox in the agribusiness-dominated regions in the USA's Mid-West, California, Southern Spain, Argentina, South Africa or where Brazilian monoculture plantations continue to expand. On 13 June 2012, the conservative Brazilian magazine *Veja* published an article with the title '*Guess Which the Land of the Indians Is*' to compare the supposed progress of agribusiness against the backwardness of indigenous lands in the State of Mato Grosso do Sul. The key message of the piece is that indigenous families and communities in the region are desolately poor because they have not accepted the 'rational' solution of agribusiness production (and the few exceptions, indigenous individuals who take part in commercial agriculture, are treated as the example to be followed). According to the magazine, poverty is consequence of their primitive way of life and in the rare cases when the land was returned to the ancestral inhabitants, the result was a total waste of land and opportunities. Replicating elements of the old ideologies brought by the Iberian conquerors after Columbus, the material misery of the native population is basically their fault (in the history of the American continent, the victims are always guilty of their fate). The only viable alternative would be to submit their domestic and professional life to market-based relations, as championed by businesspeople and large-scale landowners. The journalist made no mention to the fact that this alleged solution only benefits a small minority of the non-indigenous population and is based on increasing social and ecological degradation.

The article in the magazine also belittles the legitimate, reiterated demands of indigenous communities for land, which was illegally grabbed by farmers and the government, as well as the firm protest against the over-exploitation of the indigenous workforce. From the perspective of the hegemonic mass media, indigenous peoples continue to suffer because they insist on their arcane differences and continue to evoke past, now irrelevant, tragedies. However, contrary to this mainstream argument, a bottom-up defence of crucial socio-spatial differences is not just what guarantees the endurance of indigenous people as such, but is the cornerstone of their political struggle against a profoundly exclusionary reality that tends to dilute the most fundamental features of their existence. It is not coincidence that their reactions are primarily informed by their claim of difference to cope with contemporary trends that both disregard the other and incorporate them through subordination. Indigenous peoples have been able to creatively mobilise the treasured elements of their collective existence and take it to diverse political arenas, reaching an increasingly broader audience with the support of social media. Key demands include the devolution of grabbed land, the reversal of socio-ecological degradation and compensation for past violence. This movement is an attempt to express difference on their own terms and reveals that alterity is not merely the final outcome of social interactions, but something that exists before of, and is integral to, social relationships. Indigenous or peasant difference is, thus, not simply a qualification, but it is the subject of processes of convergence or divergence between individuals and social groups that take place in specific spatial settings and produce space anew. They precede and affect socio-spatial interactions, which are themselves motivated by the feeling of difference or the lack of it (due to previous interventions). It is, therefore, tautological to assert the immanence of interaction in difference, given that one or something is only different through another, that is, nothing and nobody can be different in isolation, but only because of the contraposition with the other (Ioris 2022).

There are specific historical and geographical reasons for this widening gap between agri-food and basic socio-ecological goals. For many centuries, going back to ancient farming and Babylonian irrigation, agriculture was a source of staple food and raw materials, with only a fraction of production commercialised or exchanged in (almost exclusively) local markets. Although some pre-capitalist societies did achieve sizeable food surpluses, this happened over long periods of time and vast territories (as in the case of Aztec, Roman, Chinese, Indian and Islamic agri-food systems). More importantly, pre-capitalist agriculture was not 'premised on a state- and market-enforced productivity model', but 'with the transition to capitalism, the new property relations propelled a process of dispossession and differentiation that enabled rising labour productivity in agriculture and rising food surplus' (Moore 2015: 242). In this process, the life-sustaining and socio-cultural properties of agriculture were progressively disregarded in favour of commodified versions of food. Capitalist agriculture changed not only in terms of the scale of production, but its fundamental qualities were also transformed as food was transmuted into commodity. This is a highly politicised phenomenon, with both local and geopolitical repercussions. For instance, the need to secure cheap, commodified food for a growing layer of non-agricultural labourers was a key feature of the early expansion of capitalism and was pivotal to the rise of the Dutch and British world hegemonies in previous centuries, as well as North American world supremacy later in the twentieth century.

It should come as no surprise that the chief consequence of commodification is the reinforcement of malnourishment and socio-ecological degradation due to combined distortions in both the production and consumption sides of agri-food systems. Shameful levels of food waste and scandalous distribution losses only aggravate the problem and betray the narrow rationality of mainstream agri-food. Every night, around 800 million people in the world still go to bed hungry and almost a billion suffer from extreme poverty, according to the highly regarded report *The State of Food and Agriculture* (FAO et al. 2015). Ironically, and sadly, most of these people live in rural areas and depend on agriculture-related jobs for most or all of their income. Nonetheless, the public debate on the causes of hunger and poverty is currently dominated by simplistic claims and technocratic—largely neo-Positivistic (i.e., basically, an argument of the equivalence between physical and social phenomena)—calls for mere increases in production and productivity. The official discourse of agricultural development and food security adopted by national governments and multilateral agencies emphasises that the aggregate number of people suffering from hunger (vaguely defined as 'chronic undernourishment') has declined in recent decades due to technological and managerial improvements. This claim is apparently supported, at least in part, by the statistical evidence. The year 2015 marked the end of the monitoring period for the two internationally agreed targets for hunger reduction included in the Millennium Development Goals (MDGs). The number of people in developing countries suffering from hunger was to be halved compared with the base years of 1990–1992 (from 23.2 to 11.6%). The recorded percentage during 2014–2016 was 12.9%, implying that the goal has almost been met. However, if we analyse the results in terms of undernourished people, the figures are rather different: since 1990, the number of undernourished people in developing regions has fallen from 991 million to 790.7 million, but the goal was 495 million (half of 991 million), meaning that the target has not in fact been met (FAO et al. 2015: 8–12).

Although any reduction in hunger should be celebrated, there are serious concerns about the reliability and significance of such statistics and, more importantly, the unwillingness to question mainstream agri-food operations. Estimates of hunger are highly problematic due to the difficulty of accurately defining the term, debatable thresholds and inconsistent methodologies, and we all know that national statistics are fraught with imprecision. Governments and multilateral agencies are constantly moving the goalposts with redefinitions of key concepts, data revisions and adjustable targets. On the whole, while the MDGs basically concentrated on eradicating extreme poverty and eliminating hunger, a major food crisis continues to affect both those with and without easily available food. The difficult conditions of those suffering from hunger and malnutrition remain concealed behind aggregate data and the persuasive discourse of agriculture modernisation advanced by agribusiness corporations. Beyond cold statistics, the crux of the matter is really whether agriculture should serve the desires of a small minority of the population and merely support economic growth, or whether the industry should be concerned with social justice and the promotion of socio-ecological sustainability. In practical terms, this is not just about changing global and national trends, but also about fostering alternative solutions and creative strategies at the local and individual levels. In any case, it is becoming clear that business-as-usual (or agribusiness-as-usual, to make an unavoidable pun) is a risky and hopeless option that is not serving the needs and expectations of farmers, consumers and wider society.

This is because mainstream agri-food approaches remain firmly within the established paradigm of market solutions and mitigatory, end-of-the-pipe measures. For example, the aforementioned report by the FAO and partners makes clear that investment in agriculture remains the single most effective way to provide opportunities to generate income and improve nutrition, especially for women and young people in rural areas, but very little is said about the underlying causes of rural deprivation (such as the unequal distribution of land or public policy support). Ultimately, documents such as these can obscure the prevailing forces of dispossession and displacement promoted by market globalisation and enacted by national governments and their allies. The influence of state policies, including environmental regulation, produces results that can either benefit wider society or, in other cases, concentrate gains in the hands of corporations or powerful landowners (for instance, in the United Kingdom, the payment of farm subsidies to billionaire Saudi princes, dukes and even to Queen Elizabeth II, who received £557,706.52 in 2015 (BBC post, 29 September 2016). Likewise, the prominence of global commodity markets and top-down rural development masks the agro-industrial and financial priorities that pervade the production, distribution and consumption of food. An emblematic example of this controversy can be found in the World Bank's attempts to improve the productivity of small farmers to integrate them into commercial chains controlled by powerful, normally foreign, players. It is also the case that, despite any progress, the vast majority of those suffering from hunger live in the Global South, where more than six decades of international development promises have not resolved the matter.

However, the negative environmental impacts and growing socio-political tensions derived from the expansion of agribusiness to many new corners of the planet are systematically played down by those who control its symbolic and material dimensions. That includes the disruption of traditional food, utilisation of mechanised deforestation and fast concentration of landed property. Consequently, there is a pending demand for interdisciplinary critical analyses able to explain the false claims of the sector amid a sustained mystification of the contribution of agribusiness to local, national and international economies. Furthermore, the controversial features of agribusiness are also relevant to help to understand the challenging risks and responsibilities of agriculture in the contemporary, increasingly urbanised world. An obvious consequence of rapidly changing production and consumption patterns is that any investigation into agri-food issues needs to consider the material, subjective and discursive dimensions of market globalisation and the multiple contradictions, as well as achievements, of contemporary capitalist agriculture. A central question of this debate is the fact that the technological and managerial practices of commercial agriculture are largely determined by the activity of mega-corporations selling agrochemical inputs, machinery and equipment, and by the complementary activity of agri-food companies controlling the purchase and distribution of goods. As observed by Clapp and Fuchs, large corporations and their commercial allies hold different and interrelated forms of power, including instrumental power (the ability to lobby governments and influence social actors), structural power (influence over the public agenda and rule setting) and discursive power (shaping the public debate and the choices presented to wider society). The power of agri-food corporations, including lobbying and pressure on governments, is never far away from the household or the dinner table. While agro-industries pursue high field productivity, and large supermarkets operate extensive delivery networks, the great majority of the population is dangerously reliant on a small number of supply chains and the narrow menu of fast-food restaurants. The colourful shelves of most shops seem to offer a range of food options and a variety of choices, but are in fact dominated by a small list of plant species and animal breeds. Mass selling of convenient, ready-to-eat options is achieved at the expense of food's nutritional value, traceability and contribution to local economies. The influence of corporate interests is particularly significant among urban populations and on the periphery of large cities, where there is a growing tendency to buy cheap sugary food or consume frozen microwaveable meals. The perverse appeals of convenience and standardisation have seriously affected not only the daily diets and health of both younger and older generations, but also the power of farmers to decide what to produce.

It is highly relevant that an expanding number of critical studies have demonstrated that conventional agri-food systems and neoliberal agribusiness activities no longer produce food that is safe and healthy, and this cannot be sustained in the long term. For McMichael (2010: 16), liberalisation and privatisation are combined 'to accelerate food circulation globally and restructure food production and retailing along corporate lines'. It is also worth remembering the observation of Derrida (1994: 52) that 'never, never in history, has the horizon of the thing whose survival is being celebrated (namely, all the old models of the capitalist and liberal world) been as dark, threatening, and threatened'. The key question therefore posed for society, academics, politicians and farmers is how to reject the configuration and repercussions of agriculture-cum-agribusiness in favour of diversified, largely localised, agricultural practices owned by producers and consumers. Those aims can only be achieved through the simultaneous construction of a new socio-economic order and novel patterns of production, consumption and ecological conservation. This agenda of radical change vividly and inescapably connects the local with the national and the international. The transition to a more sustainable agriculture and fairer agro-industry will be the result of social mobilisation at different levels, from small-scale agriculture systems to the profound transformation of national and global markets. Moreover, and contrary to the claims of many academics and activists, such changes must necessarily involve more stringent economic, health and environmental regulation and better-informed consumers, but these alone will not be sufficient. The transformation of food and agriculture must necessarily contribute to, and follow, wider politico-economic, ethical and technological reconstructions.

Considering the complex and contested evolution of agribusiness and related disputes over land, resources and social opportunities, the second version of the book expands and deepens the main issues addressed in the first edition (Ioris 2016), namely the uncertainties associated with the prevailing agribusiness model, the management of natural resources, disputes between classes and social groups, the importance of places and geographical locations, and the barriers to enhance food security and sovereignty. The series of interconnected chapter offers a critical reinterpretation of the tensions associated with the failures of mainstream regulatory regimes, land and resource grabbing, and the impacts of global agri-food chains at local, regional and inter-sectoral scales. The various chapters also examine past legacies and emerging challenges associated with agriculture modernisation, politico-spatial disputes, climate change, social movements, gender, ethnicity and education. In addition, because of the emergence or persistence of other important issues, the chapters of the new book will also address questions related to land ownership, environmental regulation, social movements (associated with organic food production, peasant struggles and consumer mobilisation), gender and ethnic minorities, food quality and public health. It likewise addresses the transformative potential of different combinations of biophysical, socio-technical and socio-spatial practices of food sovereignty. The focus of the second edition of the book is on the economic, political and environmental challenges shared by countries in the Global North and in the Global South to manage agri-food supply chains fairly and effectively, considering the strategic role of agriculture in local and national development and in the long-term protection of ecological systems. Those multiple demands cannot be easily reconciled, but require profound adjustments in public policies, technological patterns, property relations and decision-making.

Taking into account the challenges and their multi-scale ramifications summarised above—from household to global scale, and connecting past, present and future—the present discussion is located in the tradition of critical social theory, and aims to contribute to the search for radical alternatives in an attempt to reinstate food as food, farmer as food producer and markets as allocation tools (Ioris 2015). It is also part of the endeavour to remove the distortions and falsifications of the hegemonic agri-food system that feeds contemporary, market-based society (Fernandes 2016). What brings all the chapters of the book together is the common intention to explore the competition and the asymmetries between the prevailing 'green revolution' model and agro-ecological systems. Such disputes have been aggravated by the intensification of global transactions, the influence of corporate management, the intensive use of agrochemicals and genetic engineering, the grabbing of land and resources, the financialisation of production and the disregard for the rights of traditional communities, indigenous peoples and peasant family farmers. The consequence is that mainstream agri-food production (i.e. agribusiness) is now considered by many researchers, farmers and consumers to be unfair, risky and unsustainable due to mounting social-ecological impacts. At the same time, many small farmers, who champion agro-ecological systems, have developed multiple mechanisms to resist politico-economic pressures and to maintain the small production unit. The book will provide a rich comparative analysis in key agricultural regions around the world, to increase empirical, theoretical and policy-relevant knowledge about the current challenges, future perspectives and the reproducibility and transformatory potential of agro-ecological approaches, with emphasis on community farming and consumer rights.

If there are no easy answers or pre-determined routes to follow—on the contrary, the agenda of change is wide open—it should at least be asserted that these are all questions that animate the discussion and provokes in-depth reflection throughout the articles of this Special Issue. The central element of this debate is necessarily political and should operate at different levels: from the farm and community level to alliances with informed, conscious consumers, other economic sectors and changes in public policies and environmental standards. Responses to the dominance of agriculture-cum-agribusiness will fail to achieve long-term results unless local and global food systems undergo major, fundamental transformations. It is really fascinating to realise that these multi-scale changes will, more than anything else, be felt at the local scale where people work, live and dine (obviously under the influence of wider national and global pressures). Rather than being a pre-structured phenomenon, rural localities are 'reproduced, and the social relations therein recomposed, by virtue of their contemporary magnetism for relocation due to the wider discontinuities of capital activity' (Cloke et al. 1990: 15). Consequently, local and interpersonal interactions are as important as more general, structural dimensions. It is increasingly evident, at least among critical academic and activist circles, that the technologies inherited from the Green Revolution (and its more recent version, the 'Gene Revolution', associated with genetic engineering and genetically modified organism [GMOs], and closely related to digital technologies and precision agriculture) were limited in terms of their impact in reducing food supply and distribution problems; the main challenge continues to be a search for politico-ecologically viable and socio-culturally acceptable options.

The departure point of the new book is the need for critical and innovative studies to investigate the basis and consequences of the agriculture dualisms and risks unfolding at different scales and with complex repercussions for consumers, communities, ecosystems and national economies. The contributors will develop new theoretical points on agrarian and agro-ecological transformations that consider different stakeholder sectors and propose governance pathways and public policy reforms that can be constructed for facilitating sustainable, inclusive and fair processes of change. Political, economic and sociological critiques of agri-food dynamics have identified the intensification, specialisation, distancing, homogenisation and concentration of power as the key processes resulting in food insecurity and unsustainable outcomes. These trends have configured a placeless foodscape disconnected from diverse social demands and the ecological basis of distinct locations and countries. Moreover, transition studies have failed to effectively address the transformative potential of different types and combinations of biophysical, socio-technical and social practices that cut across multiple regimes and scales of agri-food activities. Harnessing this potential requires a new co-produced and scientific integration of markets and business models, governance dynamics and capacity-building perspectives. The new book will address those dilemmas by recognising social and ecological relationships between places at local and regional scales. One of the key contributions of the publication is to demonstrate that a more foundational socio-spatial approach is required for contested sustainability and for agro-food transformations. There are currently more hard questions than convincing answers, in particular, how to achieve transformations in the food system and promote a holistic and sustainable approach to food production based on local, place-based food interactions?

The authors who contributed to this edited volume were encouraged to reflect on and engage with those fundamental questions. The final result is a collection of chapters that address these from different angles and based on particular geographical experiences. Among the contributors, there are distinguished scholars and more junior colleagues, who are either based on the Global North or in the Global South. There is also a diversity of disciplinary perspectives, including geographers, historians, sociologists, agricultural engineers and economists. What is common in all chapters is the commitment to critical agri-food research and the pursuit of alternatives that respond, first of all, to the demands of the majority of food producers and consumers. It is worth mentioning that the first chapter, after this introduction, is a text left by the prestigious Brazilian intellectual, the late Milton Santos, and reproduced under the generous permission of the family. Professor Santos (1926–2001) is probably the most distinguished Brazilian geographer, with an prominent international career, recipient of the 1994 Vautrin Lud Prize (considered geography's equivalent of the Nobel Prize; to date, he is the only Latin-American scholar to ever to have won it) and a posthumous recipient of the Anísio Teixeira Prize, awarded every five years by the Brazilian agency for the improvement of higher education personnel to distinguished contributors to research and development in Brazil. Santos' views about space helped to shape radical geography and renovate the discipline in the 1970s. Chapter 2 analyses the interrelationship between money, other commodities and the production of space, in the form of the territorialisation of money. First, territory rules money; afterwards and progressively, this relation is reversed, money rules territory. Exchange constitutes the historical mediation of the relation and the reversion. By means of exchange, money and then territory become information and regulation. Money becomes national money and it is internationalised when the territorial (national) state is created. Finally, it becomes global money, the flux of the fluxes, the despotic and cruel train sustained in the velocity of circulation and in the today's informational technique.

In Chapter 3, by Marsden, Lang and Millstone, three renowned British scholars who are among the most important critical agri-food scholars in the world today, it is proposed the concept of 'Disruptive Governance' to explain the impacts of Brexit on the UK food system, and to a lesser extent across Europe. The chapter proposes that a new Disruptive Governance is now visible in politics, not just for food policy. It explores what this disruptive food governance means. Agri-food policy frameworks created after World War II are now being significantly shaken and uncertainty is being normalised, replacing the incremental development of a European financial, market and regulatory infrastructure which has underpinned food supply and management for decades. Taking a particular look at Wales as well as UK-wide territorial relations, the chapter proposes that a paradox lies at the core of the current variant of Disruptive Governance in the United Kingdom. On the one hand, there is an influential political narrative about 'taking back control', often without explanation about who will have the new control or why. In this, new powerful forces mostly off the land compete for dominance. At the same time, UK institutions that were already weakened by financial cuts and centralisation, are now shrouded by short-termism and political uncertainty. Despite the weaknesses and unstable fluidity in these food politics, there are also new possibilities for radical change to address long-term criticisms about the UK food system's unsustainability, food insecurity and impact on public health. While these opportunities were initially recognised mostly by environmental civil society interests, from 2018 ideas began to emerge from Whitehall too. This possible new strategic direction is itself subject to uncertainties, not least within the Government grappling with Brexit, but also by threats from and to Wales, Scotland and Ireland. The paper concludes that, as the policy paradox is fought over, the normalisation of disruptive governance continues and growing food insecurity is a casualty.

In Chapter 4, Borsatto and collaborators make use of the concept of 'Disruptive Governance' to explain the attacks made by the neo-fascist government of lieutenant Jair Bolsonaro against policies aimed at Brazilian family farming. Since taking power in 2018, in a turbulent and anti-democratic context, the new government has implemented a series of actions aimed at dismantling policies that benefited family farming, actions that have been supported by a populist discourse that this rupture is desirable and good for everyone. It is argued in the chapter that the Bolsonaro administration, through a disruptive governance regime, represents an intensified dismantling of public policies aimed at family and peasant farming, intending to strengthen and consolidate the agribusiness model, marked by predatory extractivism. This is demonstrated by two cases of dismantling public policies that used to benefit family and peasant farming. In Chapter 5, Fernandes and Welch deal with conflicts generated by land tenure disputes have as protagonists the families of the Landless Workers' Movement (MST), traditional landowners and agribusiness corporations in Brazil. It is a reflection on land conflicts and the spatial and territorial production of latifundia, peasant smallholders and large-scale agribusiness through studies of the landscapes of two municipalities in the state of São Paulo. Among the various tools for analysing territorial disputes, Google Earth was used for local, regional and international comparative studies of how landscapes have been transformed. The chapter includes a diversity of photographs, from satellite to field shots, as historical evidence of landscapes that express conflicting social relations in dispute over different models of social and territorial development.

Chapter 6, by Zimerman and colleagues, argue that in Brazil there are many more people living in rural areas than what is reported in official publications. Inequality in the country is not only verified in income terms, but similarly in land ownership and tenancy and how it is distributed across the rural population. As land and income are concentrated in a few hands, a significant number of farmers who do not own land would fall short of the benefits that land ownership assures them. Hence, Brazil does not differ much from the region in which it is located, Latin America. However, democracy cannot persist in such a situation. Reducing violence over land disputes involves reducing inequality via the redistribution of land and income for a more just and prosperous society. To modify such perverse trends, the authors defend that it is necessary to fight rural poverty as a priority, not as governmental policy bus as state policy, avoiding being hostage by different administrations. In the following Chapter 7, Jara and Paz scrutinise the particular characteristics of Santiago del Estero, a province in north-western Argentina, which is significant given that the region is home to the greatest proportion of rural population in the country. Also relevant is the presence of family farming run by indigenous peasants that live in agricultural holdings with undefined boundaries. Those spaces have no legal titles or deeds.

Indeed, their inhabitants are vulnerable to evictions due to the advance of agribusinesses. The chapter identifies processes of different temporalities that formed and conditioned the current agrarian question in Santiago del Estero. The analysis primarily attempts to understand capitalist development in an unequal, disorganised and, therefore, highly problematic agricultural structure, where peasantry has a strong presence, although it has not been given a significant role in rural development policies.

In Chapter 8, Mann discusses the contradictions of dominant discourses that promote Australia as 'food bowl to Asia' in the context of local food insecurity, environmental mismanagement and climate change. The recent Covid-19 pandemic vividly exposed the fragility of a food system built to rely on interconnected and complex global food supply chains that facilitate trade between nations. Rising food prices can be directly attributed to the impacts of movement restrictions and illness on local markets and the availability of agricultural labour, but the roots of the problem extend much deeper to a colonial imaginary that continues to marginalise those on the wrong side of the socio-economic divide. However, this imaginary is also coupled with emerging alternatives such as the 'eco-innovator' embracing regenerative agroecology, together with growing appreciation for Indigenous knowledges, cultures and practices regarding land management. Many farmers are focusing on land regeneration and increasing on-farm biodiversity through practices that include strategic grazing, pasture cropping and native perennial grass regeneration. Also related to the impact of agribusiness on indigenous peoples, Chapter 9, by Ioris, examines the genocide unfolding in the Brazilian State of Mato Grosso do Sul—described as Kaiowcide—which is not just a case of hyperbolic violence or widespread murdering, but it is something qualitatively different from other serious crimes committed against marginalised, subaltern communities. Kaiowcide is actually the reincarnation of old genocidal practices of agrarian capitalism employed to extend and unify the national territory. In other words, Kaiowcide has become a necessity of mainstream development, while the sanctity of regional economic growth and private rural property are excuses invoked to justify the genocidal trail. The phenomenon combines strategies and procedures based on the competition and opposition between groups of people who dispute the same land and the relatively scarce social opportunities of an agribusiness-based economy. Only the focus in recent years may have shifted from assimilation and confinement to abandonment and confrontation, but the intent to destabilise and eliminate the original inhabitants of the land through the asphyxiation of their religion, identity and, ultimately, geography seems to rage unabated.

Chapter 10, authored by Bruna and Abbas, is about access to land and food in the challenging context of climate change and in relation to rural development in Mozambique. The work was intended to further understand disturbing trends and their role in shaping rural development in the southern section of the African continent. The research's main question was related to the environmental crisis and the best policies to address it (mitigation and adaptation) to shape access to food and land rights. By answering that question, it was possible to grasp how rural development might be shaped by both trends of impacts, direct and indirect, of climate change. The investigation involved documental analysis, which was conducted including the review of the legal framework related to climate, agriculture and land topics. This chapter was also based on the doctoral research of both young academics, which basically relied on primary and secondary data about rural Mozambique, rural livelihoods and agricultural data sets. In Chapter 11, Fernandes and collaborators rely on the theory of the rent of land to analyse the forms of appropriation of different types of rent by companies controlled by the financial capital in areas of expansion of the modern agricultural frontier in Brazil, which engender territorial conflicts with peasant, *quilombola* and indigenous communities. Several companies establish relations with the financial capital in the modern agricultural frontier, called MATOPIBA, aimed at territorialisation of agribusiness. This region shows large properties with capital and technology-intensive agriculture primarily directed to exports, thus territorialising agrarian extractivist policies since the 1970s. High economic productivity also generates territorial conflicts over the ownership of land, water, labour and produce. Such conflicts constitute the 'conflictivity' that unveils the class struggle in disputes over territories and development models.

Chapter 12, by Oliveira Neto, examines and reinterprets dispossession caused by the expansion of agribusiness in the Amazon Region. Agrarian dispossession is considered a process through which capitalism expands throughout several world regions, mainly in the southern hemisphere, and features the fight of local individuals against the state politics which promotes the advance of financed capitalism aiming to take more lands and natural resources by capturing and expelling as a modern approach, both through annexation, subordination or exclusion of local individuals. Issues related to the various kinds of agriculture which opposes to the agrobusiness model, the importance of forests conservation, the climate change, the conflicts, the environmental justice and the ecological politics, the land titling issues, the social movements, the place, the territory and the land rights are some that currently surrounds the debate about dispossession. Chapter 13, by Dulci, deals with the global governance of coffee production with an emphasis on the context of Minas Gerais, in Brazil. This state concentrates small, medium and large coffee growers, the operation of warehouses, processing plants and roasting industries, in addition to brokers and exporters, both national and transnational. Because of that, it is a privileged territory for observing dynamics, limits, and why not possibilities of reversing existing inequalities along the network. By following its journey from farm to cup, it is possible to observe the coexistence of a process of valorisation and 'revival' of coffee in the countries of the North while there is an increasing devaluation of coffee growers in the South. There is a key paradox behind positive rates of expansion: despite being responsible for more than 70% of the world's coffee production, smallholders retain only 5 to 10% of the financial profit. Strategies to expand the commercialisation of green coffee through export have not affected the fact that most of the rent is being created and captured by the larger players.

In Chapter 14, James and colleagues study agroecology in relation to the evolution of the so-called agri-environmental state, making reference to the Brazilian case. The agri-environmental state is characterised as possessor of a significant set of institutions and practices that can support environmentally and socially sustainable agriculture. However, the authors identify limitations in state-centric analyses, particularly amid the emergence of authoritarian state regimes and in settler-colonial contexts, where jurisdiction and the legitimacy of the state and any state policy is unsettled and highly contested. Similarly, creating economy-ecology dichotomies can obscure the fact that any economy is always embedded in an ecology, and runs the risk of subsuming heterogeneous economic modes into a homogenous, globalised and neoliberalised market. Chapter 15, by Meek and Tarlau, develops a decolonial and feminist approaches to food systems education, which not only take place in primary and secondary schools through garden-based learning, but also include initiatives led by grassroots social movements that are most often linked to a radical political vision for more egalitarian societies. The chapter revisits the current political context, which is marked by an anti-democratic, virulent right-wing fascism that has popularised through the vilification and dehumanisation of Black, indigenous, Muslim, immigrant, landless, lower caste, women and transgender communities globally. Such brutal white supremacy that has direct implications for collective food systems and education plays an important role in terms of resistance and political reactions. Chapter 16, by Origuéla, also deals with territorial forms of resistance and the Brazilian peasant food system. It discusses socio-territorial movements, in particular the Landless Rural Workers Movement (MST). Land-related struggles depend on the specific level of organisation, scope of action and acquired political, economic, environmental and cultural features, territorial resistance can give rise to peasant food systems. At the same time, there are important commonalities and synergies between agri-food resistance around the planet and many important lessons to be learned.

The final chapter is a discussion of the crucial role of agrarian social movements and its role in present-day politico-economic and socio-spatial transformations. The author, Saturnino ('Jun') M. Borras Jr., is a professor of Agrarian Studies at the International Institute of Social Studies (ISS) in The Hague and is also Editor-In-Chief of the Journal of Peasant Studies. Borras has been deeply involved in rural social movements since the early 1980s in the Philippines. He was a founding member of the international small farmers' movement *La* Via *Campesina* and received the 2020 Ester Boserup Prize for research on development. Borras is also a professor at the College of Humanities and Development Studies of China Agricultural University in Beijing, Fellow of the Amsterdam-based Transnational Institute (TNI) and the

California-based Institute of Development and Food Policy (Food First). He is a member of the Emancipatory Rural Politics Initiatives or ERPI collective. Relying on this invaluable academic and personal experience, Chapter 17 examines the parallels, resemblances and interconnections between contemporary right-wing populism and the populism of agrarian movements are examined in this essay. It explores an agenda for political conversation and research on possible contributions to the twin efforts of splitting the ranks of right-wing populists while expanding the united front of democratic challengers. The challenge is how to transform the identified interconnections into a left-wing political project that can erode right-wing populism. This requires a reclaiming of populism. The chapter revisits the ideas and practices of right-wing populism and agrarian populism, and the awkward overlaps and fundamental differences between them. It concludes with a discussion on the challenge of forging a reformulated class-conscious left-wing populism as a countercurrent to right-wing populism, and as a possible political force against capitalism and towards a socialist and ecological-humanist future.

References

Cloke, P., R. Le Heron, and M. Roche. 1990. Towards a Geography of Political Economy Perspective on Rural Change: The Example of New Zealand. *Geografiska Annaler* 72 (1): 13–25.

Davis, J., and R. Goldberg. 1957. *A Concept of Agribusiness*. Boston: Alpine Press.

Derrida, J. 1994. *Specters of Marx*, trans. P. Kamuf. London and New York: Routledge.

FAO, IFAD, and WFP. 2015. *The State of Food Insecurity in the World 2015. Meeting the 2015 International Hunger Targets: Taking Stock of Uneven Progress*. Rome: UN-Food and Agriculture Organization.

Fernandes, B.M. 2016. Development Models for the Brazilian Countryside. *Latin American Perspectives* 43 (2): 48–59.

Friedmann, H. 2005. Feeding the Empire: The Pathologies of Globalized Agriculture. *Socialist Register* 41: 124–143.

Ioris, A.A.R. 2015. Cracking the Nut of Agribusiness and Global Food Insecurity. In Search of a Critical Agenda of Research. *Geoforum* 63: 1–4.

Ioris, A.A.R., ed. 2016. *Agriculture, Environment and Development: International Perspectives on Water, Land and Politics*. Basingstoke: Palgrave Macmillan.

Ioris, A.A.R. 2017. *Agribusiness and the Neoliberal Food System in Brazil: Frontiers and Fissures of Agro-neoliberalism*. London: Routledge.

Ioris, A.A.R. 2020a. *Frontier Making in the Amazon: Economic, Political and Socioecological Conversion*. Cham: Springer.

Ioris, A.A.R. 2020b. Controversies around Food Security: Something Difficult to Swallow. In *Routledge Handbook of Sustainable and Regenerative Food Systems*, ed. J. Duncan, M. Carolan, and J.S.C. Wiskerke, 420–435. Abingdon and New York: Routledge.

Ioris, A.A.R. 2021. *Kaiowcide: Living through the Guarani-Kaiowa Genocide*. Lanham, MD: Lexington Books.

Ioris, A.A.R. 2022. World Out of Difference: Relations and Consequences. *Philosophy and Social Criticism* (forthcoming).

Manuel, G., and M. Posluns. 2019 [1974]. *The Fourth World: An Indian Reality*. Minneapolis and London: University of Minnesota Press.

McMichael, P. 2010. The World Food Crisis in Historical Perspective. In *Agriculture and Food in Crisis: Conflict, Resistance, and Renewal*, ed. F. Magdoff and B. Tokar, 51–67. New York: Monthly Review Press.

Moore, J.W. 2015. *Capitalism in the Web of Life*. London: Verso.

Patel, R. 2013. The Long Green Revolution. *Journal of Peasant Studies* 40 (1): 1–63.

Shiva, V. 2016. *Who Really Feeds the World? The Failures of Agribusiness and the Promise of Agroecology*. Berkeley: North Atlantic Books.

2

Prolegomenon: Money and Territory

Milton Santos

Geography has reached its golden age at the end of this century, because 'geographicity' imposes itself as a historical condition, insofar as nothing considered essential today is done in the world that can avoid the knowledge of what the Territory is. The Territory is the place where all actions, all passions, all powers, all strengths, all weaknesses merge, that is, where the history of man is fully realised from the manifestations of his existence. Geography becomes that discipline that is more capable of showing the dramas of the world, of the nation, of the place.

What I bring here is an essay. It is much more an experiment of method than something finished. As a matter of fact, why should a

Originally published as Santos, M. 1999. O Dinheiro e o Território. *GEOgraphia*, *1*(1), 7–13, as the transcript, revised by the author, of a conference given on 15 March 1999 at the Fluminense Federal University (UFF), Brazil. Reproduced under permission of the copyright holder. Translated by Antonio Augusto Rossotto Ioris.

M. Santos (✉)
University of Sao Paulo, Sao Paulo, Brazil

A. A. R. Ioris and B. Mançano Fernandes (eds.), *Agriculture, Environment and Development*, https://doi.org/10.1007/978-3-031-10264-6_2

teacher address anyone with ready-made things? A class is always a set of questions, not simply answers. The class that is intended to be an answer is something almost unnecessary. The class has to be a set of questions which the teacher incompletely formulates, and which the listeners take as a guide both to accept or, after accepting, discuss and even refuse.

The question I am asking here is about these two poles of contemporary life: money, which seeks to disestablish everything, and territory, which shows that there are things that cannot be undone.

Territory and Money: Some Definitions

The first thing to do is define what we want to talk about. If I do not, it is as if I did not allow people to argue with me either. The first condition for those who start from an ideology—which is my case—is to clearly set up the terms of the debate. If I do not declare it, I avoid the discussion, I prevent it and I preclude them from debating with me. It is therefore necessary to initially define these two words: territory and money.

Territory is not just the collection of natural systems and systems of superimposed things. The territory has to be understood as the *something that is used*, not the territory in itself. The territory that is used is ground plus the identity. Identity is the feeling of belonging to what belongs to us. The territory is the foundation of labour, the place of residence, material and spiritual exchanges and the fulfilment of life. Territory itself is not an analytical category in historical disciplines, such as Geography. Only the used territory that can be a category of analysis. In fact, the very concept of the nation, and then the idea of the National State, derive from this relationship that has become profound, because one makes the other, in the manner of that famous phrase by Winston Churchill: 'first we shape our buildings, and afterwards, our buildings shape us'. Such is the territory that helps to produce the nation, so that the nation later becomes fond of it.

Money appears as a result of an economic life turned complex, when simple barter is no longer enough, and over time ends up imposing itself as a general equivalent of all things that exist and are, or will be, or could be, the object of trade. In this way, money is intended to be the measure of value that is thus attributed to labour and to its consequences.

Metamorphoses of Money and Territory

We shall take a quick tour of what we would call, to facilitate the explanation, the beginning of their history, that is, of money and territory.

At first, there is local money, expressive of limited geographic contexts and with a limited commercial horizon. It was the time of a world whose compartmentalisation produced vestibules that would be almost self-contained. A world without movement, a slow, stable world, like those numerous monads of Leibniz, but different than Leibniz's, without a general principle. It was money with only local circulation, or almost like that. In this first moment, the functioning of the territory owes a lot to its natural features, to which men and their works could adapt with little technical mediation, because the techniques then were somehow heirs of the surrounding nature, or an extension of the body. They were at the same time the result of this attachment of the body to nature, and of this command of nature over the history that was possible, in such a way that the technicity of objects manufactured beyond the body was limited.

The social relationships were few in number in this first phase, and they were also simple and not very dense. If the surroundings were fraught with mysteries, these were not due, as today, to the production of history, but to unknown natural forces, unlike today when we somehow know better how nature works. Therefore, at that time, material life somehow imposed itself on the rest of social life, and the value of each piece of land was attributed to it by the very use of that piece of ground. Existence could be interpreted from relationships that were directly felt, or as if they were direct.

In this period of history, the territory thus delineated governs money; territory was used by a localised society, as it was money.

From the Reason of Use to the Reason of Exchange

These categories metamorphose over time. With the expansion of exchanges, the amplification of commerce, with the growing interdependence between societies, with the production of a greater number of objects and a greater number of values to be exchanged, comes the complexity of money, with the expansion of its use and of its effectiveness. In order to guarantee the stability of exchanges and the production of each group, the need for regulation appears and money begins its trajectory as information and as a regulator.

It grows, expands and is identified, not only international trade but also domestic trade. Everything tends to become an object of exchange, valued more and more through exchange than through use. The role that exchange begins to take on is an enormous transformation in the history of places and the world, altering the primacy of use, and even commanding use, as opposed to the previous command of exchange by use. This is fundamental both to understand the changes that the world experienced, and to create the possibility of interpreting what exists as if it were a result of immediate mechanical relationships between the group and its surroundings, between man and what would still be called nature.

Money increasingly appears as an arena of numerous movements, founded on a law of value that owes as much to the character of production, as to the possibilities of circulation. Circulation takes over production the control of explanation, because it takes over production the command of life. The larger the complexity of external and internal relations, more this law extends to places, there is greater need of regulation, and the need for the State arises: the State and the limits, the State and production, the State and distribution, the State and the guarantee of labour, the State and the guarantee of solidarity and the State and the search for excellence in existence.

The Territorial State, the national territory, the National State are then created, which start to govern money. There is at this stage internationalised national money. It is evident that national money undergoes international modulations. It is partially an internal response to international modulations. More profoundly from the strong presence of the

State, this money is representative of the then deep relations between territorial State, national territory, national State and nation. It was relatively domesticated money, which was made within the territories.

Money and the Territory of Globalisation

The money of globalisation arrives. This end of century allowed the installation of information techniques, which are techniques that connect all other techniques, which allow the most diverse techniques to communicate. These information techniques that, after all, from the planet, produce a world (and that is why we talk about globalisation), and that lead us to the illusion of speed, as the matrix of everything, as an indispensable necessity and that certainly create a potential fluidity transformed into this effective fluidity at the service of globalised capital, in such a way that money appears as a fluid of fluids, the element that gives speed to the other elements of history. However, if the money that commands is global money, the territory still resists. We just have to mentally recreate the map of money in Brazil, and we will find a place where there are all possible types of money (São Paulo), and another where the only possible type is money-currency (an isolated point in the poorest state). In other words, the territory can also be defined in its inequalities from the idea that the existence of money in the territory does not happen in the same way. There are condensation zones and money rarefaction zones. However, the command of the financial activity is there where all the moneys can be present: São Paulo. But, above all, the command comes from global money. This fluid money, which is also invisible, money rendered practically abstract, is global and despotic money.

Never in the history of man has there been a tyrant as tough, as ruthless as this global money. It is this fluid, invisible, abstract, but also despotic, global money that plays a role in the current production of history, imposing pathways on nations. The general equivalent eventually becomes the really universal equivalent. But this money is not supported by infrastructure operations. It is money supported by an ideological system. This global money is the general equivalent of itself. And that's why it works autonomously and based on its own rules. Producing a falsification of all criteria, this autonomous and pure money would not exist, if the technical conditions used by the political conditions that dominate the historical period did not have the possibility of twisting the information.

The Role of Ideology

Our era is characterised above all by these dictatorships: the dictatorship of information and the dictatorship of money, and the dictatorship of money would not be possible without the dictatorship of information. Pure money feeds on impure information, made possible when we imagined it to be crystalline. Curiously, this formidable ideological system ends up playing a role in the production of materiality and in conformation of people's existence.

Ideology, as never happened before, starts to show itself as that metaphysics susceptible of appearing as something empirical. Twenty-five years ago, we were excited to assimilate the difference between the truthful and the untrue, between appearance and existence, between the ideological and the real. Today ideology has become reality, which complicates our analysis task, because it imposes itself on the production of the concrete human history from a single perfectly elaborated discourse, and which becomes credible from the bombardment of the media, but also from of the endorsement of the University. This is how the logic of money imposes itself on the rest of social life. Thus, money creates its law and imposes it on others, forcing mimicry, adaptations, surrenders, based on two other complementary logics: that of companies and that of world governments.

The logic of corporate money is the logic of competitiveness, which compels each company-turned-global to increase its sphere of influence and action in order to grow further. The last few years are emblematic because they are the theatre of major mergers both in the field of material production and in the production of information. These mergers reduce the number of global players and, at the same time, based on the notion of competitiveness, lead companies to compete for the smallest space, the tiniest share of the market.

When, knights errant, the ministers go out to beg in the North what they call 'understanding' of companies, it is as if they do not know that these global companies need markets, however minimal it may be, because the loss of the smallest gram of activity inflects the power of one to the benefit of the other. All markets, no matter how small, are fundamental—this is also globalisation. No matter how small a place, no matter how insignificant it seems, in the world of competitiveness this place is fundamental because global companies depend on all contributions to maintain their power. That power is blind, because it does not look around. This power is concerned with precise, individualistic, selfish, pragmatic goals, it is a blind power, since it does not look around. But it chooses places here and there, today and tomorrow, according to the answers they imagine to have, and they desert those places when they discover that they can no longer offer those answers.

Now, these individual logics need a general intelligence, and this general intelligence cannot be entrusted to the States because these may inadvertently decide to respond to the demands of their populations. Hence the global governments, represented by the International Monetary Fund, the World Bank, regional international banks such as the IDB, the Washington consensus, the main universities that produce globalisation ideals and also the subaltern universities that accept to reproduce them.

In this world of deception, the so-called national accounting appears as a central indication. However, this national accounting is, in essence, a fancy name for global accounting that chooses, among the categories to be used, those that privilege the interests of a certain type of agent, and exclude all categories of another nature. This global accounting is based on parameters inspired by global finance itself, in a world in which it is no longer capital as a whole that governs the territories, but a part of it, that is, money in its pure state.

Beforehand, the territory contained the money, which was in part regulated by the money, by the territory that was used. Today, under the influence of money, the content of the territory escapes all internal regulation, bringing to agents a feeling of instability, a systematic production of fear, which is one of the products of the perverse globalisation within which we live, this fear that paralyses, this fear that summons us to support what we do not believe, just for fear of losing even more.

Allow me to make a parenthesis here. The association that this end of the century has created between science and technology, technology and the market, this techno-market in which we live and this techno-science that drags us along and is also unfortunately dragging the human sciences, is what reduces the scope of academic work and sometimes deviates us away from the search for the truth. We must be concerned with the fate that the University may take, especially when it condemns outside criticism but rejects its own self-criticism.

Returning to the central theme, we said that before the territory contained money, regulated by the used territory, while today the content of the territory escapes all internal regulation. This is the problem in Brazil today. This fight between governors, this anger between some more vocal mayors, challenges together the work of geographers and political scientists. Because a political science that is not based on the functioning and dynamics of the territory can hardly offer an empirical contribution to the solution of national problems. The content of the territory has fundamentally changed with globalisation, whether its demographic, economic, fiscal, financial or political content. The content of each fraction of the territory changes rapidly. This current instability and nervousness of the territory is the empirical representation of the nervousness, nervosity, impatience and volcanism of the nation.

Dictatorship of Money and Deregulation of the National Territory

At this stage of national life, this extraordinary role of the dictatorship of money in its purest state has shown us the difficulty of internal regulation and also of external regulation, since each company has interests that are only exercised from the deregulation of others, it helps organise the company in question and disorganises everything else. In other words, the presence of global companies in the territory is a factor of disorganisation, of disaggregation, since they blindly impose a multitude of nexuses that are in their own interest, while other particularistic needs are relegated. Therefore, the Brazilian territory became ungovernable. As the territory is the place of all men, all companies and all institutions, the country has also become ungovernable, as a nation, as a state and as a municipality.

Ultimately, this is the result of the influence of money, in its pure state, on the territory. Finance, made international as a norm, contradicts existing structures and imposes others. When it has an autonomous existence, it does not need to consult anyone to settle in, it works in spite of the other actors, and brings to the place an existence without autonomy. There are certainly barriers to this action of pure money, ways of reorganising the territory, like the European Community. Another important element that resists this blind action of money is citizenship. In the case of Brazil, it reveals a serious problem, because we never had real citizens. It makes these fluid disorganisation forces to establish themselves with great speed.

These extremely elementary, unassuming, pretentiously unpretentious ideas are, basically, an invitation to conduct critical research and to produce a new discourse about the territory. The University exists exactly for that, because it is the place for the production of this discourse that should result from an innovative analysis.

3

Disruptive Governance in the UK Food System and the Case of Wales

Terry Marsden, Tim Lang, and Erik Millstone

How Brexit Disrupts the British Agri-Food System

There have been previous episodes and phases when radical governance disruptions shook the food systems of the British Isles, such as wars (from the Napoleonic to World War II), social upheavals (the Irish Famine, the 1930s) and the process of industrialisation (1780–1850) which de-ruralised parts of the UK. What makes the present period so important, however, is that it began in peacetime yet also when the food system is

T. Marsden (✉)
Cardiff University, Cardiff, UK
e-mail: MarsdenTK@cardiff.ac.uk

T. Lang
Centre for Food Policy, City University of London, London, UK

E. Millstone
Science Policy Research Unit, University of Sussex, Sussex, UK

A. A. R. Ioris and B. Mançano Fernandes (eds.), *Agriculture, Environment and Development*, https://doi.org/10.1007/978-3-031-10264-6_3

recognised to be in the front-line of new, massive external challenges. These include ecosystems tensions such as over climate change, water stress, biodiversity loss, also public health problems from unhealthy diets and antimicrobial resistance, all overlaid by rising economic and political tensions over trade, not least now exacerbated by the war in Ukraine and growing food nationalism. This combination of disruptions from outside, alongside and inside the UK is of a scale of impact not witnessed in the UK food system since the 1940s. This chapter proposes that the notion of disruptive governance is now a useful concept with which to understand the present situation and to examine the immediate future. We argue that it is not just something being done to the agri-food system but also a political style being actively championed and inserted within it.

The chapter considers the dynamics of food disruptive governance as a multi-level phenomenon occurring at global, continental, national and intra-national levels. It focuses on Wales-one devolved region within the UK polity, to highlight how it has barely been discussed in the UK debates about Brexit. The present authors raised the example of Wales in an early paper in 2017 to illustrate the disruptions to existing Wales-UK food dynamics, to consider what could be learned from Wales about the wider political discussion about how an English-dominated political vote could change the status of the UK's three devolved administrations—Scotland, Wales and Northern Ireland (Lang et al. 2017). While Scotland and Northern Ireland voted strongly to remain in the EU in the 2016 Referendum, Wales was more ambivalent, narrowly favouring leave, although voting preferences have subsequently shown signs of fluidity. In the May 2019 European Parliament elections, for example, the Brexit Party topped the poll with 32.5% of the vote, but votes for parties favouring a second referendum outweighed those favouring an immediate exit.

The Characteristics of the New Disruptive Governance

Leaving the European Union in 2021 created conditions which mean that UK's future is now one of upheaval and redesign, not just in relation to Europe but across and through all governance levels. The UK has, for instance, rearranged and began creating more formative global governance links. We see this process as the outcome of a massive and potentially long-running experiment in what we call 'Disruptive Governance'. This concept is employed to recognise that some political activists, mostly but not all on the neo-liberal Right, see disruption as a desirable and positive tactic. Jacob Rees-Mogg, in March 2022 was appointed to the role of Minister for Brexit opportunities, has called for Britain to be deliberately disruptive to the EU to press its Brexit 'freedoms' and opportunities, and appears to accept 'unknown unknown' risks are the new normal. Boris Johnson's 'do or die' promise is of this ilk, was intended to reassure libertarians who seek a shake-up of policy frameworks, and who reject the political settlement that had emerged within, and beyond, the UK over recent decades. They might genuflect to Mrs Thatcher's legacy but their political practice often far exceeds her political economy or her criticisms of the EU. They reject the notion of the EU as a peace project, which Mrs Thatcher had understood.

From their perspective, disruption is a necessary step towards the UK becoming a more autonomous and powerful nation state. While it partly harks back to mythic mid-twentieth century heroic 'standing alone' and partly to Empire, there are tensions within Disruptive Governance—notably over whether to favour a UK–USA nexus or to be more globalist. But it generally seeks a new, harder laissez-faire economic dynamism and government minimalism (while promising largesse to favoured corporate interests). It adopts but transcends the business language of disruption, which refers to how new entrants can reconfigure or create a market. This is thus a political credo. Partly, Brexiteers characterise their ideology as one of pursuing 'creative destruction', a term Schumpeter introduced in 1942. Schumpeter used that term to highlight capitalism's unremitting technological dynamism as a driving force for global economic evolution.

He was not referring to the political status of individual national states (Schumpeter 1942).

In business and investment circles (notably venture capitalists), the term 'disruption' is used in a mostly positive sense, but often realistically. Dominant companies can fall quickly (e.g. Xerox, Kodak, IBM), so the big money is to be made by spotting and backing the disrupters who will displace them; hence the big financial backing and betting on Uber or Deliveroo. Our use of the word 'disruption' within the term Disruptive Governance is less celebratory than the prevailing business use. We use it to denote not just winners and losers, though these clearly emerge, but as a process that changes the 'rules of engagement' in political economy. We see Disruptive Governance as a deep-rooted political tactic which sometimes becomes seen as a project in itself; a style and goal which is being both advocated and pursued. More basically, it shrouds a deepened neo-liberalist agenda built upon nationalist political means. This is not full-blown anarchic politics but something a grade or two down the scale. The intent is to create dynamics whose instability will have transformative structural consequences.

Among the architects, and some but not all, supporters of Brexit, there is a strong theme of desiring a radically different type of future, often a revised neo-liberal one where entrepreneurs are unencumbered by EU regulatory restrictions and in a policy regime allowing for the free movement of resources, goods, capital, technologies and services, but not labour. But above all, the advocates envisaged a redefined and drastically slimmed-down State for whom disruption rather than conserving the status quo is a goal. There is no Plan B. Multi-level governance arrangements slowly established over years are portrayed as 'Brussels bureaucrats', an infrastructure that is outdated and if not illegitimate, against 'British' interests. Another, not entirely distinct, variant of Disruptive Governance is a decidedly nationalistic, rather than globalist, perspective that insists on rules and regulations, just as long as they serve particular national, or at any rate nationalist, interests. UKIP (the anti-EU United Kingdom Independence Party), for example, has consistently espoused a nationalist vision for food. This is more conservative, more protection-oriented but emphasising the protection of assets rather than people.

These aspirations are very different from recent forms of conservatism, and by no means confined to the UK or Brexit, but disruption is a common core across new politics. Disruptive Governance can be seen in varied forms in North America with the election of Donald Trump, and the continuing far-right politics dominating the Republican party, the high-pressure re-writing of NAFTA trade arrangements, the weakening of US environmental and food safety protections, and in Latin America with fiscal austerity and the corporatisation of government, not least agri-business, for example in Argentina and Brazil. We see agri-food disruption too in the high-stakes US-China trade disputes, and in EU-Russian trade embargoes and especially now with the Ukrainian crisis. In this sense there are wider and multiple-layered levels of disruption to which a Brexit UK (and Wales) will be further exposed, let alone the important subsequent UK-EU negotiations associated with the post-Brexit settlement and the Northern Ireland Protocol. A common feature of Disruptive Governance is to challenge and unsettle long-standing regulatory and jurisdictional arrangements; and indeed to oppose the application of the 'precautionary principle' and their domestic and established international forms of intermediation, while proposing a new 'innovation principle' in its place (European Risk Forum 2015). The implicit intention is to enhance the appetite and legitimacy of governments and citizens to take risks.

The impact of this new governance could reshape the entire political Settlement of the British Isles, or the form it took in the late twentieth century. Wales was annexed by England in the mid-sixteenth century, and fully integrated when Scotland agreed to merge the crowns in 1707. The UK was formally created in 1801, but altered by Irish partition in 1922. Joining the EEC in 1973 occurred amidst the serious and violent 'troubles' of UK-Irish liaison until the 1997 Good Friday Agreement. Brexit now disrupts these long-evolved, multi-level governance political Settlements and, as we see with reference to Wales, disturbs and dislocates earlier political settlements between the nations of the UK.

Agri-Food as a Key Location for Disruptive Governance

With a formal UK-EU schism still fluid and contested, let alone any post-Brexit trade deals being in place to replace EU membership, Brexit is already proving disruptive. Companies are nervous and facing trading and supply chain obstacles. Low-waged jobs proliferate. Huge investments, such as in the steel and car industries are diminishing. Food companies have had to plan, then suspend, then plan again for supply disruptions; this volatility is likely to be long-running and costly. The food system's just-in-time supply logistics, which deliver food to retailers at short notice, with minimal storage, are underpinned by long-term plans and investments in land, farms, down-stream processing, retail and food service. Throughout the food system, Brexit-shaped decisions are being taken, both because of and in spite of the uncertainties. Investments are variously being relocated or delayed. Currencies are recalibrated, making food imports more expensive whether paid in Euros or US dollars. EU-based Workers were essential to the UK food system but most have moved back to mainland Europe (Duncan 2022). Efforts to recruit from beyond Europe have been largely unsuccessful, while the rhetoric (and hope) is that automation will massively reduce labour requirements. The impact of disruption on food—a fragile but malleable biological product—was initially sidelined by the government. Yet with Covid and now the Ukrainian war additional effects on food supply, national food security is beginning, if reluctantly being raised as a key political issue.[1]

Given the historical centrality for the EC/EU of its agri-food sector, which still accounts for around 40% of the EU's annual budget, and the extent to which the UK has relied on EU member states to supply around

[1] While farmers unions and many commentators have been arguing for some time to develop food self-sufficiency targets for the UK given growing concern over declines in domestic production and growing dependence upon overseas imports, these have only recently begun to be taken seriously by the British government in the light of the Ukrainian war, and wider international concerns over food supply and prices. This is now an issue being actively raised in the upcoming and long-awaited food strategy as of June 2022.

45% of our food supply, Brexit is having a huge impact on the UK agri-food sector. Brexit is hugely important both to farmers and to urban consumers. Enduring with the effects of austerity coupled with a rapid rise of food poverty, entails that the current attempt by the UK government to disrupt many aspects of its food policy framework, is aggravating the already-unacceptable levels of food insecurity, in the UK especially for low-income household, and increasing the economic vulnerability for farming communities. There are also regional dimensions to this food system disruption. One area seriously scrutinised, of course, is the Irish border and the threat to the peace agreement (of 1997) from competing political perspectives. But few have acknowledged the huge flows of food in and out of Ireland. While Ireland's exports, largely of meat and dairy, pass through England—mainly via Holyhead on Anglesey (the second highest UK port for throughput)—less recognition has been given to how the same tonnage of foods, mostly for direct human consumption, flows in the other direction from Britain to Ireland. Scotland actively threatens secession and it, like Ireland and Northern Ireland, has major agri-food flows through England, notably of whisky (McFarlane et al. 2018).

The silence on agri-food matters after the Referendum was deafening. Industry began to get nervous. Civil Society began to organise, conscious of the political void. It was not until 18 months after the Referendum that puffs of smoke began to emerge from the UK government, albeit under duress from the cross-party House of Commons Committee on Exiting the EU. On 21 December 2017, that Committee published 39 sectoral 'impact statements' on the effects of Brexit. One was on agriculture, animal health and food and drink. In January 2018, there was a cross-Whitehall briefing. Those initial sectoral impact statements were vague and slight. The Agri-Food document just summarised the state of the UK food system, failing to spell out how any variants of Brexit might affect it; it failed to identify likely impacts. Then, seemingly stung into action by the rising prospect of a no-deal Brexit (which some in the Government party actively want but few others do), a series of Guidance Notes began to emerge in August 2018. The choice of topics was strange, seemingly randomly chosen pin-points of light in a darkened policy space, but ones which did little to clarify what might happen to the UK's food sector. The public was left almost entirely in the dark.

In February 2018, England's Department for Environment, Food and Rural Affairs (Defra) produced a Consultation Paper optimistically titled 'Health and Harmony', which did not mention food or human health at all, but did chart a post-Brexit vision for agriculture as centred on subsidies for environmental land management (Defra 2018). This turned into the September Agriculture Bill (HM Government 2018), which barely mentioned food but did say useful things about a fairer financial distribution across the food chain. Above all, it prepared farmers for tapering subsidies two years after Brexit. This is now ongoing with a transition period which is phasing out former EU subsidies for farmers based upon 'area-payments' and replacing them with 'public-good' support for a range of environmental management actions. All the signs are that the new environmental management schemes (ELMS) will fall far short of the levels of basic payment income support which they are replacing; further exacerbating the removal of full-time farmers and the further amalgamation of holdings.

In addition, much heralded new trade deals involving food goods with Japan, Australia and New Zealand have been formalised potentially allowing cheaper produced and lower standard foods to be imported. An independent English food strategy was published in July 2021, but its conclusions, especially on the need to improve food security and food diets have been sidelined. A new food white paper is planned but has been delayed as we write. At the same time the devolved nations are attempting to develop their own bespoke food and post-Brexit agricultural policies. As of June 2022, this is still to be enacted. Here we consider one devolved region—the case of Wales—in the context of these unfolding shifts in overall UK policy.

Genetic Technology (Precision Breeding) Bill has been published ...

Wales: A Case Study of Imposed Disruptive Food Governance?

What was Wales to do in the face of the arrival of these disjointed policies and governance? One urgent item was to challenge the simplistic narrative that 'Brexit means Brexit'. Wales's political classes are acutely aware that Brexit is nothing less than a total restructuring of the UK's relationship with its continental neighbours, i.e. far more than the simplification implied by the tautological mantra. The impact on Wales's food and farming system is immense. No-one describes the current state of Welsh agri-food as perfect. It is characterised by a dominant and traditional ruralism, centred on the extensive production of sheep and cattle, and linked closely to the cultural traditions of family farming and the Welsh language. Since devolution in 1998 and with the EU as its reference point (and significant financer through both CAP payments and regional structural funds), the Welsh Government has been developing a mixed strategy based upon part rural development and family farming and part emphasising the promotion of sustainable Welsh food exports (such as Welsh lamb, beef and a variety of cheeses). But this is being undermined by uncertainties regarding post-Brexit Westminster relations and financing. Publication of the Welsh Government's Consultation paper: 'Brexit and our Land: securing the future of Welsh Farming' is still ongoing (details to be published in July 2022), and a revised Welsh Food Strategy delayed. There are no clear guarantees regarding future UK funding for Welsh farming and food after Brexit; only the assumption about a medium-term 'transition period'. The question in Wales as elsewhere is 'transition to what?'.

Table 3.1 provides some key summary statistics on the Welsh Agri-food sector. These are grouped under four headings: people, products, flows and funding. In rural Wales, and in the upland areas of Scotland and Northern England, Brexit significantly disrupted all four of those spheres in interconnected ways. Farms and farming families are likely to continue to go out of business; uncertainty over subsidies and blockages in markets (especially for red meat and fisheries) are likely both to constrain existing modes of production as well as undermining the adoption of more sustainable agro-ecological farming practices. Welsh farmers are highly dependent on complex interconnected markets, not only with continental EU countries but also with Ireland, and Brexit could well disrupt those connections, and territorial quality standards (such as Protected Area status for Welsh Beef and Lamb) challenged and potentially devalued. Because of Russia's invasion of Ukraine, we have rapidly rising prices for fuel, energy and fertilisers. While many retail prices have risen, the supermarkets are competing with each other on keeping retail prices low. So, while there should be incentives to encourage UK farmers to increase their production, many are cutting back because they do not expect farm-gate prices will increase sufficiently to cover their higher costs.

The future of funding mechanisms in Wales, not just for agriculture, but also for regional and rural development, is now unpredictable (we return to this, discussing Whitehall thinking, later). Welsh agri-food businesses are highly integrated with English processors and markets. Despite significant proportions of EU rural and regional development funding being invested in Welsh food businesses, there are real dangers of further spatial concentration of the processing and retailing sectors, as the 'friction of distance' increases. If processors relocate away from Wales, there is a risk that Welsh farmers would become less competitive, as suppliers of raw materials. Those multiple and interconnected vulnerabilities are not of course unique to Wales. Similar challenges afflict much of Northern and Western England and most of Scotland. Above all, for Wales, a small country with limited devolved powers, Brexit disrupts established ways in which it can operate within the multi-level governance frameworks that apply in the UK and the EU.

The agronomic data in Table 3.1 provide a sketch of what, in the Welsh context, is now being disrupted. It is actually extensive—finance (much farming exists only due to Common Agricultural Policy [CAP]-funding; and it is likely that post-Brexit farming support will be much reduced); people (labour shortages have resulted in food processing and veterinary), and work. Woven into this are the ecologies and culture which matter so much in Wales, not just for tourism. Brexit has offered a myth that 'borders' can be re-imposed, when Wales needs to export and buy in food via and from England, as does Northern Ireland. Birds which

Table 3.1 Wales' agri-food sector: some key facts and flows

People

- Total farmers: 39,903 (2016), full time 18,564; part-time 21,339
- Total number of workers: 12,000 (2017), full time workers: 4000 (2016)
- 1038 Vets in Wales, of whom 25% are EU nationals
- Farms: Large number of farms with low turnover: 21,200 farms (86%) produce 29% of agricultural output but use 57% of farmed land
- There are 500 farms with a high turnover that contribute 33% of output but only use 8% of the farmed area
- 78% of farmland is classified as 'disadvantaged'
- Total income from farming per hectare: England £331, Scotland £116, Northern Ireland £246, Wales £102

Products

- Sheep and lambs 10 million (2017)
- Cattle and calves 1.1 million (2017)
- 180,000 Farmed hectares (excluding common land)
- 64% permanent grassland, 15% rough grazing, 5% arable, other grassland 9%, Land down to Horticulture: 0.1%, Other land 7%
- 306,000 hectares of forest: 49% broadleaf: 51% commercial conifers (15% of land area) 2000 jobs
- Fisheries lands 25,300 tonnes of sea fish per year worth £38.8 million, 851 employed

Agri-food flows (pre-Brexit, 2019)

- 96% of sheep meat exports to EU27: a third of Welsh lamb 63,000 tonnes of sheep meat per year
- 47,700 tonnes of beef produced a year some exports to EU (15–17%) some 90% of beef exports to EU. Republic of Ireland represents 70% of imports to UK (including Wales)
- 1.8 billion litres of Milk per year from 1700 dairy farms (50% processed in Wales, mainly for cheese) 50% of milk is exported to UK (some EU)
- There are only 389 (1%) horticultural farms in Wales; 54% vegetables, 43% orchards
- Tourism accounts annually for £2870 million creating over 100,000 jobs
- Agriculture generated £385 million
- Food and drink supply chain employs 240,000 people in 2016 in 27,575 business units, turnover 91.1 billion, Gross Value Added of 4.5 billion
- Rising food exports from Wales: 113% growth in value 1999–2013, mainly (90%) to EU
- 18% of all employment in food, 22% of all business units
- 85% of food businesses are micro, employing fewer than 10 people
- Some Welsh agri-food products have EU PGI status

(continued)

Table 3.1 (continued)

Funding (pre-Brexit, 2019) • Wales received £760 million per annum from the EU • The annual overall 'Barnett' Welsh budget is £15.5 billion; 6 billion goes on health • EU funding (for agriculture and regional development) is less than 10% of Welsh health budget • Wales receives more per capita for these funding mechanisms than English regions • Support payments for agriculture (Pillars 1 and 2) is the largest amount in Wales of EU funding amounting to 2.8 billion between 2014 and 2020 • Welsh government also 'match funds' EU funds (50% East Wales, 32.6% Rural Development (Pillar 2) and 30.9% West Wales and the Valleys • The average Welsh (dairy) farmer receives £32,300 per annum and each sheep farmer an average of £19,300 per annum

Source Welsh Agricultural Statistics (2017), published by the Welsh Government

winter on the farmland or shores of Wales often come from Scandinavia. Those which spend the summer in Wales (e.g. terns, swallows) will have flown from Africa or further. Biological isolation is entirely unrealistic; avian 'flu' for example readily crosses borders and even oceans. The cultural messages that entice people to eat junk foods and unhealthy diets are bought by giant corporations, not by the EU. In fact, the EU is one of the few power sources which can help consumers resist the onslaught of US junk food. Wales's horticulture is tiny. Without the EU, Wales's capacity to transform its obesogenic environment will be weakened, and the same is true for all parts of the UK. More costly fresh foods—already hit by sterling's fall—will have a deleterious impact on Welsh diet-related ill health, already wanting.

Like others, Welsh food processors and retailers have been stockpiling. While farmers might continue their daily and annual practices, hoping (and/or praying) for the best, food interests off the land face more immediate and urgent concerns. Labour flows have become uncertain. The UK-wide Seasonal Agricultural Workers Scheme was closed before the Referendum on the assumption that EU migrants' right to relocate made it obsolete. Immediately after the Referendum, the Senedd of the Wales Assembly began hearings (we appeared before them) but the general view from Cardiff was that the politics centred on London and Brussels. Even within the dominant Labour Party, there was and remains some disarray. The limits of devolution of powers were being exposed.

London First Abandons Then Reclaims Leadership: But Will It Work?

Nowhere have the limitations of the Wales Settlement become more apparent than over the funding on which most of Wales's farmers depend. The Conservative's Coalition Government with the Liberal Democrats in 2010 closed the UK-wide Labour-led *Food 2030* strategy developed after the 2007–2008 global commodity crisis which had exposed the vulnerabilities of agri-food systems. The Conservatives re-imposed walls between policies on food, agriculture and environment. Two Plans on Food and Environment were drafted but dropped. In 2016, approaching the Referendum, there was nothing. The priority had been to cut the Department for Environment, Food and Rural Affairs (Defra). Only when Michael Gove was appointed to Defra did any thinking emerge. Defra staffing doubled. Bills on Agriculture and Environment were drafted. The former severed food from agriculture. The purpose of farming was now to be environmental land management. Food, the single biggest driver of ecosystems damage globally, was apparently unimportant. Not until summer 2018, did the Guidance Notes preparing for a hard Brexit emerge. These were short term, including hiring ships to by-pass expected chaos at Dover (but not Holyhead, note, the second heaviest-used port). Then belatedly, from London—but without including Wales or Scotland in the preparation—Defra in London trickled out (via a letter to the chair of a Select Committee) that a new Food Strategy process would begin in 2019 after the March Brexit date! Then the Strategy was published in 2021 and subsequently 'shelved'.

For example, when the Food Research Collaboration (FRC) published its Feeding Britain report in July 2018, drawing on the Government's own figures on UK reliance on food supplies from other EU Member States, its concerns were dismissed by Defra; then just 48 h later, another Secretary of State admitted there was contingency planning underway, and a few weeks later, the astonishing leak that the Transport Secretary of State was planning to hire a flotilla of ships to bring in food supplies! This is the sixth richest economy on the planet. No wonder food industries were alarmed.

The Disruptive Governance Paradox: Whose Version of Control and Regulation?

Amidst these multiple disruptions and discontinuities, we also detect a Disruptive Governance paradox: hope amidst the turmoil. This epitomises paradigm shifts. On the one hand, Brexiteers are creating uncertainty as a new norm, a state whose duration and territorial extent nobody knows. What else is it when Ministers protest their commitment to retaining 'high' food standards, while also trying to pre-negotiate trade deals with countries, such as the USA, which have far lower food standards (see Lang et al. 2017)? On the other hand, severance from the EU could offer new opportunities to all, or parts, of the UK to forge a radically improved direction for our agri-food systems. Simultaneously, knowing the demise of EU CAP farm subsidy system, farmer and environmental interests compete for new subsidy and support-payment systems, while food consumer groups and some trade unions press for higher welfare and food quality standards.

We see already significant public counterforces arguing over future agri-food governance, which could eventually undermine its plausibility. The paradox of Disruptive Governance is that it *could lead to opposite effects from those originally intended*: more public and institutional regulation rather than less, more stress for food producers and consumers concerned about weakened food standards. This paradox is currently mostly pitched as UK outwards rather than UK inwards, towards US or Australian food deals rather than London-Cardiff-Edinburgh negotiations. But now that HMG is de-regulating gene edited crops and animals, we know the direction in which they intend to travel.

What is clear is that, in the key UK 'Europeanised' policy areas of environment, agri-food and trade, Brexit has created an urgent need in Westminster and the devolved regions to develop new coordinated institutional capabilities and to implement new policies. A new politics of agri-food is emerging with dramatically different visions: Atlanticist, globalist, neo-imperial (let others feed us) and neo-European. Trade deals are being progressed by Westminster politicians with the USA, the former colonies (such as Australia and New Zealand)—almost anywhere except Wales or Scotland. Yet clarity is urgently needed about new trans-boundary and UK cross-border cooperation for the devolved regions of the UK as well as for the very divided and unequal English regions. If Brexiteers were serious about allowing people to 'take back control', they would be proposing enhanced devolved powers and consulting on their proposals. But HMG has introduced the (Internal Market) Bill that gives Westminster power to impose a unified market within the UK (except N Ireland).

Such a prospect sits rather uncomfortably, of course, with the aspirations of some of the most enthusiastic Brexiteers who advocated leaving the EU as a step towards dismantling regulations, which they portray as burdensome, to liberate entrepreneurship, commerce and trade. This is why the issues FRC's Brexit briefing papers have raised, concerning the risk from Brexit to UK food security, including the threats of declining standards, reduced access to decent diets, lack of affordability and diminished animal welfare, have resonated with the British public and media. The UK public did not vote for a 'race to the bottom' in trading and food standards, nor continued declines in rural and farm businesses, nor reductions in environmental protection (Cracknell and Rose 2021).

A socially just and environment-protecting Brexit would necessitate smarter re-regulation not de-regulation. Rural Wales, which currently relies on animal exports to the EU (e.g. lamb for French school meals), needs to know whether the promises from London to maintain high standards will really be kept, when faced with demands from the US Secretary of State for Commerce, Wilbur Ross, that the UK must break with the EU's high food and environment standards.

The New Containment and New Dynamics of Devolution

Wales's options have been diminished by Brexit; and Wales knows it. Not being offered an opportunity to vote on any final deal, Wales must make the best of it. But it is unlikely to be able to do so, unless a new type of order replaces Disruptive Governance. Plans and structures are urgently needed but are not forthcoming. The current machinery for coordinating policies and practices across the UK and especially between the devolved governments (such as the Joint Ministerial Committees) is 'not fit for purpose'.

One might have hoped that Whitehall would by now have developed new bespoke coordinating structures and frameworks, which are transparent and accountable to both legislatures and citizens across all devolved regions. That has not happened. There is a distinct lack of trust between the devolved authorities and Westminster, and reduced levels of communication between different executives. It is difficult to avoid the conclusion that Brexit is fuelling constitutional fragmentation (as well as a variety of English nationalism) just when the UK ought to be displaying the pragmatic, clear and integrated thinking, for which it is used to be celebrated (Burns et al. 2018). UK and EU public sector funding has been far the largest provider of resources enabling the devolved administration in Wales over the twenty years since it was established. Levels of funding have been governed by the so-called 'Barnett formula'.[2] Rural policies have also historically relied

[2] The Barnett formula allocates UK government funding to the devolved regions as an annual block grant. The annual Welsh budget for 2018–2019 was £15.5 billion. This is allocated with respect to population and some assessment of needs. It is largely up to the Welsh Government how it allocates its block grant. EU funding alternatively, for both regional development and agriculture is a ring-fenced component. Because this is allocated on a needs-basis, and because Wales has qualified as the most disadvantaged region in the UK for these funds it has historically received higher proportions of these funds (see Table 3.1). Measured on a *per capita* basis, for instance, structural regional development funds have been more than double those of any other part of the UK. Again, with agricultural EU funding, given Wales' Less favoured Area status it has received higher proportions of these funds as subsidies. Once these preferential allocation systems for Wales come to an end with Brexit, a key question is how the UK government will allocate public funds.

heavily on ring-fenced EU funding associated with Regional Development ('Convergence funding') and CAP programmes. Those latter EU programmes (representing annually something like six per cent of all annual public funding in Wales) were based upon a 'needs-based' set of priorities built by reference to the EU's goal of reducing regional economic disparities. It has been used to build regional capacities and support farming families to deliver a combination of public and market goods. They have provided important ring-fenced EU funding for Wales in its relatively stable first twenty years of devolution.

Wales's first decade of devolved powers occurred under Labour governments and the second decade has mostly been under Conservative-dominated coalitions in Westminster; and then a conservative government (with a large majority from 2019). Yet across those apparent politically different government periods, there was a general consensus between EU, Westminster and Cardiff governments that the distinctive features of Wales' post-industrial and sparsely populated rural landscape and economy required considerable state investment. That consensus is now undermined by Brexit. Commitments to reducing regional economic disparities in the UK are now vague and unreliable, despite the considerable 'levelling-up' rhetoric aimed mainly at (post-2019 election) northern-English conservative constituencies.

It remains to be seen whether the intra-UK rupture is deep or permanent. Many in the politically dominant Wales Left circles were taken aback by the overall UK Referendum result. The 52.5% vote to leave mirrored 53.8% in England but differed from 44.2% in Northern Ireland and only 38% in Scotland. Ironically, the post-industrial Welsh valleys and the large and dispersed rural areas had both been the major recipients of EU investments over the past 20 years. Like Cornwall, the largest per capita UK recipient of EU funds, Wales apparently voted to bite the hand that helped to feed it. In fact, EU funds were more like compensation payments than core funding for Wales. The resources often came from EU regional funds, set up to act as counterweight to the centralisation and concentration trends unleashed by the completion of the Single Market, initially championed but later denounced by Mrs. Thatcher.

Uncomfortably, the dominant Welsh Labour administration, which was and is strongly pro-Remain, found it difficult to deny that this was a major act of collective 'self-harm' by the poorest and most vulnerable Welsh communities. But, as other analysts have argued, the Referendum coalesced many sentiments—deserved and undeserved—into one stark yes/no vote. Much of the societal marginalisation from de-industrialisation and fiscal austerity was the responsibility of policy-makers in London rather than in Brussels, of concentration and cost reductions for competitiveness in larger markets rather than just those of Europe. The inequalities in, and neglect of, the regions emerged despite, not because of, the efforts of the EU Committee of the Regions. Whatever the palliative benefits to Wales from the EU's continued support for Wales's agri-food and environmental economy, those benefits were overshadowed by the wider political rhetoric stressing immigration and 'red-tape'. While the European Commission is often derided in the UK as a 'bloated bureaucracy', few acknowledge that there are in fact more civil servants in Edinburgh running Scotland than there are in the entire European Commission. Agri-food and eco-economy concerns were ignored or discounted by (what Adorno called) an apolitical 'context of delusion', exhibiting 'false consciousness', (which in German he termed *Verblendungzusammenhang*).

The problematic status quo in food governance has been reliant upon Westminster; and the current state of devolved regional politics has recently been described as a case of a policy 'without map or compass' (Douglas-Scott 2018). England continues to be highly centralised, and arguably has become more centralised since 2010. The Regional Development Authorities and Regional Government Offices created after 1997 by Labour have been replaced by smaller, under-funded and weaker Local Enterprise Partnerships. A competition over which authorities can or cannot have a mayor gives semblance to a concern for the regions. A short-lived burst of enthusiasm for a 'Northern Powerhouse', championed in 2010 by Chancellor of the Exchequer George Osborne was an attempt at big thinking but it was ineffective and ignored in Whitehall and the rest of England.

In Wales, England and Scotland the relationship of agri-food and rural policy to the regional development agendas has been marginalised by the dominant issue of Brexit. If one positive long-term development were to emerge from Brexit, it could well be a revitalised regionalisation and de-centralisation of the UK state. The rigidity of Whitehall and its reluctance to re-distribute to the devolved administrations any powers, which might be repatriated from Brussels, are not faults that can be attributed to the EU. Other EU member states are far more devolved internally—Germany was deliberately constructed to have relatively strong regions and relatively weak central government by the victorious Allies after 1945. Meanwhile Scotland, which voted strongly to remain, is enthusiastically manoeuvring for greater independence and contesting any suggestions of further centralisation to Westminster. The Scottish National party is trying to exploit the Disruptive Governance paradox more than Wales, and argues that competencies repatriated from Brussels should be transferred to Edinburgh not to London. Scotland is not covered by Defra's Agriculture Bill that was tabled in Westminster in October 2018.

All the while, it must be stressed, the Northern Ireland Assembly currently remains suspended, yet on the border question Northern Irish politicians, especially the DUP MPs, receive the highest (and disproportionate) political attention of any section of the UK, even more than the City of London with all its financial clout. Northern Ireland perhaps best illustrates the risks of the new Disruptive Governance. The Brexit settlement which left the EU Northern Ireland border intact as part of the need to continue to protect the 20-year peace agreement for Northern Ireland but necessitated new border restrictions in the North Sea between mainland Britain and Northern Ireland is creating significant food supply problems. Currently the protocol which was agreed by the Johnson government as part of the EU withdrawal agreement in 2021 is now not accepted by the Ulster Unionist party who are refusing to form the devolved government in Northern Ireland (as of May 2022). Meanwhile the UK government want modifications to the protocol, not least around the food trade. This is continuing disruptive politics between the EU and UK, as well as with the devolved territories.

Unlike Scotland (and Northern Ireland for obvious reasons), Wales has nominally agreed new devolved arrangements with Westminster about the scope of 'devolved competences', though this is still contested as the effects of Brexit unfolds in ways that are exceptionally difficult to predict. For many parts of the devolved administrations Westminster seems ironically a far more (politically) remote and less trustworthy negotiating partner than Brussels ever had. Several observers and committees have called for greater cooperation among civil servants and departments across the UK's governments, while also wanting to see greater autonomy for the devolved administrations. Inevitably Brexit has provoked calls for longer-term and more root-and-branch reforms to coordinating mechanisms across, at least the 153 areas where EU law currently intersects with devolved competences. 41 of those are closely related to the environment, agriculture and fisheries.

It should be noted that this unfolding and contested process of disruptive governance does not have its roots in the political arenas of agri-food or environment nor regional economic development. Yet because these policy fields have been, and were by far, the most Europeanised, they are the policy domains (together with trade) over which the disruptive UK government now seeks most control. Consequently, they are becoming a key policy fulcrum around which the new and more disrupted domestic UK devolved relations are most likely to be recast. As we see temporal disruptive governance is also leading to territorial disruptions, which indeed in turn could be long-standing.

Conclusions: Embedding or Tackling Disunity (A Postscript from 2022 and Beyond)

From 2016 to 2022 in the UK we saw the emergence of Disruptive Governance and its presence in food as opening new fissures while threatening existing political settlements. This juncture is as profound a transformational moment for agri-food and rural development policy and its long-term governance in the UK as was 1947. UK food systems have during this more recent period become far more destabilised, just at a time when scientific opinion calls for ordered and radical systemic change in European food systems, and indeed in countries like Wales (see SAM 2020; Sanderson-Bellamy and Marsden 2020). Politically, the dynamic and dialectical interplay between conservative right-wing, economic libertarian culture on the one hand, and the more socially liberal political cultures (as experienced in Scotland and Wales) on the other hand, are now highly relevant in seeing how the more entrenched systems of disruptive governance are playing out in various part of the UK, not least with regard to agri-food. In this sense agri-food has become more politicised since the onset of the Brexit process.

The post-Brexit politically dominant Westminster UK government is hoping to seize the opportunities that Brexit has provided to further centralise powers as well as policy approaches in the areas of food and agriculture, and indeed more broadly environmental policy. In the past most of these policy areas were located under the more socially liberal EU policy frameworks of both its regional, environmental and common agricultural policies and programmes. Now the UK Westminster government is applying far more economically liberal approaches associated with:

- Further de-regulation of food laws and standards, such as the opening up of GM and GE technologies and their commercialisation;
- Creating new 'free-trade' agreements in other Non-EU parts of the world; namely Australia and New Zealand, Japan and the USA; while also allowing inspection light food EU imports into the UK while abiding by stricter food EU export rules;
- Developing more limited new agricultural environmental schemes which do not support farm production (as seen as a private/market good), but focusses upon 'public funds for public environmental goods';
- Resisting attempts to proactively tackle growing food poverty and food security issues through separating food policy from welfare policy; as a result, food inflation is creating rises in food poverty, with 7.3 million estimated to be food insecure in the UK (May 2022), up from 4.7 million in January 2022;
- Reducing the devolved powers of the devolved regions of Scotland, Northern Ireland and Wales by implementing a revised UK-wide internal market act which constrains devolved powers on the basis of prioritising 'level-playing fields' in competition policy in for food and farming businesses.

- Overall trade with the EU collapsed in the wake of 2021's exit, with leaving the EU single market and customs union reducing trade by 15%; and
- The employment of EU nationals in the hospitality and production sectors has declined dramatically with emerging labour shortages, falling by 171,000 (6%) between June 2019 and June 2021; there was a 25% fall in EU nationals working in the hospitality sector.

These policy initiatives and impacts are being implemented on top of severe post-Brexit (and covid) disruption in supply chains as a result of significant restrictions in the import and export of food goods both between the EU and the UK, but also between Northern Ireland and Great Britain (see below). Also severe labour shortages are occurring not least in the seasonal horticulture and veterinary sectors as many European workers and professionals return home. In addition to this the current government's food 'policy' is strongly to rely upon the concentrated and corporate-controlled retailer, catering and food processing sectors to deliver national food security goals. 'Free-markets' lower taxes, balanced budgets and cheap disposable labour are the mantra of the Westminster government, with a continuous reliance upon the concentrated corporate retail sector to deliver 'cheaper' food options to low-income consumers.

The tactic of de-stabilisation could lead to a new paradigm of sustainable rural and agri-food development. At the same time, there are already signs such opportunities could be lost. It could set different, potential-collaborating interest groups against each other, and lead to a nominally eco-focussed system propped up by a wave of imported food products from far beyond the EU. For Wales this is vital. Wales's rural heartland is (currently) about meat and dairy from relatively small, often upland farms. The macro politics will continue, potentially embedding disruptive governance. For Wales—like Scotland and perhaps Northern Ireland-—this is the time to develop a new and collective vision of sustainable rural and regional development in collaboration with other UK regions and territories. England's lack of regional governance becomes a block. But Wales, for example, already has its own ambitious, inspiring Future Generations Act and Environment Act, with its devolved statutory obligations. These provide a statutory framework that no other part of the UK has matched.

While Scotland is preparing its Good Food Nation Act, Wales is already well-placed to use these ambitious Acts as the basis to renegotiate intra-UK links. Unlike England, where the Agriculture Bill barely mentions food and where an Environment Bill has not even been tabled, Wales can build on its commitment to develop an agri-food economy that takes future generations' needs and interests as its core focus. Wales has an opportunity both to participate in new UK-wide frameworks after the Brexit process, and to develop its own distinctive vision and strategy for agri-food, rural and regional economy based upon a strong ecological-economic approach. As so often with the unevenness of power within the UK, the problem is England, which cannot be reduced to the whims of London or Southeast centrism. In all these dynamics, agri-food systems will be important simply because it is where the public eats politics, even if they sometimes do not know it.

This chapter has analytically focussed upon the evolution of disruptive governance in the UK and Wales following the Brexit referendum of 2016. It demonstrates that disruptive governance is perhaps a symptom of deeper structural forces associated with the emerging failures of globalised and financialised capitalism to sustain its carbonised growth model especially since the financial crash of 2008 and what followed; and indeed the even deeper global resource and climate crisis which is leading to rising nationalist security of resources (as now depicted in Russia's invasion of Ukraine and China's new mercantilist route of developing its 'Belt road' initiative). For the neo-liberal right in order to gain some level of political legitimacy out of these structural changes it has resorted to forms of disruptive governance based upon distorted views of nationalism and protectionism. The EU becomes one of its scapegoats, as do increasing levels of migration and public sector inefficiencies. What becomes clear in all this is a growing level of ecological, economic and social vulnerability for many communities in the UK, not least in Wales. Moreover, the early 2000s reliance upon relatively cheap imports from China and elsewhere, and food imports from around the world, can no longer rely upon a cheap 'just-in-time' delivery and logistics system which continually provided wider choices for consumers.

In this sense the food system in the UK, as indeed in most places currently, is caught up in a geo-political conundrum not entirely of its own making and indeed subject to what might be termed external nexus systems effects—energy, transport, mobility and wider resource scarcity perturbations. Needless to say this makes sustainable transitions or transformations all that more imperative, but at the same time, more and more difficult in the face of the immediacy of political reactions to the multiple crises, and the capacity for the conventional and intensive food system to continue to 'lock-in' asymmetrical and dysfunctional market and power mechanisms. Until the current government accepts that sustainable food security cannot simply be left 'to the market' it is difficult to envisage how the UK emerges out of the Danteian depths of this conundrum.

References

Burns, C., N. Carter, R. Cowell, P. Eckersley, F. Farstad, V. Gravey, A. Jordan, B. Moore, and C. Reid. 2018. *Environmental Policy in a Devolved United Kingdom: Challenges and Opportunities after Brexit*. Swindon: ESRC Briefing Paper.

Cracknell, J. and E. Rose, 2021. *A View from the 'Red Wall: Swing Voters' Attitudes to Regulation and Food Standards*. Retrieved from https://redwall.unchecked.uk/wp-content/uploads/sites/2/2020/12/A-view-from-the-Red-Wall.pdf

Defra. 2018. *Health and Harmony: The Future for Food, Farming and the Environment in a Green Brexit*. London: Department of Environment, Food and Rural Affairs.

Douglas-Scott, A. 2018. Without Map or Compass. *London Review of Books*, May 2–5.

Duncan, G. 2022. UK Farms Missing up to 75% of Seasonal Workers Needed for Harvest. *The Grocer*, May 19. Retrieved from https://www.thegrocer.co.uk/fruit-and-veg/uk-farms-missing-up-to-75-of-seasonal-workers-needed-for-harvest/667635.article?utm

European Risk Forum. 2015. *Better Framework for Innovation: Fuelling EU Policies with an Innovation Principle*. Retrieved from https://eriforum.eu/uploads/2/5/7/1/25710097/businesseurope-erf-ert_innovation_principle_joint_statement.pdf; see also: Ridley, M. 2017. *Britain Should Adopt the Innovation Principle*. Retrieved from https://www.rationaloptimist.com/blog/the-innovation-principle

HM Government. 2018. *Agriculture Bill 2017–2019*. London: Defra.

Lang, T., E. Millstone, and T. Marsden. 2017. *A Food Brexit: Time to Get Real-A Brexit Briefing*. Brighton: SPRU University of Sussex, Cardiff University and City University.

McFarlane, G., T. Lewis, and T. Lang. 2018. *Food Brexit and Northern Ireland: Critical Issues*. London: Chartered Institute of Environment Health and Food Research Collaboration.

SAM. 2020. *Towards a Sustainable Food System: Moving from Food as a Commodity to Food as More of a Common Good*. Brussels: Independent Expert Report, European Commission, Directorate-General for Research and Innovation, Group of Chief Scientific Advisors.

Sanderson-Bellamy, A., and T. Marsden. 2020. *A Welsh Food System Fit for Future Generations*. A Report by the Sustainable Places Research Institute at Cardiff University, commissioned by WWF Cymru. Cardiff.

Schumpeter, J. 1942. *Capitalism, Socialism and Democracy*. London: Routledge.

Welsh Agricultural Statistics. 2017. *Welsh Government Statistics*. Cardiff: Wales, UK.

4

Back to the Past: Authoritarian Populism, Disruptive Governance and Policy Dismantling in Rural Brazil

Ricardo S. Borsatto, André de Camargo Macedo, Wolney Felippe Antunes Junior, and Vanilde Ferreira Souza-Esquerdo

Introduction

Populism is not something new, but the debates on populism gained momentum with the rise of right-wing populist leaders both in the Global North and South, like Donald Trump in the United States, Jair Bolsonaro in Brazil, Viktor Órban in Hungary, Narendra Modi in India, Boris Johnson in the United Kingdom, among others. Despite the growing recent literature on populism, the differences between right-wing populist governments in the Global North and the Global South have only been minimally explored. This tide of right-wing populist

R. S. Borsatto (✉)
Federal University of São Carlos (UFSCar), São Carlos, Brazil
e-mail: ricardo.borsatto@ufscar.br

A. de Camargo Macedo · W. F. Antunes Junior · V. F. Souza-Esquerdo
State University of Campinas (Unicamp), Campinas, Brazil
e-mail: vanilde@unicamp.br

A. A. R. Ioris and B. Mançano Fernandes (eds.), *Agriculture, Environment and Development*, https://doi.org/10.1007/978-3-031-10264-6_4

governments worldwide has roots in the severe and combined financial, fiscal, food and fuel crisis that disrupted the world markets in 2007–8. In response to the crisis, many national governments adopted an austerity posture, introducing severe public sector financial cuts in their welfare policies, directly affecting the lower socioeconomic classes that were already suffering from the price increase of fuel and food. With the inability of both traditional right- and left-wing governments to offer responses to the crisis that did not penalise their most vulnerable populations, right-wing authoritarian populist politicians found an environment conducive to their rise.

Differences between right-wing populist governments in the South and North are particularly striking in food and rural policies. Evidence shows that right-wing populist governments in Global North adopt a nationalist and protectionist discourse promising to defend the farmers' interests, which are threatened by neoliberal policies (Mamonova and Franquesa 2020). Conversely, the opposite is the rule in the Global South. Right-wing populist governments have adopted agri-food and rural policies that extol the power of the free market and aligned with the neoliberal agenda, promoting deregulations and dismantling rural development policies. The 2020 peasant protests against pro-market agricultural policies by the government of Narendra Modi in India is perhaps the most emblematic reaction to this process (Singh et al. 2021).

In this text, we intend to analyse the dismantlement of two Brazilian policies committed to a more socially engaged path of rural development, intending to demonstrate that the authoritarian populism in Brazil is disrupting earlier socioenvironmental political assumptions and supporting a predatory agro-extractivism model of rural development, that recalls the time when Brazil was a mere Portuguese colony. We use the concept of disruptive governance to explain the attacks made by the populist government of Jair Bolsonaro against policies aimed at a more social-environmentally friendly model of rural development. Since taking power in 2018, the Bolsonaro government have used a populist discourse advocating that this rupture is desirable and good for everyone and implemented a series of actions to dismantle previous rural development policies (Sabourin et al. 2020a). According to Marsden et al. (2020a), disruptive governance is an intentional strategy to deregulate established structures and processes that were originally put in place to protect the public interest. Under the disruptive governance promoted by authoritarian populist governments, rational policy options and developments are neglected, and private interest bodies become more influential inside the government.

Given this scenario, we argue that the Bolsonaro administration, through a disruptive governance regime, represents an intensified dismantling of public policies aimed at a more socially and environmentally friendly rural development, intending to strengthen and consolidate an agribusiness model marked by predatory extractivism. We explore two cases of dismantling of public policies to support our arguments that the alliances of right-wing populist governments in the Global South are with large landowners and agribusiness transnational corporations, to the detriment of family and peasant farming.

Brief Context of Rural Brazil

Some considerations and a brief review of the country's political events are first necessary to inform our discussion of the populism employed by Jair Bolsonaro[1] and its impacts on rural areas, especially for family and peasant farming. In Brazil, as in most countries of the Global South with a history of colonization, family and peasant farming has historically been on the margins of development processes. Relegated to marginal spaces between large colonial estates, family farming history is characterised by miscegenation between peoples, permanent instability and constant dynamics of internal migration (Wanderley 1996). Significantly, the Brazilian State first developed public policies aimed at family and peasant farming only in the 1990s. Based on an agricultural credit programme specific to this sector, policies aimed at family farmers in Brazil began, which experienced, between 2003 and 2016, a

[1] Elected in 2018 for the post of president, Jair Bolsonaro is a former military man whose political trajectory took place, mostly, in the Legislative Power, holding the post of federal deputy (congressman) for almost thirty years. With a populist, authoritarian and conservative profile, Bolsonaro defends the legacy of the military dictatorship that took place in Brazil (1964–1985) and, in his government program, allied himself with the neoliberal agenda.

promising period of investments and innovations (Grisa and Schneider 2014; Medina et al. 2021).

From 2003 to 2016, a centre-left coalition led by the Workers' Party (PT—*Partido dos Trabalhadores*) was in charge of the Brazilian government. This period is divided between the terms of Luiz Inácio Lula da Silva (2003–2010) and Dilma Vana Rousseff (2011–2016). Although their social and political base included rural social movements and other family and peasant farming supporters, the policies aimed at this sector differed slightly during these two administrations. Lula da Silva's government represented a moment towards innovation and expansion, in which new forms of governance were experimented with, strengthening the exchange of ideas between government, academics and social movements (Abers et al. 2014; Grisa 2010; Grisa and Schneider 2014). Dilma Rousseff, whose profile was more technical and distant from social movements, embraced a less innovative agenda and focused more on the volume and delivery of actions (Lotta and Favareto 2016; Pires and Gomide 2016). In this period, important policies aimed at family and peasant farming emerged. For example, the Food Acquisition Programme (PAA—*Programa de Aquisição de Alimentos*) used public funds to purchase food directly from family farmers (Borsatto et al. 2021). Other programmes aimed to promote sustainable territorial development by financing projects to support activities of family farmers in their own territories (Antunes Jr et al. 2021).

The governments of Lula da Silva and Dilma Rousseff make up the so-called 'pink tide' of Latin America at the beginning of the century, in which different countries in this region, such as Argentina, Bolivia, Venezuela, Uruguay and Ecuador, experienced progressive and left-wing governments. Although the rise of these governments has been supported by social movements of peasants, family farmers and indigenous communities, they have not broken with the model of agribusiness and mineral extraction. Furthermore, it was a period in Latin America in which agribusiness and mining sectors expanded and accumulated power, mainly driven by the growth in world demand for primary commodities (Vergara-Camus and Kay 2017). Having made this consideration, it is still possible to state that this was the unique period in

Brazilian history, in which the State recognised the importance of family farming and implemented a comprehensive agenda of policies for this segment.

In 2016, with the removal of President Dilma Rousseff—in a hybrid, parliamentary coup[2]—the new President, Michel Elias Temer, implemented a neoliberal agenda, breaking the dialogue with rural social movements and removing the main ministry that supported family and peasant farms—the Ministry of Agrarian Development (MDA—*Ministério de Desenvolvimento Agrário*). The process of dismantling public policies aimed at family farmers gained momentum (Sabourin et al. 2020a). However, the worst was yet to come. In the 2018 presidential elections, the far-right candidate Jair Bolsonaro emerged victorious. In one of the first acts of his administration, Jair Bolsonaro eliminated the main instrument of governance for food policies in Brazil—the National Food and Nutrition Security Council (CONSEA—*Conselho Nacional de Segurança Alimentar e Nutricional*)—which was also responsible for public procurement programmes for family and peasant farming. Authoritarian and populist, the government of Jair Bolsonaro uses the so-called disruptive governance to challenge institutions constantly, deregulate historically achieved and established laws, norms, and agreements, and dismantle the list of policies aimed at family farming.

On the other hand, to paraphrase the words of the then minister of the environment, they '*passam a boiada*', which metaphorically means to open the gate for a stampede of actions that directly benefit large landowners and agribusiness corporations. Examples include making legislation to combat deforestation more flexible; deregulating the control of mineral extraction activities; reclassifying and liberating numerous active pesticide components (many that are banned in the European Union); reducing inspection services, and relaxing fines for environmental crimes; in addition to the complete abandonment of the

[2] In August 2016, then-President Dilma Rousseff was ousted from her post through an articulation between right-wing parliamentarians—largely supported by the hegemonic media, as well as by industrial and agribusiness sectors—, who accused her of committing a crime of responsibility. However, jurists, academics, leaders and politicians demonstrate that this accusation only worked as a parliamentary manoeuvre to break with Dilma's mandate.

policy of rural settlements of agrarian reform, and of demarcation of indigenous lands (Pereira et al. 2021b; Vale et al. 2021).

Populism and Disruptive Governance

Many scholars argue that a political consequence of the 2007–8 financial crisis is the growth of more unstable and disruptive forms of governance (Marsden et al. 2020b). The concept of disruptive governance is still under construction and was coined by Terry Marsden when analysing the gradual decline of the once stable UK food and rural regulation model in the context of Brexit. The concept of disruptive governance is increasingly used to recognise that some populist political activists, mostly but not all on the neoliberal right, see disruption as a desirable and positive thing (Marsden et al. 2020b). Disruptive governance, thus, is often defended as an end in itself. However, this is almost always a misrepresentation. Most frequently, disruptive governance is intentionally pursued, with the aim of shifting the governance system to a radically different type of future (Marsden et al. 2020b). In many cases, it is an expression of radically neoliberal strategies, which attempt to free entrepreneurs from regulatory restrictions so that they can enjoy an unrestricted opportunity to manage resources, goods, capital and services. Another not entirely distinct variant of disruptive governance is a decidedly nationalist rather than a globalist perspective that insists on rules and regulations as long as they serve nationalist agendas. Under disruptive governance, the new order is disorder (Marsden et al. 2020b).

In addition to keeping their true objectives veiled, processes of disruptive governance seek high visibility (Table 4.1), which populist governments use as a way of demonstrating to their popular base that the government is fulfilling its promises to reform the system. The Bolsonaro administration is an emblematic case in using disruptive governance to dismantle policies focused on promoting more sustainable rural development. A list of institutions and policies was quickly dismantled under the justification that it was necessary to destroy the past and build something new. However, in the Brazilian case, the new benefits only large farmers and corporations linked to export agribusiness to the detriment

Table 4.1 Types of dismantling strategies and their characteristics

Type of strategy	Characteristics	Visibility
Dismantling by default	Absence of formal political decision, aiming to circumvent opposition to dismantling and the high political costs; this type of dismantling promotes the ineffectiveness of the policy through budgetary restrictions and/or reductions in services and/or activities	Low visibility
Dismantling by arena shifting	Deliberate transfer of one public policy to another political arena. The change of arena can cause demobilisation of the organisational bases of a policy and thus affect its intensity	Low visibility
Dismantling by symbolic action	Situations where political actors publicly declare their intention to dismantle a public policy—affecting intensity and density. However, the declarations often do not lead to dismantling, possibly due to high institutional restrictions, but they are symbolically important	High visibility
Active dismantling	Political actors have clear and deliberate action to dismantle a policy and/or political instruments. In this strategy, political administrations can understand dismantling as rewarding due to ideological positions. Therefore, political actors openly assume the eventual costs and institutional constraints. This type of dismantling affects the intensity of public policy and consequently its density	High visibility

Based on Bauer and Knill (2012), Grisa and Niederle (2021), Sabourin et al. (2020b)

of family and peasant farming. The dismantling within the regime of disruptive governance allows populist governments to hide their true intentions (e.g. supporting predatory agro-extractivism) and simultaneously demonstrate to their popular base of support that they are acting as promised, rupturing the structures 'that prevented the development of the nation'.

Disruptive governance is undoubtedly a new strategy used by far-right populist governments to dismantle rational policies. However, at least in the Brazilian case, the novelty is restricted to the form the dismantling is occurring, because the content of the dismantling (support for predatory agro-extractivism and suppression of the rights of family and peasant farming) is the resumption of a long-standing public agenda—to go back to the past!

Policy Dismantling

The approach of dismantling public policies—policy dismantling—has been well established in recent years, especially in the post-2008 crisis period, with the adoption of fiscal austerity policies in different countries of the Global North and South (Bauer et al. 2012a; Jordan et al. 2013; Jordana 2014; Sabourin et al. 2020a). The construction of this framework comes from the interaction with neo-institutionalist theorists (Mahoney and Thelen 2010) on changes in public policies. Starting from the field of historical neo-institutionalism, Mahoney and Thelen (2010) question the theory that institutions change only from critical situations and ruptures (e.g. wars, economic and political crises) and characterise the existence of subtle and gradual changes over time. However, in the debates about gradual institutional changes, few elucidations were made about the dismantling of public policies—a specific type of institutional change (Grisa and Niederle 2021; Jordan et al. 2013). The policy dismantling approach seeks to understand the decision-making process of dismantling (or not) a public policy, focused on identifying political actors, external factors, opportunities and institutional restrictions, as well as the strategies that are designed in the dismantling process (Bauer et al. 2012b; Jordan et al. 2013). Bauer and Knill (2012: 35) define policy dismantling as: 'A change of a direct, indirect, hidden or symbolic nature that either diminishes the number of policies in a particular area, reduces the number of policy instruments used and/or lowers their intensity'.

In analysis of dismantling, two dimensions are essentially mobilised: (1) intensity, which is the degree of restrictions on public policy contents, and (2) density, which refers to changes in the number of programmes and instruments of a public policy (Bauer and Knill 2012; Jordana, 2014). Furthermore, authors that support this approach identify two main scenarios that support both perception and action of political actors for dismantling: (1) the political benefits of dismantling are more advantageous to political actors than the costs of dismantling, and (2) the maintenance costs of a policy are higher than the dismantling costs (Bauer et al. 2012b; Bauer and Knill 2012). In both scenarios, political actors are influenced by external factors and by institutional opportunities and constraints, which can constrain the behaviour of political actors when making decisions (Jordan et al. 2013). Bauer et al. (2012b) point out that external factors can be supranational pressures; changes in economic doctrines; stability (or not) in the financial system; political-ideological changes; the integration of new agendas in the political scenario (e.g. climate change; Covid-19 pandemic); etc. In addition, the authors state that eventual external shocks (such as economic, political, and climate crises, as well as the emergence of new technologies) must be taken into account when analysing the dismantling of public policies.

The dismantling of policies can be done in various ways, and Table 4.1 presents a brief summary of the strategies behind the dismantling. The visibility adopted varies according to the perception of political actors about when dismantling occurs; for example, politicians generally prefer highly visible dismantling when they have just been elected. Furthermore, the different dismantling strategies can be mobilised together to achieve this objective, or they can vary along a dismantling trajectory (Bauer et al. 2012b; Sabourin et al. 2020b). In the next section, we will present emblematic cases of public policies dismantled under the context of disruptive governance: Food Acquisition Programme (PAA—*Programa de Aquisição de Alimentos*) and two programmes aimed at territorial development—National Programme for Sustainable Development of Rural Territories (PRONAT—*Programa Nacional de Desenvolvimento*

Sustentável dos Territórios Rurais) and Citizenship Territories Programme (PTC—*Programa Territórios da Cidadania*).

Case One: Food Acquisition Programme (PAA)—Expansion, Dismantling and Ending

The Food Acquisition Programme (PAA—*Programa de Aquisição de Alimentos*) emerged in 2003 after two debates that occurred in Brazil during the 1990s: the fight against hunger and the recognition of family and peasant farming as a relevant social category. Innovative and multisectoral, the PAA used public money to purchase food produced by family farming, at previously defined prices, and distributed it to populations suffering from food and nutritional insecurity (Grisa and Zimmermann 2015; Hespanhol 2013; Maluf et al. 2015). Under this programme, public procurement processes were simplified and adjusted to the organisational reality of family and peasant farmers. Guarantee of food and nutrition security was one of the central agendas from 2003 to 2012; therefore, the budget of the PAA continually increased during this period (Fig. 4.1). Until 2012, there were no signs of dismantling the PAA. Only incremental changes were made to the programme to improve and expand its implementation in this period.

In 2014, the PAA and other public policies aimed at family and peasant farming in Brazil began to suffer from the political and economic instability in the country. The slowdown in agricultural and mineral exports, the fall in tax revenue, the statements of politicians and businessmen opposed to the government, among other factors, prevented the maintenance of previously implemented social policies and paved the way for an internal crisis in the Brazilian State (Boschi and Pinho 2019). Faced with an unstable political and economic situation, Dilma Rousseff's second term (from 2015) embraced an economic agenda of austerity. At that time, the Dilma Rousseff's government was facing an economic crisis, which motivated it to adopt a contractionary fiscal policy to contain the growing dissatisfaction of the financial groups (Bastos 2017). Vasconcelos et al. (2019) noted that in 2015, BRL 70

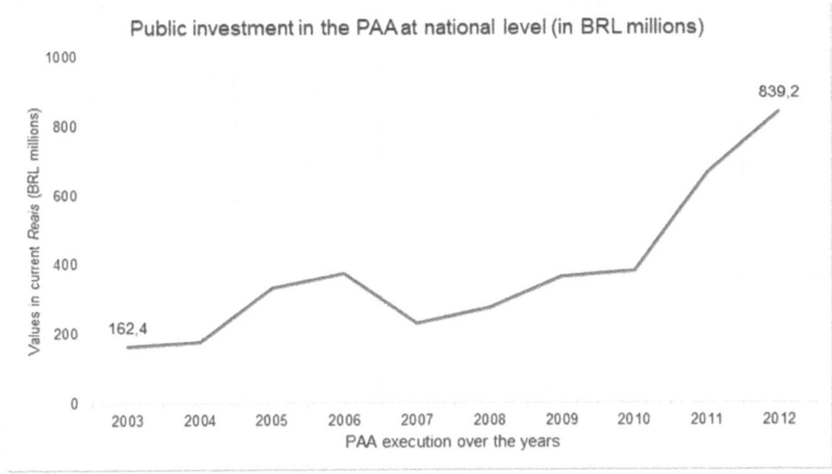

Fig. 4.1 Public investment in the PAA at national level (2003–2012) (Based on data from CONAB [2021]; Delgado et al. [2005]; Hespanhol [2013]; Porto [2014])

billion (≈ USD 20 billion) was cut in government spending and a significant portion of this cut was from policies aimed at family and peasant farmers. Hence, a classic process of dismantling public policies began (dismantling by default—budget restrictions that lead to the ineffectiveness of public policy), but not a radical break with what had been implemented (Bauer et al. 2012a; Sabourin et al. 2020b).

With the eruption of the political and economic crisis in the country in 2016, the government of Dilma Rousseff was interrupted by a parliamentary coup. Michel Elias Temer ascended to the presidency in an administration dominated by conservative and neoliberal agents deepening the austerity agenda. In this context, public policies aimed at family and peasant farming began to face an accelerated dismantling process in favour of agribusiness exports (Sabourin et al. 2020a, b). The main structures of the State that worked in favour of this category [Ministry of Social Development and Fight against Hunger (MDS) and the Ministry of Agrarian Development (MDA)] were eliminated at the beginning of the new government. The agendas of these ministries were successively transferred to sectors with a lower hierarchical position in the government until they were part of a lower-level agency called the Special Secretariat for Family Agriculture and Agrarian Development (SEAD—*Secretaria Especial de Agricultura Familiar e do Desenvolvimento Agrário*) (Sabourin et al. 2020b). These changes are characterised as dismantling by arena shifting, which symbolises the weakening of the institutional support given to family and peasant farming. The change in the arena of policies aimed at family farmers eliminated the political forums previously occupied, especially by social movements and civil society organisations linked to this category. Considering that the PAA obtained negligible investments during Michel Temer's government (2017–2018) compared to during the programme's peak period, the dismantling by arena shifting adds to the dismantling by default (Fig. 4.2).

In the Temer administration, the social category of family and peasant farming began to be discredited. Political actors linked to agribusiness pushed for the incorporation of this sector into the hegemonic agro-export model—a movement that was further evident in the government of Jair Bolsonaro—disregarding the profound differences between plantation farming and family-based agriculture, as well as regressing the institutional advances made in recent decades (Grisa and Niederle 2021;

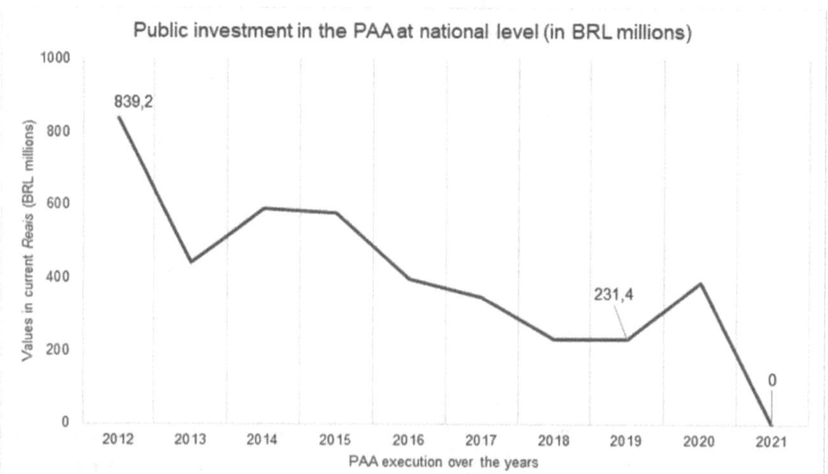

Fig. 4.2 Public Investment in the PAA at the National Level (2012–2021) (*Source* Prepared based on data from CONAB [2021]; Salgado et al. [2020]; SAGI [2021])

Wanderley 1996). The election of Jair Bolsonaro in 2018 brought a qualitative change in the dismantling process. This government mainly adopted high visibility dismantling strategies (Dismantling by symbolic action and Active dismantling), using dismantling as a political tool to strengthen its relationship with its social base.

An example of Active Dismantling occurred on the first day of Bolsonaro's government when the National Food and Nutrition Security Council (CONSEA) was extinguished. CONSEA was an arena of social participation that played an essential role in the fight against hunger and food insecurity in the country and in articulating policies aimed at family and peasant farming. In addition to CONSEA, other areas for social participation aimed at family and peasant farming were eliminated by the government of Jair Bolsonaro, including the National Council for Sustainable Rural Development (CONDRAF—*Conselho Nacional de Desenvolvimento Rural Sustentável*) and the National Commission for Agroecology and Organic Production (CNAPO—*Comissão Nacional de Agroecologia e Produção Orgânica*), both for ideological reasons (Sabourin et al. 2020a). As discussed by Scoones et al. (2018—and intrinsic to the concept of disruptive governance—the populist government of Jair Bolsonaro eliminates or dismantles all channels of social participation or governance, which could design strategies to resist its pro-dismantling of public policies.

In the organisational structure of the Bolsonaro administration, the agendas of family and peasant farming were incorporated into the very ministry responsible for the large agribusiness sector. Hence, the movement to integrate family farming into agribusiness, which began under Michel Temer, was expanded under Bolsonaro. Furthermore, the PAA continued to suffer dismantling through default. In 2019, the programme budget was the lowest in its entire history (Graph 2). The initiative was only not completely dismantled due to pressure from social movements and family farmer organisations (Niederle et al. 2021).

In 2020, with the emergence of the Covid-19 pandemic, social movements and civil society organisations emphasised that the PAA should be used as an instrument to combat the socioeconomic effects caused by the pandemic, especially due to the programme's ability to guarantee both food and nutrition security of vulnerable populations and the income of family farmers (ABA 2020; Borsatto et al. 2020; Sambuichi et al. 2020). Therefore, the federal government made an emergency allocation of BRL 500 million ($\approx$ USD 100 million) for the PAA in that year. However, only R$ 387.5 million were spent (SAGI 2021), an amount lower than in the 2016 budget, when the programme was already undergoing the process of dismantling by default (Fig. 4.2). The final blow to the PAA comes in 2021. A non-statutory measure by the federal government ended the PAA and created a new programme to replace it, the *Programa Alimenta Brasil* (Feed Brazil Programme). Under the government of Jair Bolsonaro, backed by the support of his base, dismantling strategies with high visibility have been commonly employed in the areas of family farming and the environment, education, culture and health (Sabourin et al. 2020a).

Case Two: Abandoning the Agenda of Sustainable Territorial Development and Embracing Predatory Extractivism

In Brazil, the debate about social participation in constructing development strategies began in the 1990s (Wanderley, 2014). The country, which had recently transitioned from a dictatorial regime to democracy, was experimenting with a new Federal Constitution (1988), which began to grant subnational federative entities greater political and fiscal autonomy. These elements, along with the resistance posed by socio-territorial movements (indigenous, peasants and landless rural workers) of the neoliberal policies that were advancing in Latin America in that period (Veltmeyer 2020), contributed, in Brazil, to the emergence of some initiatives of the federal government to promote local development. With the beginning of the government of Lula da Silva in 2003, these initiatives led to the institutionalisation of a territorial approach to development, culminating in the creation of the Secretariat for Territorial Development (SDT—*Secretaria de Desenvolvimento Territorial*), as part of the MDA (Antunes Jr et al. 2021).

Under the operationalisation of the SDT, the National Programme for Sustainable Development of Rural Territories (PRONAT—*Programa Nacional de Desenvolvimento Sustentável dos Territórios Rurais*) and the Citizenship Territories Programme (PTC—*Programa Territórios da Cidadania*) were instituted. The primary objectives of PRONAT and PTC were to strengthen social capital; promote sustainable development; stimulate local economies, primarily through activities of family and peasant farming; and provide access to essential citizenship services (Brazil 2005, 2008). These programmes started from the premise that the diversity of actors and elements present in the territories (expressed in their economic, social, environmental, political and cultural dimensions) could be important agents to stimulate development. To this end, these programmes established (in the territories served) governance spaces in which public authorities and civil society actors coordinated actions for territorial development. Such actions were mainly financed by the Programme for Infrastructure and Services in Rural Territories (PROINF—*Programa de Infraestrutura e Serviços em Territórios Rurais*).

PRONAT and PTC were very clear examples of the perspective of the Lula da Silva (2003–2010) and Dilma Rousseff (2011–2016) on rural development: the recognition of the social, cultural and environmental diversity of different territories and the attempt to reconcile it with the advancement of agribusiness and mineral extraction (Soyer and Barbosa Jr 2020). This understanding (and action) was not only present in progressive Brazilian governments but also in other Latin American countries that had centre-left governments in the transition from the twentieth to the twenty-first century. These governments—called post-neoliberal (Veltmeyer 2020)—witnessed the commodity boom and used it to implement social policies and/or policies aimed at historically marginalised groups, which some authors call the period of neo-extractivism (Pereira et al. 2021a, b; Soyer and Barbosa Jr 2020). Attributed mainly to Latin American countries, neo-extractivism is a development model rooted in the appropriation of natural resources (mineral resources and land exploitation) aimed at exporting primary commodities and achieving economic growth. This model intensifies the primarisation of the economy of the countries, as well as inserting them in a subordinate way in the international division of labour. However, the moments of great global demand for commodities and the increase in the prices of these products, as occurred at the beginning of this century, may favour governments to implement their public policies.

Brazilian policies aimed at territorial development (based on the idea of coexistence between agribusiness and family farming, and unaccompanied by structuring measures, such as agrarian reform) have had their limitations problematised both in paradigmatic debate (Oliveira Neto 2020) and empirical research (Antunes Jr et al. 2021). Nevertheless, they remained on the government agenda from 2003 to 2016 and financed a significant number of projects, which certainly achieved important results. However, beginning under Dilma Rousseff, from 2013 onwards, public investments in territorial policies dropped significantly (Fig. 4.3), characterising a dismantling by default, with loss of intensity (Bauer et al. 2012a; Sabourin et al. 2020a). As discussed in the previous section, the coup that ousted Dilma Rousseff in 2016 led to Michel Temer taking office, who promptly eliminated the MDA and, consequently, the SDT. At that time, agrarian policies in Brazil experienced complete uncertainty, in which the transfer of the political arena, the absence of decision-making agents and the suspension of investments left programmes, such as PRONAT and PTC, in a state of inaction. Articulating strategies of dismantling by default and by changing the arena (Bauer et al. 2012a), Michel Temer showed that, from then on, the understanding of the rural Brazilian as a diverse area and with multiple demands and interests ceased to have strength on the government agenda (Soyer and Barbosa Jr 2020).

As in the case of the PAA, there was a qualitative change in the dismantling process. High visibility dismantling strategies (Dismantling by symbolic action and Active dismantling) intensified. If policies aimed at territorial development were not officially defunct at first, with the rise of Jair Bolsonaro, they were actively eliminated (Bauer et al. 2012a). Through a non-statutory measure issued by the federal government, the National Management Committee of the PTC was dissolved. Furthermore, the agenda for sustainable territorial development, which reconciles and promotes coexistence between family and peasant farming and the agribusiness model, came under attack and was abandoned entirely (Fig. 4.3).

Fig. 4.3 Path of policies aimed at territorial development in Brazil (*Source* Secretaria Especial da Agricultura Familiar e Desenvolvimento Agrário [SEAD])

With a messianic and ultra right-wing profile and wholly aligned with the interests of agribusiness agents, Jair Bolsonaro destroyed everything that could become an obstacle to the advancement of an extractive and predatory agricultural model (Pereira et al. 2021a, b; Soyer and Barbosa Jr 2020). In line with the typical nationalist and authoritarian discourse of right-wing, populist governments (Scoones et al. 2018), Jair Bolsonaro argues that the only possible model for rural development is the agribusiness model. In this sense, programmes that incorporate the demands of family and peasant farmers, such as PRONAT and PTC, were not the only obstacles encountered. The entire legal and institutional framework for environmental preservation and conservation, guarantee of indigenous territories and traditional communities, and the establishment of rural settlements for landless families must be disrupted (Kluck 2020; Pereira et al. 2021a, b; Soyer and Barbosa Jr 2020).

Conclusions

The Jair Bolsonaro government can be considered an emblematic case of how far-right populist governments in the Global South use disruptive governance to dismantle policies that do not meet the interests of predatory agro-extractivism. Unlike the Global North, family and peasant farming is one of the great victims of the far-right populist governments of the Global South. In the Brazilian case, policies that directly benefited family and peasant farming were quickly and widely dismantled within an atmosphere of disruptive governance. The dismantling was justified by a discourse that radical changes were necessary, used high visibility strategies and demanded only non-statutory changes (not requiring congressional approval to be effective).

In March 2019, at a meeting with supporters, Jair Bolsonaro made the following statement: 'Brazil is not an open terrain where we intend to build things for our people. We have to dismantle many things' (Lavalle and Bezerra 2021). The dismantling within the regime of disruptive governance allows populist governments to hide their true intentions (e.g. supporting predatory agro-extractivism) and simultaneously demonstrate to their popular base of support that they are acting as promised, rupturing the structures 'that prevented the development of the nation'. This rupturing basically means: flexibilising the appropriation by private enterprises of public lands; loosening instruments that control deforestation; militarising public bodies responsible for environmental and land issues; subjugating environmental and agrarian agendas to the agricultural agenda; dismantling channels of participation and social control; etc. Although such disruptive actions, undertaken by the government of Jair Bolsonaro, can be seen as a 'generalised mess' (Kluck 2020: 683), they are reality-ordered actions, very well-articulated and with well-defined interests: the consolidation of agribusiness as the only model for rural development.

Disruptive governance is undoubtedly a new strategy used by far-right populist governments to dismantle rational policies. However, at least in the Brazilian case, the novelty is restricted to the form the dismantling is occurring, because the content of the dismantling (support for predatory agro-extractivism and suppression of the rights of family and peasant farming) is the resumption of a long-standing public agenda—to go back to the past!

References

ABA. 2020. Abaixo Assinado pela Retomada Urgente do PAA. Retrieved from https://agroecologia.org.br/abaixo-assinado-pela-retomada-urgente-do-paa.

Abers, R., L. Serafim, and L. Tatagiba. 2014. Repertórios de Interação Estado-Sociedade em um Estado Heterogêneo: A Experiência na Era Lula. *DADOS—Revista de Ciências Sociais* 57 (2): 325–357.

Antunes Jr, W.F., R.S. Borsatto, and V.F. Souza-Esquerdo. 2021. Why Is It So Difficult to Promote Territorial Development Through Public Policies? The Obstacles Faced by a Brazilian Experience. *World Development Perspectives* 24. https://doi.org/10.1016/J.WDP.2021.100367.

Bastos, P.P.Z. 2017. Ascensão e crise do governo Dilma Rousseff e o golpe de 2016: poder estrutural, contradição e ideologia. *Revista de Economia Contemporânea* 21 (2). https://doi.org/10.1590/198055272129.

Bauer, M.W., et al. 2012a. *Dismantling Public Policy: Preferences, Strategies, and Effects*. Oxford: Oxford University Press.

Bauer, M.W., et al. 2012b. Dismantling Public Policy: Preferences, Strategies, and Effects. In *Dismantling Public Policy: Preferences, Strategies, and Effects*, ed. M.W. Bauer et al., 203–225. Oxford: Oxford University Press.

Bauer, M.W., and C. Knill. 2012. Understanding Policy Dismantling: An Analytical Framework. In *Dismantling Public Policy: Preferences, Strategies, and Effects*, ed. M.W. Bauer, et al., 30–51. Oxford: Oxford University Press.

Borsatto, R.S., F. Grigoletto, A.C. Macedo, and A.C. Martensen. 2020. Respostas dos Municípios para Garantir Segurança Alimentar e Nutricional em Tempo de Pandemia. In *COVID-19: Crises Entremeadas no Contexto de Pandemia (Antecedentes, Cenários e Recomendações)*, ed. N. Valêncio, and C.M. Oliveira, 165–184. São Carlos: UFSCar.

Borsatto, R.S., A.C. Macedo, L.L. Santos, W.F. Antunes, and V.F. Souza-Esquerdo. 2021. Food Procurement as an Instrument to Promote Local Food Systems: Exploring a Brazilian Experience. *International Journal on Food System Dynamics* 12 (2): 177–191.

Boschi, R.R., and C.E.S. Pinho. 2019. Crisis and Austerity: The Recent Trajectory of Capitalist Development in Brazil. *Contemporary Politics* 25 (3): 292–312.

Brazil. 2005. *Marco Referencial para Apoio ao Desenvolvimento de Territórios Rurais*. Série Documentos Institucionais 02–2005. Brasília: Ministério de Desenvolvimento Agrário (MDA).

Brazil. 2008. *Territórios da Cidadania: Integração de Políticas Públicas Para Reduzir Desigualdades*. Brasília: Ministério de Desenvolvimento Agrário (MDA).

CONAB. 2021. Transparência Pública do PAA. Retrieved from https://consultaweb.conab.gov.br/consultas/consultatransparenciapaa.do?method=abrirConsulta.

Delgado, G.C., J.C.P.R. Conceição, and J.J. Oliveira. 2005. Avaliação do Programa de Aquisição de Alimentos (PAA). *Texto para Discussão*, 1145. Brasília: Ipea.

Gravey, V., and A.J. Jordan. 2020. Policy Dismantling at EU Level: Reaching the Limits of 'An Ever-closer Ecological Union'? *Public Administration* 98 (2): 349–362.

Grisa, C. 2010. Public Policies for Family Farming in Brazil: An Essay from a Cognitive Approach. *Desenvolvimento Em Debate* 1 (2): 82–111.

Grisa, C., and P.A. Niederle. 2021. Paradigms, Institutional Changes and Policy Dismantling in the Mercosur Specialized Meeting of Family Farming. *Lua Nova* 112: 251–284.

Grisa, C., and S. Schneider. 2014. Três Gerações de Políticas Públicas para a Agricultura Familiar e Formas de Interação entre Sociedade e Estado no Brasil. *Revista De Economia e Sociologia Rural* 52: 125–146.

Grisa, C., and S.A. Zimmermann. 2015. Estado e Sociedade Civil na Promoção da Segurança Alimentar e Nutricional no Brasil: A Construção do Programa de Aquisição de Alimentos da Agricultura Familiar (PAA). *Agroalimentaria* 21: 17–36.

Hespanhol, R.A.M. 2013. Programa de Aquisição de Alimentos: Limites e Potencialidades de Políticas de Segurança Alimentar para a Agricultura Familiar. *Sociedade e Natureza* 25 (3): 469–483.

Jordan, A., M.W. Bauer, and C. Green-Pedersen. 2013. Policy Dismantling. *Journal of European Public Policy* 20 (5): 795–805.

Jordana, J. 2014. Multiple Crises and Policy Dismantling in Spain: Political Strategies and Distributive Implications. *Political Studies Review* 12 (2): 224–238.

Kluck, E.G.J. 2020. Terra e Floresta em Foco: Percursos Recentes da Regularização da Terra e Anistia a Crimes Ambientais. *Revista Da Anpege* 16: 681–711.

Lavalle, A.G., and Bezerra, C.P. 2021. Por que 'desconstruir' a participação social, *Nexo: Políticas Públicas*, https://pp.nexojornal.com.br/opiniao/2021/Por-que-%E2%80%98desconstruir%E2%80%99-a-participa%C3%A7%C3%A3osocial.

Lotta, G., and A. Favareto. 2016. Desafios da Integração nos Novos Arranjos Institucionais de Políticas Públicas no Brasil. *Revista De Sociologia e Politica* 24: 49–65.

Mahoney, J., and K. Thelen. 2010. *Explaining Institutional Change: Ambiguity, Agency and Power*. Cambridge: Cambridge University Press.

Maluf, R.S., et al. 2015. Nutrition-sensitive Agriculture and the Promotion of Food and Nutrition Sovereignty and Security in Brazil. *Ciência e Saúde Coletiva* 20: 2303–2312.

Mamonova, N., and Franquesa, J. 2020. Right-Wing Populism in Rural Europe. Introduction to the Special Issue. *Sociologia Ruralis* 60 (4): 702–9. https://doi.org/10.1111/SORU.12306.

Marsden, T., C. Lamine, and S. Schneider. 2020a. *A Research Agenda for Global Rural Development*. Chaltenham: Edward Elgar.

Marsden, T., T. Lang, and E. Millstone. 2020b. *Brexit, Wales and Agri-food: Can we Manage Disruptive Governance?* Working Paper, Cardiff.

Medina, G.S., M.S. Gosch, and M.E. Delgrossi. 2021. Development Pathways for Family Farmers: Lessons from Brazil on the Need for Targeted Structural Reforms as a Means to Address Regional Heterogeneity. *Geoforum* 118: 14–22.

Muddle, C., and C.R. Kaltwasser. 2017. *Populism: A Very Short Introduction*. New York and Oxford: Oxford University Press.

Niederle, P. et al. 2021. *Rupturas na Transição Agroecológica: Mudança Institucional e Desmantelamento das Políticas Públicas no Brasil*. Working Paper, Porto Alegre.

Oliveira Neto, A.C. 2020. O Desenvolvimento Terrirorial Rural no Brasil: Uma Análise a Partir do Debate Paradigmático. *Ateliê Geográfico* 14 (2): 189–207.

Pereira, L.I., E.L.F. Coca, and C.F. Origuéla. 2021a. O "Passar a Boiada" na Questão Agrária Brasileira em Tempos de Pandemia. *Revista Nera* 24: 8–23.

Pereira, L.I., C.F. Origuéla, and E.L.F. Coca. 2021b. A Política Agrária no Governo Bolsonaro: As Contradições entre a Expansão do Agronegócio, o Avanço da Fome e o Antiambientalismo. *Revista Nera* 24: 8–27.

Pires, R.R.C., and A.Á. Gomide. 2016. Governança e Capacidades Estatais: Uma Análise Comparativa de Programas Federais. *Revista De Sociologia e Política* 24: 121–143.

Porto, S.I. 2014. *Programa de Aquisição de Alimentos (PAA): Política Pública de Fortalecimento da Agricultura Familiar e da Agroecologia no Brasil*. MSc dissertation, Universidad Internacional de Andalucía, Baeza.

Sabourin, E., C. Grisa, P. Niederle, S.P. Leite, C. Milhorance, A.D. Ferreira, S. Sauer, and J.M. Andriguetto-Filho. 2020a. Le Démantèlement des Politiques Publiques Rurales et Environnementales au Brésil. *Cahiers Agricultures* 29: 1–8.

Sabourin, E., C. Craviotti, and C. Milhorance. 2020b. The Dismantling of Family Farming Policies in Brazil and Argentina. *International Review of Public Policy* 2 (1): 45–67.

SAGI. 2021. Programa de Aquisição de Alimentos (PAA)—Execução Geral. Retrieved from https://aplicacoes.mds.gov.br/sagi/vis/data3/data-explorer.php.

Salgado, R.J.S.F., et al. 2020. A Natureza Teórico-epistemológica das Publicações Brasileiras no Programa de Aquisição de Alimentos (PAA). *Brazilian Journal of Development* 6 (3): 9998–10025.

Sambuichi, R.H.R., et al. 2020. O Programa de Aquisição de Alimentos (PAA) como Estratégia de Enfrentamento aos Desafios da COVID-19. *Revista De Administração Pública* 54 (4): 1079–1096.

Scoones, I., et al. 2018. Emancipatory Rural Politics: Confronting Authoritarian Populism. *Journal of Peasant Studies* 45 (1): 1–20.

Singh, T., P. Singh, and M. Dhanda. 2021. Resisting a "Digital Green Revolution": Agri-logistics, India's New Farm Laws and the Regional Politics of Protest. *Capitalism, Nature, Socialism* 32 (2): 1–21.

Soyer, G., and R. Barbosa Jr. 2020. O Extrativismo Agrário do Governo Bolsonaro a Partir das Relações Estado-Sociedade. *Revista Da Anpege* 16: 522–554.

Vale, M.M. et al. 2021. The COVID-19 Pandemic as an Opportunity to Weaken Environmental Protection in Brazil. *Biological Conservation* 255. https://doi.org/10.1016/J.BIOCON.2021.108994.

Vasconcelos, F.A.G. et al. 2019. Public Policies of Food and Nutrition in Brazil: From Lula to Temer. *Revista de Nutrição* 32.

Veltmeyer, H. 2020. Latin America in the Vortex of Social Change: Development and Social Movement Dynamics. *World Development* 130: 104916.

Vergara-CamuS, L., and C. Kay. 2017. Agribusiness, Peasants, Left-wing Governments, and the State in Latin America: An Overview and Theoretical Reflections. *Journal of Agrarian Change* 17 (2): 239–257.

Wanderley, M.N.B. 1996. Raízes Históricas do Campesinato Brasileiro. *XX Encontro anual da ANPOCS*, Caxambu, MG.

Wanderley, M.N.B. 2014. Gênese da Abordagem Territorial no Brasil. In *Participação, Território e Cidadania: Um Olhar sobre a Política de Desenvolvimento Territorial no Brasil*, ed. J.S.B. Cavalcanti, M.N.B. Wanderley, and P.A. Niederle, 80–103. Recife: Editora UFPE.

5

Contested Landscapes: Territorial Conflicts and the Production of Different Ruralities in Brazil

Bernardo Mançano Fernandes
and Clifford Andrew Welch

Introduction[1]

It is through landscape that we see and feel the complexity of the relations and objects that make up space, territory, region, and place. Landscape is the expression of the multidimensionality and multiscalarity of territory and its ruralities (Marsden 1989). In rural studies landscape is made manifest by rurality. Rurality is the configuration of social life in relation

[1] This chapter is a revised version of B.M. Fernandes, and C. Welch. 2019. Contested Landscapes: Territorial Conflicts and the Production of Different Ruralities in Brazil. *Landscape Research* 44 (7): 892–907, available online https://www.tandfonline.com/doi/abs/10.1080/01426397.2019.1595549?journalCode=clar20.

B. M. Fernandes (✉)
São Paulo State University (Unesp), São Paulo, Brazil
e-mail: mancano.fernandes@unesp.br

C. A. Welch
Federal University of São Paulo (Unifesp), São Paulo, Brazil

A. A. R. Ioris and B. Mançano Fernandes (eds.), *Agriculture, Environment and Development*, https://doi.org/10.1007/978-3-031-10264-6_5

to natural resources, goods, services, cultural values, and social categories like class—the social formation, as it were—of the territories that make up rural landscapes (Hecht 2010; Wanderley 2000). In this article, the diversity of landscapes produced by different ruralities is understood as the product of territorial disputes between different models of development. The varied territories produced generate diverse social relations that, in turn, produce different ruralities. Territorial disputes are conflicts arising from several factors. Here we analyse disputes based on conflicts between capitalist and non-capitalist social relations of production, such as the use of family labour in peasant agriculture. The territorial production of these two ruralities produces distinct landscapes that are evidenced here by terrestrial photography, aerial photography, and satellite images, at different geographic scales. The analysis is based on an examination of territorial disputes in the municipalities[2] of Mirante do Paranapanema and Ribeirão Preto in the state of São Paulo, Brazil. By analysing the different landscapes created by disputes among agrarian reform[3] supporters like *Movimento dos Trabalhadores Rurais Sem Terra* (MST) peasants, cattle ranchers, and sugarcane agribusinesses, we demonstrate through landscape how different actors, advocating for different models of development, produce different ruralities.

In this article, we combine vertical and conflict analysis to the study of rurality, that is to say, our objects are locations that have been the subject of disputes and our analysis is based on field research as well as terrestrial, aerial, and satellite images, the latter from Google Earth.[4] Relationships and conflicts in the two specific cases examined include

[2] In Brazil, states are subdivided into municipal administrative districts. These municipalities include one closely populated headquarters city and numerous sub-districts, which are often sparsely populated and rural.

[3] The term agrarian reform is used in Brazil to distinguish the country's project from land reform. Land reform is seen as a simpler question of land distribution, whereas agrarian reform adds to that low-cost credit, technical assistance, schools and other services to ensure the new small-sale agriculturalists have a chance to prosper.

[4] We owe a debt to geographer Messias Modesto Passos (2004), whose study inspired our use of photographs and satellite images to describe the region's landscapes. With an environmental focus, Passos analysed tropical deforestation in the Pontal do Paranapanema, not agrarian issues.

those between latifundia,[5] and agribusiness,[6] agribusiness and peasants,[7] and latifundia and peasants. Informed by ample empirical research, including interviews with participants, the study of relevant literature and documents, emphasis has been placed on the three levels of imagery mentioned. We selected certain shots and perspectives to express differences in the ruralities and landscapes produced by agrarian capitalism and the agrarian question,[8] which we see as two distinct paradigms for understanding the development of the countryside. These paradigms represent divergent economic models that seek different policies for the use of territories, the first favouring agribusiness and the second peasant agriculture. Fernandes (2009b) developed a typology to help understand the multi-scale organisation of spaces and territories in disputes by socio-territorial organisations like the MST. Using the multi-scalar approach, it can be useful for interpreting contested landscapes as territorial fractions expressed in layers of analysis scales. In this typology, the first territory is the space of governance at all scales; the second territory is property of different types and scales, while the third territory is space in movement.

---

[5] Traditionally in Latin America, latifundia is defined as a large, unproductive rural property, owned by an individual or family of political or economic import. More recently in Brazil, the word has also been applied to large properties that might be economically active, as in the case of cattle grazing or rental for use by planters of monocultures like sugarcane, but serve little societal utility other than to enrich their owners. In many senses, they define the worst characteristics of agrarian capitalism, in which private property rights are more valuable than human rights and productivity more highly valued than sustainability (Fernandes 2009a).

[6] Agribusiness is a set of systems linking farms and ranches to industries, financial services, technology, and markets. Families may operate agribusiness corporations, but their large-scale and profit-orientation requires hiring non-family members. These agricultural companies are part and parcel of the hegemonic agrarian capitalist model of development (Fernandes 2008).

[7] The peasantry is a sub-set of family farmers who depend on family labor to produce for themselves and the market on relatively small sections of land that may or may not be their own. Their focus is on reproducing a way of life in the countryside, more than on producing profits. Their self-provisioning provides them with limited autonomy, but their relative poverty maintains them in a dependent position *vis-a-vis* the hegemonic capitalist model. Historically, peasants have rebelled to defend their interests and strengthen their territories (Welch 2009).

[8] Our references to the agrarian question are informed by Karl Kautsky's 1899 study *The Agrarian Question* (1988). His central argument is based on the discovery that agribusiness is not a totality, destined to eliminate the peasantry. He found, to the contrary, that small-scale agriculture persisted and even grew with capitalist agriculture, playing roles in the production of foodstuffs, reproduction of seasonal labour and source of surplus land. Thus, to defend the agrarian question is to believe in the persistence of the peasantry.

These territorial forms influence landscape, as shown in the examples to follow.

About Mirante Do Paranapanema and Ribeirão Preto

Locating the municipalities of Mirante do Paranapanema and Ribeirão Preto from a global perspective, as shown in Fig. 5.1, underscores the nature and scale of reach, depth, and grasp of the territories disputed by national and international capitalists (Cubas 2017; Pereira 2015; Pitta and Mendonça 2015). As locations where sugarcane has been grown for many years, both attracted the interest of international capital in the early twenty-first century. First came financial interest in 'flex' crops, that is commodities that offered diverse opportunities for profits. Sugarcane is a flex crop because it can be sold as a sweetener and as a fuel (ethanol). Moreover, ethanol's framing as a low pollution fuel and sugarcane's framing as a renewable resource made both the crop and the industry attractive internationally as 'socially conscious' investments. Second, following the financial meltdown of 2008, even far-off land came to be seen as offering a relatively secure return on invested funds. Foreign investment by central countries like the United States and China in the land of developing countries like Brazil soon became criticised as 'land grabbing,' essentially an international threat to national rights. The *Journal of Peasant Studies* has dedicated several thematic issues to each of these trends. The competing territorial development interests represented by international, national, state, and municipal institutions express the multi-scalar dynamics of the first territory.

The two localities also illustrate the second form of territory in the typology. These two territories were created and given socioeconomic form by the hard work first of peasant-squatters and later by wage-workers controlled by propertied capitalists. Ribeirão Preto became the centre of Brazil's late nineteenth-century coffee export boom; Mirante do Paranapanema was reached by railroad some decades later, developing relatively slowly as a cattle grazing area until it participated in the sugarcane boom that started in the 1960s. Ribeirão Preto's rural

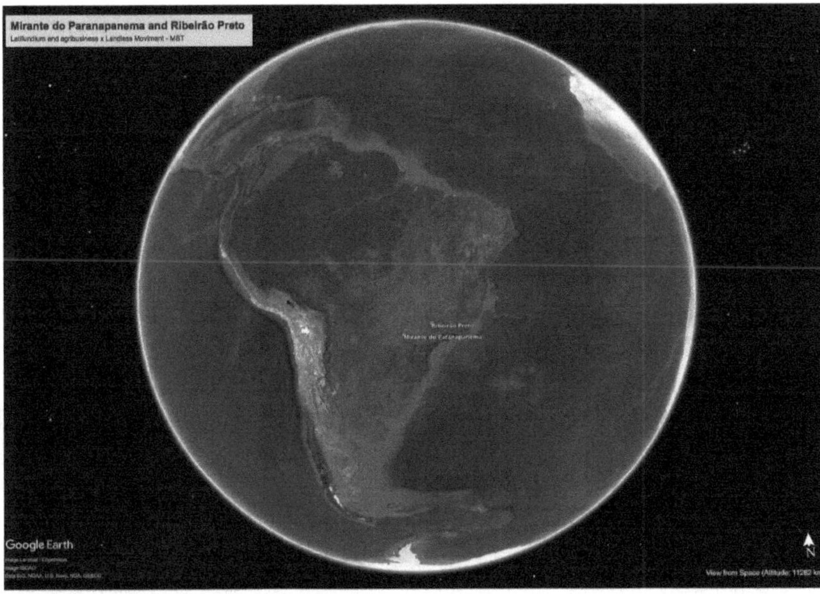

Fig. 5.1 Mirante do Paranapanema and Ribeirão Preto at planetary-scale

areas were bought and sold from the late nineteenth century on. Half a century later, politicians played politics with the lands of the Pontal do Paranapanema, the micro-region where Mirante is located, as if it were a football. Considered public lands, both appointed and elected governors gave huge fragments of it away to supporters (Leite 1998). In short order, both municipalities were largely denuded and their rural areas dedicated to coffee or cattle and cotton. Many hundreds of peasant families were dislocated as a consequence, making the areas ripe for agrarian reform. In Mirante, the struggle over property rights took off in the 1970s. In Ribeirão, where wage-labour provided an alternative for displaced workers, the contemporary struggle for land took shape in the 1990s.

In Brazil, the meaning of 'rural' provokes considerable debate. The official meaning defines rural as the spaces that begin with urban perimeters. In addition to this spatial definition there is another one based on changes in social relations that refer to the old and the new rural in a conservative interpretation of the agrarian capitalism paradigm. The rural concept is often used to refer to development, land use, labour relations, and production (Buainain et al. 2003; Graziano da Silva 1997). It is related to public policies, inequalities, and conflicts (Fernandes 2008), that is, closer to the ideas of Kautsky (1988), who underscored the fruitful coexistence of small-scale and large-scale agriculture. This is where the third territorial form becomes useful, as it helps to orient a geographical analysis of the dispute between small-scale and large-scale agriculture. Each has its own space, determined by distinct but mutable social relations. In the case studies at hand, each form of agriculture, represented by opposing groups, tried to superimpose itself on the other, pushing and pulling boundaries to territorialise opponents. Varied social relations produced varied ruralities, creating diverse rural territories with their distinct territorialities. The geographical method allows for a selection of elements to interpret the significance of the landscapes of different rural territories.

On Contested Landscapes

The principle point of the landscape images examined here is the materialisation of alternative realities. Convention has it that a municipality has to be divided between city and countryside, taking shapes that are either residential (the space of reproduction) or worked (spaces of production, like factories or farms on the outer margins). In Brazil, it is the rare municipality that reserves areas to build and maintain in-between places, like public parks. This is evident in the residential areas shown in the images of Ribeirão Preto, especially in Fig. 5.10—houses about one another and where marginal open space remains, historical patterns suggest that it will soon be occupied for residential construction purposes. These are 'contested landscapes,' a concept that appeared at least a generation ago in debates over land use in the western United States, where conservationists advocated for wilderness designations and cattle ranchers sought grazing rights (Goodman and McCool 1999).

The concept has since matured theoretically. In 2013, the University of Stirling, Scotland, hosted the 'Landscapes of Conflict Conference.' Through a multidisciplinary approach, not unlike our own, 'three distinct themes' emerged in the relationship between landscape and conflict. They involved disputes over: (1) 'access to resources'; (2) 'cultural identity, heritage, and place within landscapes' and (3) 'the impact of military conflict upon landscape' (Adderley and Mills 2014: 99–100).

Each of these themes is relevant to our examination. The land itself is the principal resource the landless fight to control. They have built their movements and organisations on the idea that the earth and its bounty can and should not be monopolised. Their identity as militants of landless movements, characterises them as survivors of a violent, unjust past, one that dispossessed and displaced their ancestors, but gives them the courage and sense of rights that compels their struggle. While strictly military conflict did not play as profound a role in our case studies as in those of the Stirling conference, the battles against indigenous peoples, enslavement of Africans, exploited workers, hired gunmen, and the brutality of Brazil's contemporary military police have penetrated the soil with blood. Furthermore, the contested landscapes analysed below are also ideological and political battlegrounds. Movement leaders (who are but more experienced landless workers themselves) associate their struggle with those of socialist guerrillas like Cuba's Che Guevara, pitting themselves as heroic warriors against capitalist armies. They seek to empower the poor and undermine the rich, explicitly defining their confrontations as class conflicts.

Analytically, a host of recent literature on 'contested landscapes' illuminates the political facets of the concept. A 2018 'contested landscape' article in the *Journal of Peasant Studies* examines examples from discourse to farmland to demonstrate how a favourite phrase among the international community's global planners obscures traditional mechanisms of domination (Newell and Taylor 2018). The authors unmask the corporate, monopolising advocates of 'Climate-Smart Agriculture' (CSA), showing the concept to be another attempt to concentrate wealth and undermine terms like 'agro-ecology' rather than any sort of truly innovative approach to producing more food at less cost to the environment and society. With neo-Gramscian theoretical references as their guide, the landscape is mainly cultural and contestation a battle of wits, but all of this is rooted in concrete entities like the World Bank and agro-food giants like Walmart. Thus, the authors demonstrate how the 'new' CSA fights to take territory away from the old 'CSA'—Community Supported Agriculture—by falsely claiming to help small farmers produce more with less water and fewer toxic chemicals. The authors write to sound an alarm against an emerging paradigm that gradually excludes small producers.

In Brazil, the agrarian reform settlements are emblematic of the agrarian question paradigm. They seek to offer an alternative to living on the margins of society, either in the cities or the countryside, where many farming people are reduced to poorly compensated migrant workers or contract farmers. The settlements have resulted from pressuring local, state, and national governments to develop public policies that make it possible for small-scale agriculture to prosper. Advocates argue that 70 per cent of the food is produced by smallholders and work to expand and fortify this sector, placing a higher value on reproducing the sector than on production (Paulino 2014).

The MST and other socio-territorial movements are chiefly responsible for the thousands of agrarian reform settlements established in the country. Founded in 1984, the MST united many local struggles for land in one vigorous and innovative socio-territorial movement. The MST helped disperse the direct action practice of land occupations in order to force authorities to implement agrarian reform law (Welch 2006). According to Brazil's 1988 Constitution, the federal government is responsible for appropriating and redistributing land that is left unproductive or where proprietors have been found guilty of violating labour or environmental laws. Following a massacre of unarmed landless protesters in 1996, the federal government was shamed into creating an Agrarian Development Ministry (MDA), free from the control of the Agricultural Ministry, which has advanced and defended large-scale farming since its inception in the 1850s (Coletti 2002). The MDA's budget and influence grew under the Workers Party (PT) presidency of Luis Ignacio Lula da Silva (2003–2010) Still, the government appropriated relatively few private properties, meaning that while the agrarian reform landscape grew in area, the size of the agrarian capitalist landscape also grew. Most of the area dedicated to agrarian reform was unproductive state-held land. By 2015, nearly 9500 agrarian reform settlements had been established, incorporating more than 82 million hectares, which provided farmland for more than 1.1 million families (Girardi et al. 2017). In the meantime, aggressive landlords, financiers and agribusiness corporations, both domestic and foreign, bought up as much forest and farmland as possible. Under Lula's successor, the PT's Dilma Rousseff (2011–2016), the agrarian reform agenda ground to a halt. Dethroned by a parliamentary coup d'etat, the new government eliminated the MDA and worked to dismantle the settlements, as related above. In other words, in practice, agrarian reform has done little to change the balance of power in Brazil (Castilho 2012).

Mirante Do Paranapanema, Encampments, and Settlements

In terms of the agrarian reform struggle in Brazil, Mirante do Paranapanema is the epicentre of what became one of the most conflict-ridden regions in Brazil, the Pontal do Paranapanema. Territorial disputes are epical; this region exemplifies the permanence of conflict in a historical process that renews the agrarian question on a regular basis (Girardi et al. 2017; Leite 1998). The continuity of this question partially explains how the old and the new connect latifundia and agribusiness in their expropriation of the peasantry and drive for territorial domination. The territorial configuration of the Pontal produces a landscape whose texture reveals how social classes utilise their territories in forms that are different enough for observers to take note of their specific ruralities and rugosities.[9] From world to local scales, from theoretical to empirical knowledge, an examination of the Pontal's landscapes helps explain these conflicts.

---

[9] Rugosity may mean 'full of wrinkles' in English, but it is used here as the concept developed by Brazilian geographer Milton Santos (1996) to describe the physical or visual geographic features that reveal the uneven accumulation of time in space. Not unlike wrinkles, one finds the furrows of the past in landscape. In the Pontal do Paranapanema these signs include tree trunks burned more than half a century ago, structural ruins that were once railroad thruways and stations, latifundia rented for agribusiness or turned into settlements, while the original names live on like spectres.

The Pontal do Paranapanema is a region formed by conflict (Feliciano 2009; Fernandes 1996; Leite 1998; Welch 2009). From the middle of the nineteenth century until the beginning of the twenty-first century, large land brokers have achieved territorial control of the region through land grabbing. *Grileiros* are would be landlords who forge documents to takeover an area whose occupants lack proper titles and the power to defend their holdings. In the Pontal, grileiros grabbed more than a million hectares of land, deterritorialising indigenous peoples and he peasantry, reducing them to exploited labourers compelled to uproot Atlantic Forest trees and brush to create large estates and ranches. It is evident that this '*grilagem*,' as the grileiro's work is called, involved a great deal of violence, as those who lived on the land often resisted being expelled from their homes. The process generated a fight for the land that remains a vital memory in the region until today. Significant changes occurred, producing new spaces, territories and landscapes, from the alliance between latifundia and agribusiness, with the leasing of land for the production of sugarcane, to the formation of socio-territorial movements. By the twenty-first century, the MST and other peasant movements had appropriated several territorial fractions in the Pontal do Paranapanema region, as shown in Fig. 5.2, with emphasis on the municipality of Mirante do Paranapanema.

We highlight Mirante do Paranapanema as the municipality with the largest number of agrarian reform settlements in the region. Each yellow polygon is a fraction of the peasant territory conquered and produced through land occupations, encampments, demonstrations, confrontations, and negotiations. This territorial production defines the landscape and territoriality of the settlements. Figure 5.3 demonstrates this territoriality using as examples the settlements of Che Guevara, Antonio Conselheiro and Paulo Freire. We can observe the differences between the territories of the settlements demarcated by the yellow polygons and the surrounding territories to the north and west that are agribusiness areas of sugarcane production. In the three polygons one observes lines and points that are actually roads and farm-family homes and outbuildings. The surrounding sugarcane *fazendas* (Portuguese for plantations) are homogeneous geometric forms typical of monoculture and documenting the lifeless forms that the Via Campesina (2006) criticises as

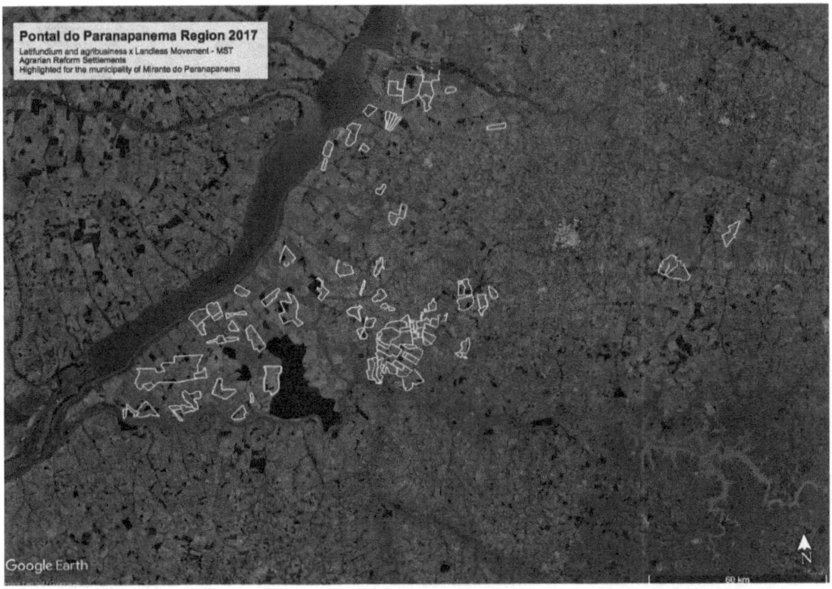

Fig. 5.2 In yellow, the map shows agrarian reform settlements in the Pontal do Paranapanema region, which is marked by the larger red boundary line. The smaller red line demarcates the boundaries of Mirante do Paranapanema

'green deserts.' The criticism points to the dearth of human life as well as the environmental hazards presented by sugarcane monoculture. For centuries, the production of this natural sweetener has demanded a high toll from land and labour. Almost no-one lives on agribusiness land as much of the work is mechanised, such that mostly seasonal, migrant workers are exploited at harvest time. Today's sugarcane is also problematic for using huge quantities of fresh water and toxic agrochemicals, which pollute streams and aquifers. These are essential differences between peasant and agribusiness territories.

As Fig. 5.3 shows, the boundaries between the territories are very well defined and represent diverse landscapes because of the social relations that produce them and by them are produced. Thirty years ago, in the Pontal, all of these territories were ranches of extensive pasturelands for breeding and fattening beef cattle. The arrival of the MST in the region changed an uninhabited landscape with the construction

Fig. 5.3 The yellow lines demarcate the Che Guevara, Antonio Conselheiro and Paulo Freire agrarian reform settlements

of thriving communities. More recently, the high price of fossil fuels brought agribusiness corporations, sugarcane and ethanol mills to the area, intensifying conflict.

The struggle for land carried out by the MST is alive with the memories of struggling families. As an example, note the names given to the conquered territories: the settlements. Che Guevara, Antonio Conselheiro, and Paulo Freire are well-known names in the fight for land, for human rights and against capitalism. The construction of settlements changes names and landscapes of latifundia as well as agribusiness territories. The latifundia usually have names of saints. Figure 5.4 shows a billboard that reveals the dispute over territorial nomenclature. After the conquest of the Fazenda Santa Clara in Mirante do Paranapanema, the landless decided to rename the settlement for Che Guevara, but faced resistance from government authorities. The negotiation led to a compromise in which the settlement was denominated 'Santa Clara "Che Guevara".'

Fig. 5.4 A 2008 billboard announces a construction project on an elementary school in the Assentamento Santa Clara Che Guevara. The name reflects land struggle tensions

Historical awareness sharpens during the experience of struggle to change the dominant structure and established order. The families shown in Fig. 5.5 fought for nearly ten years to win the land. Gaining use of the land is often referred to as a 'conquest,' reflecting the hard-fought process of agrarian reform. The occupation of the latifundia created a space of continued struggle and resistance to transform the old agribusiness territory into a new peasant territory.

The foreground of Fig. 5.6 shows rows of beans, while in the background the peasant's home and other infrastructure are visible. This is the same Antonio Conselheiro settlement that is shown in Fig. 5.3. From a street-level perspective, Fig. 5.6 focuses on a fraction of peasant territory the conquest of which resulted from the spatialisation of the struggle for land. Years of conflict have generated popular knowledge important to the process of creating spaces of struggle like those shown in Figs. 5.7 and 5.8 (Fernandes 1996; Origuela 2011). In the case of these figures,

Fig. 5.5 In 1993, the MST organised occupations by landless peasants and other workers in Mirante do Paranapanema, where many agrarian reform projects are now situated

Google Earth depicts a space of struggle—the MST's Dorcelina Folador encampment in Mirante do Paranapanema—from two different angles. As a set of moving spaces, the encampment is a good example of the third territory in the typology referenced above. It is a moveable space of political socialisation open to all families who want to fight for land. It is an interactive space, a space of exchange of experiences of struggles, culture, and tasks necessary to implement agrarian reform.

The landless families began their socio-spatial movement with a process of changing the landscape of Fazenda São Domingos, the latifundia occupying the top portion of Fig. 5.7 and the left side of Fig. 5.8. Rather than celebrate a saint, the MST commemorates Dorcelina Folador, a member of the MST who grileiro gunmen murdered in 1999. The São Domingos latifundia, an area eligible for agrarian reform due to its long-time unproductive use to breed cattle as well as its questionable purchase history, was leased for sugarcane production was the struggle began in 2011. Its transformation from pasture to plantation

Fig. 5.6 Bean production on a farm in the Antonio Conselheiro Settlement in 2002

Fig. 5.7 Aerial view of the Dorcelina Folador encampment in Mirante do Paranapanema

Fig. 5.8 Street view of the Dorcelina Folador encampment in Mirante do Paranapanema

was meant to protect it from appropriation. Even after eight years of struggle, agrarian reform had not been achieved. In Fig. 5.7 one observes the camp as a white horizontal line that borders a protected conservation reserve. This helps militants avoid direct conflict with the landholder, who has already been accused of ordering gunmen to shoot at them, leading some to be wounded over the years. If the families were to conquer São Domingos, they would produce a new peasant territory fraction, increasing by one the number of yellow outlined polygons in Mirante.

Fazenda Da Barra, Ribeirão Preto

In 2009, a road separated the Mario Lago agrarian reform settlement from the remnants of the Fazenda da Barra. Figure 5.9 represents the new status quo of one contested landscape. In June 2007, the settlement was created out of an appropriated fragment of the fazenda after a lengthy struggle. One can see two trees and 6–10 homesteads with gardens; on the right side, monoculture, a monotonous yellowish green field of sugarcane, demonstrating the homogeneity of agribusiness territory and the heterogeneity of peasant territory. The road curves to the right, projecting the homesteaders' occupation into the cane fields. The image captures precisely the contemporary struggle for agrarian reform in Brazil. On one side of the conflict, corporate and crony agribusiness interests push an agenda of large-scale agriculture, land grabbing, and monopoly. On the other side, a number of farmworker and poor people's movements struggle for agrarian reform. This side of the image shows how peasant reterritorialisation impacts landscape. In this fragment of peasant territory, families reproduce themselves and produce coffee, beans, rice, papaya, passion fruit, bananas, tomatoes, lettuce, cucumbers, and animals such as dairy cattle, chickens and pigs. In contrast, sugarcane is the single immediate product of agribusiness territory.

If we take a step back we note a long history of dispute over the occupation of the region where the city of Ribeirão Preto is located (Welch 1999). This large commercial and residential centre had a population of over 600,000 people in 2015. More than 500,000 additional people

Fig. 5.9 A partial view of the Mario Lago Settlement in 2009 (Photo credit Douglas Mansur)

live in Ribeirão's micro-region, which includes another 15 municipalities. The Kayapó people first farmed there, producing maize, manioc, and diverse fruits. Brazilians began their aggressive colonisation of the area into the nineteenth century, encouraging their cattle to trample the Amerindian crops. As demand for coffee grew, the region saw coffee plantations built on the backs of enslaved African and Afro-Brazilian workers until they devised a form of exploitation called the *colonato*, which depended on indentured European immigrant families. These so-called colonists grew counted on self-provisioning to eat, cultivating vegetables and raising animals, such as hogs and chickens.

Into the 1960s, coffee predominated on the land, government incentives encouraged planters to transition to sugarcane to respond to the U.S. decision to import sugar from Brazil after blockading revolutionary Cuba. The oil crisis of the 1970s and the development of ethanol powered motors quickly completed the transformation of the region from coffee to sugarcane. In the years to come, sugarcane would come to substitute almost every other form of farming in the micro-region. During the 1990s, as machines gradually displaced migrant labour—one cane harvester did the work of 100 human cane cutters—the landless workers movement grew in the region. By the twenty-first century, with 98 per cent of Ribeirão Preto's agricultural land dedicated to cane production, dozens of land occupations challenged the government to implement agrarian reform (Welch 2006).

Out of this history, urban-based mediators joined together with the sons and daughters of displaced colonists and migrant harvesters to wrest control of a fraction of the Ribeirão Preto sugarcane industry's territory by forging the Fazenda da Barra Sustainable Development Project, including the Mario Lago settlement. A street-level view of the landscape revealed the conflict—as Fig. 5.9 demonstrates—while missing a good bit of its complexity. Other scales were exposed from the bird's eye view now easily accessible through Google Earth. We have selected two views (Figs. 5.10 and 5.11) from these sources to further examine this contested landscape.

The first selection shows both the larger outline of the area and several internal divisions. While Fig. 5.9 emphasises the rural context, Fig. 5.10 illuminates the urban context. To the left (just outside the image) a

Fig. 5.10 A Google Earth image of the Fazenda da Barra Sustainable Development Project. The colour outlines added by geographer Tiago Cubas (2017) show the boundaries of segments where certain organisations predominate in the support of settler families. Red indicates the Mario Lago Settlement, where the MST is hegemonic. Blue indicates the Santo Dias Settlement, where the Landless Liberation Movement is dominant. Yellow shows the Índio Galdino portion, in which no single organisation holds hegemony

Fig. 5.11 Google Earth image shows details of the Mario Lago settlement, including lot configurations, agro-forestry, and street names

major highway separates this area of Ribeirão Preto from the urban perimeter. Another barrier, this one natural, is evident in the upper right corner where a curve in the Prado River contains the agrarian reform project. To the right, shown above, a sugarcane field bounds Mario Lago. Thereafter, the boundaries are mixed among residential areas, forests, and areas apparently stripped bare for construction. Suddenly, the landscape conflict gains complexity. Housing's territorial competition with the agrarian reform project appears more pressing than its conflicts with plantations. Indeed, urbanisation is a fierce competitor in the land use contest.

Figure 5.10 shows a vast difference in residential density between nearby neighbourhoods and the settlements. Each rectangular area in the project represents the homestead of one to two families and their crops (see also Fig. 5.11). The same area in nearby residential districts accommodates around 20 houses, each supporting at least one nuclear family. Another obvious contrast is that of green space. While this characteristic predominates in the agrarian project, it is nearly non-existent in the residential areas. 'Green space' is evident in the plantation areas, but there is no forestland as found in the settlements. Among sustainable development models, one of the most sustainable is agro-forestry. This is the alternative settlers opted for in the organised parts of the project (red and blue demarcations). The woods attract and retain more water, moderate temperatures, refresh the air, provide fuel as well as windbreaks to slow topsoil erosion and, depending on the variety of trees, offer income through harvesting fruits and nuts (Cubas 2017).

In their fight to create the settlement, landless movements skilfully used sustainability concerns, finding that current urban residents opposed the aggressive elimination of precious open space and diminishing land use diversity (Borelli Filho 2009). Of special concern was at subterranean scale: the Guarani Aquifer, one of the world's largest sources of subterranean fresh water. Spreading out from under the Ribeirão Preto area, a huge region of South America depends on for its fresh water. Campaigners made an issue of the damage toxic agricultural fertilisers and chemicals cause to underground water. These are some of the reasons why the final agrarian reform project emphasised 'sustainable development.' Townspeople, exhausted by decades of 'cane-ash rain,' responded positively to these arguments, tipping public opinion towards support for establishing an agrarian reform landscape.

Figure 5.11 provides a closer look inside Mario Lago. In this image, the landscape of family lots takes fuller form, agro-forestry, roads, and commercial enterprises (note the fish and carry business [*pesqueiro*] in the upper left corner), vastly different from the monoculture that was the Fazenda da Barra. The smaller patches of cultivated land at each homestead testify to the diversity of the project. The large forested area and numerous trees planted in the fields and around the buildings confirm both the diversity and adherence to the agro-forestry model. The location of the buildings also underscores the autonomy of settlers, each determining the relationship between their homestead and the roads, fields, and neighbours. The lot sizes speak to the equalitarian nature of agrarian reform projects, which also contrast sharply with the dramatically different living conditions of a planter or sugar/ethanol mill owner and his employees.

The names of the roads are clear commentaries of the ideological perspective of the organisations involved. On Mario Lago—named for the popular leftist Brazilian poet and *sambista*—the roads that facilitate transportation, communication, and community building within the entire settlement are named for other popular heroes of the MST, like Antonio Conselheiro and leftist intellectuals, like Josué de Castro and fallen MST militants, like Roseli Nunes. In contrast, Ribeirão Preto, a city the elite celebrate as Brazil's 'agribusiness capital,' has streets named for the killers of indigenous peoples, grileiros and planters who built their fortunes on the backs of subaltern classes.

Conclusion

The smiling men in Fig. 5.12, as well as the fruit, lettuce and other produce shown, remind us that human beings are a fundamental part of the contested landscape narrative. After years of struggle, MST militants like the Ferreiras helped turn an old sugarcane latifundia into a productive agro-ecological farm cooperative. For four years, they fought not only their own fears, but the police, the landowner's gunmen, government agencies, the courts, and bad weather to establish an agrarian reform settlement that provided 264 families with land on which to build homes and livelihoods. Their struggle for the land and on the land turned the Fazenda da Barra from a green desert into a thriving, innovative farm community contributing cutting edge knowledge to an agro-forestry model of territorial development (Borelli Filho 2009; Cubas 2017). While street-level photos like this one show the human face of struggle, aerial and satellite photography has helped demonstrate other dimensions and scales of contested landscapes.

Internationally, the contest over landscape between small-scale agriculturalists (peasants) and large-scale agriculture (agribusiness) has a long visual history (Burke 2001; Kimbrell 1996). But conflicting representations became more common in the twentieth century, especially during periods of economic crisis, like the Great Depression of the 1930s, when the plight of rural folk received renewed attention from artists and

Fig. 5.12 The sales booth of a Mario Lago settler Lucinei Ferreira (in apron) and his sons at the Rural Producer's Fair at Curupira Park in Ribeirão Preto in October 2017

states concerned with instability. But a new wave of agricultural industrialisation followed World War II—resulting in the 'green revolution' of the 1960s, which stimulated a massive rural exodus, increased food shortages, and environmental deterioration. More recently, the genetic revolution in agriculture has deepened rural poverty and increased the influence of capital (Wirzba 2004), but many are the representations of the power of technology over nature. All of these factors have influenced landscape, as peasants sought to stay on the land and grileiros sought to expel them. In general, the countryside became more and more dehumanised, as people left it in a frustrated search for stability, prosperity, and happiness in towns and cities. Pictures like Fig. 5.12 shows how the Brazilian agrarian reform movement has worked to reconfigure rural spaces with life, embracing both the reproduction of families and the productivity of the land.

These contemporary landscapes reflect processes Brazilian geographers have described as territorialisation, desterritorialisation, and reterritorialisation. This dynamic, non-linear, inconclusive process suggests a cyclical process of occupying land, being pushed off it and retaking it. In some settings, such as the United States and Western Europe in the 1980s, analysts noted a 'rural renaissance' (Kayser 1988). In Latin America, a 'new rurality' was identified in the 1990s (Wanderley 2000). These authors commented on statistics documenting a certain reduction in urban growth rates and a tendency towards a 'return' of people to the countryside. In the developed world, these trends were most notable among urban professionals, who seemed to seek a spiritual revival in nature (Jacob 1997). Taking advantage of fire-sale prices, they bought the homes and land of indebted farmers, took over the houses and rented the land to 'real' farmers.

This trend was not agrarian reform, but it did contribute to valorising rural places, tarnishing the brilliance of big city life as a universal goal. The agrarian reform struggles have benefited from these tendencies as they were also accompanied by greater demand for the consumption of cleaner, healthier, and less wasteful foods, like those produced by Mario Lago settlers. In other words, an increasing number of consumers sought to reduce their consumption of toxic pesticides, hormones, antibiotics, and other technological and chemical interventions in food production. Demand for locally produced, seasonal foods also grew with these changes and the food sovereignty movement responded to this demand by broadening criticism of industrial and corporate agriculture in an attempt to diminish its capacity to territorialise geographical and political 'space.' In the contest over landscape, there is little question as to the predominance of agrarian capitalism, but there is also little question as to the perseverance of the agrarian question.

References

Adderley, P., and C. Mills. 2014. Landscapes of Conflict. *Landscapes* 15(2): 98–103.

Borelli Filho, D. 2009. *A Luta pela Posse da Terra na Região de Ribeirão Preto: O Processo de Formação e Organização do Assentamento Mário Lago*. MSc dissertation, Araraquara, UNIARA.

Buainain, A., A. Romeiro, and C. Guanziroli. 2003. Agricultura Familiar e o Novo Mundo Rural. *Sociologias* 10: 312–347.

Burke, P. 2001. *Eyewitnessing: The Uses of Images as Historical Evidence*. London: Reaktion Books.

Castilho, A.L. 2012. *Partido da Terra: Como os Políticos Conqusitam o Território Brasileiro*. Editora Contexto: São Paulo.

Coletti, C. 2002. Ascensão e Refluxo do MST da Luta pela Terra na Década Neoliberal. *Idéias* 9 (1): 49–104.

Cubas, T.E.A. 2017. O São Paulo Agrário na Era da Globalização. Ph.D. thesis, Presidente Prudente, UNESP.

Girardi, E. et al. 2017. *Relatório DATALUTA Brasil 2016*. NERA—Núcleo de Estudos, Pesquisas e Projetos de Reforma Agrária: Presidente Prudente.

Feliciano, C.A. 2009. Território em Disputa: Terras (Re)tomadas no Pontal do Paranapanema—São Paulo. PhD thesis, USP, São Paulo.

Fernandes, B.M. 1996. *MST: Formação e Territorialização*. Hucitec: São Paulo.

Fernandes, B.M. 2008. Questão Agrária: Conflitualidade e Desenvolvimento Territorial. In *Luta pela Terra, Reforma Agrária e Gestão de Conflitos no Brasil*, ed. A.M. Buainain, 173–224. Campinas: Unicamp.

Fernandes, B.M. 2009a. Reforma Agrária no Governo Cardoso e no Governo Lula: Desafios da Geografia Agrária. In *Tradição Versus Tecnologia: as Novas*

Territorialidades do Espaço Agrário Brasileiro, ed. I. Falcade and R.M.V. Medeiros, 171–188. Porto Alegre: UFRGS.

Fernandes, B.M. 2009b. Sobre a Tipologia de Territórios. In *Territórios e Territorialidades: Teorias, Processos e Conflitos*, ed. M.A. Saquet and E.S. Sposito, 197–215. São Paulo: Expressão Popular.

Goodman, D., and D. McCool. 1999. *Contested Landscape: The Politics of Wilderness in Utah and the West*. Salt Lake City: University of Utah Press.

Graziano da Silva, J. 1997. O Novo Rural Brasileiro. *Nova Economia* 7 (1): 43–81.

Hecht, S. 2010. The New Rurality: Globalization, Peasants and the Paradoxes of Landscapes. *Land Use Policy* 27: 161–169.

Jacob, J. 1997. *New Pioneers: The Back-to-the-Land Movement and the Search for a Sustainable Future*. University Park: Pennsylvania State University Press.

Kautsky, K. 1988. *The Agrarian Question.* Trans. P. Burgess. London: Zwan Publications.

Kayser, B. 1988. Rural Renaissance in the United States: The Viewpoint of a French Geographer. *Geographical Perspectives* 61: 77–88.

Kimbrell, A. 1996. *Fatal Harvest: The Tragedy of Industrial Agriculture*. Sausalito, CA: Foundation for Deep Ecology.

Leite, J.F. 1998. *A Ocupação do Pontal do Paranapanema*. São Paulo: Hucitec.

Marsden, T. 1989. Restructuring Rurality: From Order to Disorder in Agrarian Political Economy. *Sociologia Ruralis* 29 (3/4): 312–317.

Newell, P., and O. Taylor. 2018. Contested Landscapes: The Global Political Economy of Climate-smart Agriculture. *Journal of Peasant Studies* 45 (1): 108–129.

Origuela, C. F. 2011. Paradigma e metodologias da questão agrária: Uma análise das ocupações de terras no Brasil com ênfase para o Pontal do Paranapanema [Agrarian question paradigm and methodologies: Analysis of land occupations in Brazil with emphasis on the Pontal do Paranapanema region]. (Unpublished Senior thesis). Universidade Estadual Paulista, Presidente Prudente, Brazil.

Passos, M.M. 2004. A Paisagem do Pontal do Paranapanema: Uma Apreensão Geofotográfica. *Maringá: Acta Scientiarum* 1 (1): 177–189.

Paulino, E.T. 2014. The Agricultural, Environmental and Socio-political Repercussions of Brazil's Land Governance System. *Land Use Policy* 36: 134–144.

Pereira, L.I. 2015. *O Processo de Estrangeirização da Terra no Brasil: Estudo de Caso da Empresa Umoe Bioenergy no Município de Sandovalina-SP. Undergraduate dissertation.* Presidente Prudente: UNESP.

Pitta, F., and M.L. Mendonça. 2015. *A Empresa Radar S/A e a Especulação com Terras no Brasil*. São Paulo: Editora Outras Expressões.
Santos, M. 1996. *A Natureza do Espaço*. São Paulo: Hucitec.
Via Campesina. 2006. *O Latifúndio dos Eucaliptos*. Porto Alegre: Pontocom Editora.
Wanderley, M.N.B. 2000. A Emergência de uma Nova Ruralidade nas Sociedades Modernas Avançadas: O 'Rural' como Espaço Singular e Ator Coletivo. *Estudos Sociedade e Agricultura* 15: 87–145.
Welch, C.A. 1999. *The Seed was Planted: The São Paulo Roots of Brazil's Rural Labor Movement, 1924–1964*. Pennsylvania State University Press: University Park.
Welch, C.A. 2006. Movement Histories: A Preliminary Historiography of Brazil's Landless Laborers' Movement (MST). *Latin American Research Review* 41 (1): 198–210.
Welch, C.A. 2009. Camponeses: Brazil's Peasant Movement in Historical Perspective (1946–2004). *Latin American Perspectives* 36 (4): 126–155.
Wirzba, N. (ed.). 2004. *The Essential Agrarian Reader: The Future of Culture*. Washington, DC: Community and the Land. Shoemaker and Hoard.

6

Land Inequality in Brazil: Conflicts and Violence in the Countryside

Artur Zimerman, Kevin Campos Correia, and Marina Pereira Silva

Introduction[1]

When Brazil became ended the long two decades of military dictatorship in 1985, there was an expectation that inequality among its population, whether in cities or rural areas, could have been eventually addressed, with the poor being represented in the political spheres and have their demands for better living conditions vindicated in a country, until then, very unequal. Concerning the rural poor, there was an opportunity often to improve their precarious condition was a deep agrarian

[1] This chapter is an updated and expanded version of the chapter *Desigualdade de Terra* (Land Inequality) published in *Os 'Brasis' e Suas Desigualdades* ('*Brazils*' and its Inequalities), UFABC University Press, 2022. The first author would like to acknowledge the support received from the Sao Paulo Research Foundation (FAPESP), grant #2020/12847-1.

A. Zimerman (✉) · K. C. Correia · M. P. Silva
Federal University of ABC (UFABC), Santo André, Brazil
e-mail: artur@ufabc.edu.br

A. A. R. Ioris and B. Mançano Fernandes (eds.), *Agriculture, Environment and Development*, https://doi.org/10.1007/978-3-031-10264-6_6

reform that would allow the possibility to feed themselves properly and even sell their surplus. Almost four decades later, there has been no significative reduction in inequality and frustration has taken over the neediest, causing eruptions of collective violence in confrontation with powerful landowners to change this zero-sum game in their favour, with the support of social movements.

In the course of this chapter, we shall see that in Brazil there are many more people living in rural areas than what is reported in official publications. In this way, Brazilian inequality is not only verified in income terms, but similarly in land ownership and tenancy and how it is distributed across the rural population. As land and income are concentrated in a few hands, a significant number of farmers who do not own land would fall short of the benefits that land ownership assures them. Hence, Brazil does not differ much from the region in which it is located, Latin America. If the country is now one of the world's largest producers and exporters of agricultural commodities (Sondergaard and Silva 2020; Souza et al. 2020; Calil and Ribera 2019), this production originates basically from agribusiness farms and large-scale production (Sondergaard 2020), probably different from the 'old style large landowner', or *latifundiário* concerning food production in scale, but similar in principles, keeping land inequality very present in the region as it was before.

The manifested intentions of the Brazilian State to redistribute land through agrarian reform did not bear fruit and, as much as Cardoso's (1995–2002) and Lula's (2003–2010) administrations altered the distribution of land in quantitative terms compared to their predecessors and successors, the type of the agrarian reform performed in their governments was insufficient, making it clear that the preferential national policy would be disproportionately privileging agribusiness to the detriment of family farming since the redemocratisation (Sencebe et al. 2020). There had been a contradiction between the Ministry of Agriculture (MA) and the Ministry of Agrarian Development (MDA), with one clashing towards the other. It was a way for different governments to try to settle conflicts in the rural areas that did not work to reduce tensions between agribusiness and family farms or even reduce the inequality in the countryside.

Thus, the difference between the expectations that the less favoured segments of the rural population had in a democratic country was that the State would be carrying out a significant agrarian reform. However, the political and economic elite, largely interested in preserving the land status quo, managed to keep any and all governments away from carrying out the agrarian reform dreamed of by the rural marginalised. We will show the agrarian reform that did not actually occur (at least not in the imagined and necessary dimensions to backtrack on inequality), in addition to trying to explain the reasons for this not happening and the consequences in terms of violence generated by the bad performance of the Brazilian State in public policies that should be aimed at the rural population as a whole in order to reduce inequalities and poverty that has persisted for many decades, perhaps centuries. Land inequality is the villain for the poorest in the interior of Brazil and one of the main roles that democratic governments should fulfil would be to reduce this gap between layers of the population.

The Brazilian Agrarian Reform

Faced with the context of accumulated and ingrained inequalities in terms of access to land, several actions were carried out by the federal government, in its various administrations, made up of different political parties and leaders. Here, we only deal with the period of redemocratisation after 1985, since only democracy considers citizens in their full meaning, a definition that includes the rights inherent to citizenship. Some attitudes in relation to giving access to land to the most vulnerable who live in the countryside were taken, even with necessary additional conditions besides land redistribution (Fitz 2018). Among them was the creation of the Extraordinary Ministry for Land Affairs (MEAF) in 1982, which opened a way to the Ministry of Agrarian Reform and Development (MIRAD) in 1985—the former only being extinguished in 1989, when its competencies were incorporated into the Ministry of Agriculture (Zimerman 2017). In that respect, it is worth observing that the Ministry of Agriculture had as its main responsibilities the agrarian reform and the development of agriculture. Being directly aligned with the objectives of the Land Statute—created in 1964, at the beginning of the military period—and used by the rulers to appease the peasants and reassure the large landowners, it was an artifice to stop the expansion of communism. in Latin America a few years after the Cuban Revolution (1959), and at the height of the Cold War, with the support of the United States. Therefore, the Land Statute has been managed since its creation only to develop agriculture, that is, not achieving a serious and comprehensive agrarian reform, which would embrace those who needed it most (Silva 1996, 2015).

Major changes in terms of agrarian reform only took place after half of the first term of Cardoso (FHC), when in 1996 the Extraordinary Ministry of Land Policy (MEPF) was created, which later changed its name to the Ministry of Land Policy and Family Agriculture (1999) and later on came to be called the Ministry of Agrarian Development (MDA), existing until mid-2016 and whose structure was, to a large extent, preserved even in governments of different and opposition parties, for example, the Workers Party (*Partido dos Trabalhadores*—PT), and the Party of the Brazilian Social Democracy (*Partido da Social Democracia Brasileira*—PSDB). During the Temer government (in May 2016), this ministry (MDA) was reduced to a secretariat and linked to the already existing Ministry of Social Development (MDS). Finally, in the Bolsonaro government (since 2019), this secretariat was emptied of its functions and transferred to the National Institute of Colonisation and Agrarian Reform (INCRA) several of its responsibilities and changed its name to Secretariat of Family Agriculture and Cooperatives, falling under the responsibility of the Ministry of Agriculture, Livestock and Supply (MAPA) which until 2016 explicitly defended the interests of agribusiness and counterbalanced the now non-existent MDA. Therefore, from that moment on, the interests of family farming were completely left aside by the federal administration in a clear statement of what the government's priority would be.

In the seventeen years (1999–2016) of its existence, the MDA's objectives focused on promoting agrarian reform; in addition to ensuring the sustainable development of family farming (Presidency of the Republic 2000); and identify, recognise, delimit, demarcate and title lands occupied by the remnants of *quilombola* communities (Presidency of the

Republic 2003). It is in this institutional scenario that there was a certain congruence in the creation and development of public policies for settlements, reaching a high number of agricultural units for the low-income population, as never seen before, and which can be observed in the table below (Zimerman 2017). In the Cardoso's and Lula's governments, the agrarian settlement grew from four to eight times when compared to the Collor/Franco's and Sarney's administrations with regard to the number of families settled per year according to the period of their mandates (Robles and Veltmeyer 2015) (Table 6.1).

Since the end of the Rousseff's administration, there has been a notable reduction in settlement policies. The phenomenon is justified by the cutback of social demand for land, motivated by the rural exodus and the growing public policies of income transfer implemented in the country (called *Bolsa-Família*), in addition to the lack of a strategic vision for the countryside and its effects of the political alliance of the federal government with the agribusiness and technocracy (Human Development 2016). At Temer and Bolsonaro's governments, the reduction was even higher than before, and the agrarian reform practically became derisory. During this period, mainly in Bolsonaro's government, we had a process of dismantling the regulatory apparatus regarding the registration of pesticides, in addition '[...] Bolsonaro says that there will be no agrarian reform and is maintaining the FPA'S "Positive Agenda", especially the denationalisation of land that constitutes a

Table 6.1 Agrarian reform and democratic governments, 1985–2020

Democratic governments	Settled families	Average number of settled families/year
Sarney (1985–90)	82,689	16,538
Collor/Franco (1991–4)	35,600	8900
Cardoso (1995–2002)	584,655	73,082
Lula (2003–10)	614,088	76,761
Rousseff (2011–15)	133,635	26,727
Temer (2016–2018)	10,077	5038
Bolsonaro (2019–2020)	9222	4611
Total (1985–2020)	**1,465,355**	**40,704**

Source Area ministries of each government

direct attack on the demarcation of indigenous and quilombo territories' (Barbosa 2020: 107). The Bolsonaro government has antagonised agrarian movements, especially *Movimento dos Trabalhadores Sem Terra* (acronym MST in Portuguese, for Landless Workers Movement), given that '[…] the option of the MST and the other peasant social movements is to strengthen alliances for the unity of popular forces in the defence of democracy and in active resistance to the intensified violence in countryside and the criminalisation of social struggle' (Barbosa 2020: 110).

On a major scale, despite advances in numbers and, respectively, in guaranteeing access to land for the poorest and most vulnerable—especially in the Cardoso's and Lula's governments—it is important to highlight that the challenge of providing the conditions for these families to be able to earn their livelihood, in fact, generate food for their subsistence and potentially to profit in this market, witnessed a setback. The agrarian reform foresees the reduction of hunger in the places where there is more poverty, which are the rural areas, as previously mentioned (Wittman and Blesh 2017) together with other additional public policies to reduce hunger and extreme poverty, such as *Fome Zero* (Zero Hunger) did in Brazil, created in 2003 by the Brazilian federal government. Especially during the food crises that increasingly affect the most vulnerable in the countryside, they could be alleviated with the redistribution of land accompanied by access to credit and technical agricultural knowledge. Families living in and out of the countryside could subsist and eat as a result of what they produce.

The food crises of 2007/8, 2011/2 and now with the Covid-19 pandemic, where countless families completely lost their livelihood, would be able to have fresh food and a decent living condition. However, unfortunately, the signal that recent governments have given is that this path is no longer open. In addition, with the decrease in the numbers concerning settled families, and with the worsening of the recent economic crises in Brazil, we perceive a strong orientation for investments in public policies in the area to continue to decline. However, contrary to what official data from Latin American countries (including Brazil) publish, there are probably many more people being affected by the step backward of agrarian policies than perceived, as we will deal in the next section.

Rural Population vs. Urban Population in Latin America and Land Inequality

In Brazil, the definition of what a city is, and the differentiation from rural to urban area, follows the parameters of the Brazilian Institute of Geography and Statistics (*Instituto Brasileiro de Geografia e Estatística*—IBGE in Portuguese), which have been used since the period known as *Estado Novo* (New State) between 1937 and 1945 (Silva 1946). Not too far apart, other Latin American countries adopt very similar definitions. In this way, other criteria to understand and measure urban versus rural areas are not taken into account, for example, the definitions of international organisations such as the Organisation for Economic Cooperation and Development (OECD) and the World Bank, that shed light on the issue of population density—comprising up to 150 inhabitants per square kilometre; the infrastructure conditions and the distance from another city with more than 100,000 inhabitants (Abramovay 2000; Girardi 2008; Gomes 2013; Rodrigues 2014).[2] In general, urban areas show a smaller, less unequal disparity than rural areas in terms of income and land area ownership/tenancy (Long et al. 2021). Certainly, different definitions of what is urban and what is rural produce different statistic results (Wineman et al. 2020), and apparently that was what happened to many Latin American countries regarding the proportion of its urban and rural populations.

This distinction is relevant when we think in terms of social inequality: in the world, nowadays, three quarters of the poor live in rural areas or at least the majority of them (Borras and Franco 2012; Souza et al. 2020) and this situation will not change unless more progress on rural development occurs (OECD 2016), or a massive rural exodus happens. Thus, when assumed as part of this measurement, infrastructure, populational density and distance to a bigger city make a considerable difference for the area to be rural or urban. Latin American countries that, according to their own biased definitions, have around 24% of the peasant population, which would, respectively, shift to 46% of their population as rural inhabitants, if we take the reference recommended by the World Bank

[2] See the following chart for more details.

and OECD into consideration. In other words, it would almost double the rural population (Ferranti et al. 2005) by updating the definition of what is urban and rural.

The tendency to use outdated, inappropriate definitions regarding the demographic issue of developing countries is that they do not want to be labelled archaic or backward (Barbier 2020). In most of the collective imagination, to present a large population that lives in the countryside is to be out of step with modernity concerning non-developed or developing countries. This view proved to be wrong, since there are developed countries with a large population living in the countryside, such as Liechtenstein with 86%, Portugal with 34% and Italy with 29% (World Bank 2018), even many just live but not work there. Therefore, the real rural population is often disregarded in the official data of several countries, and in Latin American countries specifically. According to Veiga (2004), if the characteristics determined by the OECD were considered, approximately 33% of Brazil's population would live in rural areas, as opposed to the 15% defined by Brazilian IBGE. This discrepancy generates greater inequality and wrongly influence policymaking, as it is not possible to think of public policies for a population who, statistically, does not exist.

That said, the Latin American dichotomous view of the urban as the representation of modernity and the development of countries, while the rural represents underdevelopment and lack of perspectives, is incorrect (Perez 2001). The evidence is that a large part of the exports of the countries in the region is from agricultural production, which, according to data from the World Bank (Chaherli and Nash 2013), represents a substantial portion of the region's GDP. Paradoxically, agribusiness involves fewer rural workers due to the technological revolution that reached the countryside and the already consolidated agricultural machinery (Rizov 2020). Those who manage agribusiness in the Amazon region, for example, live in the cities and move to the countryside more in times of sowing and harvesting, following a logic typical of globalisation (Ioris 2018). Looking at Chart 6.1, we can see that, apart from Guatemala, the analysed countries disregard relevant data in their censuses, which once again points to an impact on public

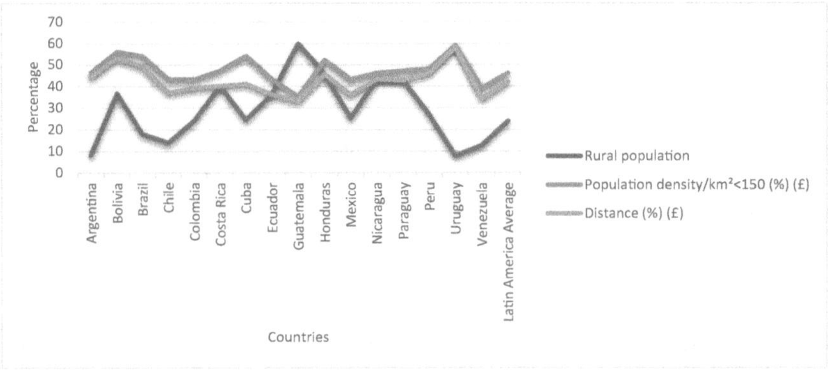

Chart 6.1 Contrasts of the Rural Population in Latin America in 2002 (*Source* (*) Indicators of world development of rural population, based on World Bank (2002). *Source* (£) Population density and distance to major centres, extracted from CIESIN/GPW [2002])

policies designed for the rural population of Latin America, especially to the poor living in the countryside (Zimerman 2014).

In general, Latin American democracies are quite recent (from the mid-80s) and have repressive remnants of previous dictatorships' governments. However, when social movement leaders start to confront elites and local power in search of rights and greater equality in the land distribution, the results are often violent conflicts (Zimerman 2014). The Latin American region has the highest levels of land inequality in the world (see Table 6.2). Whether it is concentrated in the hands of exporters associated with large companies and agribusiness, or in the hands of local elites and landowners—exploitation colonisation legacy—land inequality favours the perpetuation of a large mass of impoverished people who subsist in terrible conditions. For Frankema (2010: 418):

> The concentration of land in the hands of small political elite is widely regarded as an unfavorable initial condition of economic development in Latin America, as compared to the more egalitarian distribution in many East Asian countries, where profound rural reforms and supportive agrarian development policies created favorable conditions for balanced and sustained long-run growth.

Table 6.2 Descriptive statistics of the Land Gini divided into 13 regions of the world

Region	min	max	mdn	mean	St.dev	cv	obs
South America	63.9	86.3	80.4	79.9	6.3	0.08	11
Central America	60.7	78.3	73.9	72.3	6.0	0.08	7
Caribbean countries	46.2	81.6	69.9	68.1	11.8	0.17	7
East-Asia	30.7	43.8	39.5	38.4	5.5	0.14	4
South-Asia	41.8	62.3	55.4	53.7	8.7	0.16	6
Southeast-Asia	29.1	68.0	47.3	47.9	11.7	0.24	8
North Africa/Middle East	56.3	82.0	63.8	65.1	7.3	0.11	12
South and East Africa	36.8	83.5	66.7	62.7	17.4	0.28	12
Central and West Africa	31.2	68.1	45.2	45.2	9.1	0.20	14
Western Ramifications	47.0	78.6	61.1	61.9	16.4	0.26	4
Western Europe	47.0	79.1	63.4	63.9	10.1	0.16	14
Eastern Europe	39.2	60.0	52.4	51.0	95.0	0.19	4
Scandinavian countries	42.1	63.3	47.2	49.3	7.5	0.15	8
World	29.1	86.3	60.0	59.7	15.0	0.25	111

Source Frankema (2010: 425)

Therefore, from these inequalities in access to land, allied to globalisation, and also to the period of democratisation, which most of these countries went through in the middle of the 80s of the last century, the emergence of organisations and movements that fight against the governments and transnational agricultural conglomerates for the defence of low-income people who live in the countryside, such as the case of La Via Campesina, which operates in about 70 countries, advocating sustainable small-scale agriculture (Zimerman 2016). Added to this movement are others who work punctually on the issue of agrarian reform and whose main method of action is the occupation of land. Relevant examples, in this sense, are the MST and the Confederation of Agricultural Workers (*Confederação Nacional dos Trabalhadores na Agricultura*, acronym CONTAG in Portuguese), both based in Brazil.

Such social movements emerged basically with the aim of making access to land for the less favoured residents of rural areas more equitable, allowing the poor to ascend socially and have at least the minimum conditions of livelihood and food, reducing the gap between large landowners/agribusiness and the other workers/small landowners/tenants and, with that, decrease the inequality prevailing in the countryside.

According to Frankema (2009), there is clear inequality of land distribution, among the twenty most unequal countries, sixteen are located in Latin America. Thus, the claim that there is a type of Latino inequality can be analysed in the existing data.

In the above table, we can see the land inequality ranking, which presents South and Central America (Latin America at large), in addition to the Caribbean, as the three most unequal regions on the planet (FAO 2017). If we consider that most of the poor people live in rural areas, we realize that '[...] the fight against land inequality is the first step towards reducing poverty and income inequality [...]' (Zimerman 2016: 13), at least in Latin America. In this scenario, it is necessary to implement agrarian reform for the region. But not a reform that is done just to divide and redistribute land, but a reform that includes financing, access to technical knowledge and equipment, in order to produce and generate wealth (Kay 2019). Only a comprehensive agrarian reform in countries where there are many poor people living in rural areas would provide decent conditions and reduce the persistent inequality that affects the countryside in Latin America. It's politically necessary in democracies, where the demand of the less favoured strata must be considered and heard as they are represented in political forums (Ansell and Samuels 2014; Carter 2010). It is also socially necessary, since in the twenty-first-century countries cannot have this social disparity, with first- and second-class citizens living side by side (Oyvat 2016).

In terms of inequalities in access to land in the Brazilian case, what we can identify is that effective and extensive agrarian reforms seem to be politically improbable, as previously discussed. However, analysing the evolution of the Gini Index regarding the distribution of land according to Federation Units, as illustrated in the table below, we can see that, in more than three decades of democratic governments, inequality has not been reduced in Brazil as a whole no matter which government ideological spectrum is in power. Oyvat (2016: 220) carried out a study on agrarian structures in income inequality, reaching the conclusion that as there are about half of the world's population living in rural areas, policies that favour small farmers can have a more positive impact, with development of more egalitarian models emerging. Therefore, the Gini presented here of Brazilian states may be relevant to land tenure and its level of inequality, as both are interconnected, following Oyvat discussion (Table 6.3).

Looking at the data, it is possible to verify, that the great concern is not the scarcity of land in the country, since Brazil has continental dimensions, but its distribution. About half of Brazilian arable land is concentrated in the hands of only one per cent of landowners (Robles 2018; Wolford 2010), which is one of the highest rates of land concentration in the world. The extremely unequal division of this natural

Table 6.3 Evolution of the Gini Index, by Brazilian federation units—1985–2017

Federation units	Evolution of the Gini Index			
	1985	1995	2006	2017
Brazil	0.86	0.86	0.86	0.86
Rondônia	0.66	0.77	0.72	0.74
Acre	0.62	0.72	0.72	0.75
Amazonas	0.82	0.81	0.84	0.84
Roraima	0.76	0.81	0.67	0.81
Pará	0.83	0.81	0.82	0.85
Amapá	0.86	0.84	0.85	0.88
Tocantins	0.71	0.73	0.79	0.78
Maranhão	0.92	0.90	0.86	0.88
Piauí	0.90	0.87	0.86	0.85
Ceará	0.82	0.85	0.86	0.85
Rio Grande do Norte	0.85	0.85	0.82	0.79
Paraíba	0.84	0.83	0.82	0.82
Pernambuco	0.83	0.82	0.83	0.80
Alagoas	0.86	0.86	0.87	0.86
Sergipe	0.86	0.85	0.82	0.81
Bahia	0.84	0.83	0.84	0.85
Minas Gerais	0.77	0.77	0.80	0.79
Espírito Santo	0.67	0.69	0.73	0.74
Rio de Janeiro	0.82	0.79	0.80	0.81
São Paulo	0.77	0.76	0.80	0.83
Paraná	0.75	0.74	0.77	0.79
Santa Catarina	0.68	0.67	0.68	0.70
Rio Grande do Sul	0.76	0.76	0.77	0.79
Mato Grosso do Sul	0.86	0.82	0.86	0.87
Mato Grosso	0.91	0.87	0.87	0.88
Goiás	0.77	0.74	0.78	0.79
Distrito Federal	0.77	0.80	0.82	0.86

Source IBGE, agricultural census, 1985–2017

resource, crucial for the livelihood of the population living in and from the countryside, leads to violent disputes and deaths of rural residents, as we will see throughout this chapter. It is important to highlight that in the years of the Cardoso's and Lula's governments, there was a favourable scenario for the improvement of land distribution. Although, as a characteristic of Brazilian presidentialism, often the alliances formed for the necessary governance are not, and in this case were not, ideologically coherent. One of the biggest challenges faced by governments for the implementation of public policies in this context in the country is the 'Ruralist Block', as the group of parliamentarians who defend the interests of large landowners and agro-industries is known. According to Zimerman (2014: 12),

> These parliamentarians belong to numerous parties, including many from the governing base, both in the Cardoso, Lula, [Rousseff, Temer and Bolsonaro] administrations, indicating ambiguities in votes in both houses of the Brazilian congress on issues related to land distribution and agricultural activity. In the 1994 and 1998 elections alone, there were about 150 parliamentarians who formed the *Bancada Ruralista* [Ruralist Block] and its ideological range passed from the right to the center-left, making the distribution of land in the country politically unfeasible (Hammond 2009: 164).

Another important issue in opening a propitious scenario for a better distribution of land after redemocratisation was the enactment of the Federal Constitution of 1988, which, in its chapter III, Articles 184 and 185, establishes that land must fulfil its *social function* and be productive. Unproductive lands are subject to expropriation for agrarian reform purposes. This measure ends up curbing the existence of a group of large landowners that is still common in Brazil, who are country estate owners without capital to invest in the productivity of their lands. Among these (but not only), we can include the "squatters" (known as *posseiros* in Portuguese), and the *grileiros*, who falsify documents to make it appear they are the real landowners. The existence of these large unproductive properties already increases the risk of deaths from agrarian disputes, also among landowners and squatters (Zimerman 2014). Even with all these instrumental changes in the Constitution, which favoured the new democratic atmosphere to reduce land inequality in the countryside, several obstacles such as the existence of this Ruralist Bloc in Parliament continued to be an impediment to advancing agrarian reform and reducing inequality in the countryside.

Regarding the total Gini index of agricultural establishments by municipality in Brazil (according to IBGE), throughout the period 1995–1996, the first years of Cardoso's administration, and in 2006, the last year of Lula's first term in power, we noticed that, while the first period shows a greater concentration of land in the Northeast region, in a strip from the north of the southeast region, and in the Midwest region to the southern stretch of the Northern region (including part to the north of the state of Amapá), the second period shows an increase in inequality in the country as a whole. In addition, there is a subtle decrease in land concentration in the Northeast region, however, a large increase in the North region, mainly in areas of the Amazon Forest, with more deforestation and agricultural expansion, probably for large rural producers. This Amazonian deforestation is intrinsically related with inequality (Brown et al. 2016; Oliveira and Silva 2021; Zimerman 2018) due to the expansion of production area to the detriment of the clearing of the forest, causing even more inequality among those who inhabit the Amazon region.

The land ownership structure in Brazil leads to two controversial models, one that is widely privileged by the establishment with measures to encourage agribusiness characterised by the accumulation of land, and the other that values the survival of the humblest social groups in small pieces of land, which was left to its own fate due to the negligence of governments, but with great support from social movements. We will, in the next section, verify the abrupt growth of agribusiness in the country in recent decades, showing the model that was chosen to prosper by the state and what was despised and left in the background, resulting in conflicts over land disputes and the governments incompetence to manage both rural poverty and inequality reduction and increasing violence in the countryside.

Food Production, Land Grabbing and Violence Resulting from Land Disputes

Agriculture plays an important role in terms of GDP generation in Brazil and several other countries in the region, and it has expanded substantially over the years (Bojanic 2017) (Table 6.4).

Bojanic (2017) says that the main responsible for the technology behind production growth was the Brazilian Agricultural Research Corporation (Embrapa), whose recommendation was to increase the use of fertilisers, machinery and pesticides, the improvement in logistics infrastructure and research and extension helped to reach this level of growth and quality. This large-scale production success most likely comes as a result of a phenomenon called land grabbing, besides other elements as well. It is nothing more than the acquisition of cheap and fertile land in non-developed and/or developing countries, mainly by investors from other countries that do not have enough fertile land (such as the rich oil-exporter countries of the Persian Gulf) and/or climatic conditions suitable for agricultural production (Nordic countries) and/or population too large to feed (China), or a combination of these phenomena (Rulli et al. 2013), aiming at food security and sovereignty for the purchasing countries, or their representatives. This scenario contributes to small farmers abandoning their land and living in unemployment or underemployment as rural workers and, in many cases, they start to migrate to large urban centres in search of better opportunities, which

Table 6.4 Change in agricultural production in 1996 and 2015, by main products

Product	Production in 1996 (in millions of tons)	Production in 2015 (in millions of tons)	Percentage increase 1996/2015
Sugar	13.6	37	172
Coffee	1.3	2.6	100
Maize	25.5	76.2	200
Soy	21.5	95.6	344
Meet	207.0	310	50
Chicken	54.6	107	96

Source Bojanic (2017: 5), with data from the Agriculture Ministry, CONAB, IBGE and FAO

in most cases does not occur successfully. Thus, it is worth noting that inequalities in terms of access to land are intensified.

This very lucrative results can be obtained in the production of food for humans or livestock, and inputs for the production of bioenergy (Garcia-Arias et al. 2021). In addition, land values in undeveloped and developing countries are comparatively low with those in developed countries and target countries generally have few restrictions on selling land to foreigners besides to the market transforming it into a capital commodity and increasing its value, making it even more difficult for landless workers, squatters, tenants or small landowners to acquire it, increasing the gap from large conglomerates to family farming units that have been reducing in quantity and area size (Reydon and Fernandes 2017a, b; Varrotti and Gras 2021). A high volume of land is purchased so that production can be carried out in scale, with modern machinery, introducing high technology and little manpower. Such undertakings do not lead to the hiring of many local workers, nor to gains for the destination country, in addition to implying fluctuations in food prices, since much of what is produced is destined for export. The greatest loss is for the small producers who may find them under pressure to sell the land or just live it if not legally titled, losing their livelihood, and many times it happens through coercive means and violence (Dell'Angelo et al. 2021).

Another interesting aspect to point out, which directly affects the livelihood of small producers and rural and landless workers, in addition to the environment, is the replacement of biodiversity by the monoculture model in huge areas of fertile land (Ahmad 2017; Hunsberger et al. 2017). A large part of the agribusiness production is headed for bioenergy, such as sugarcane to produce ethanol, a very spread type of energy for vehicles and industry. In terms of the environment, there are researchers who have discovered that this generation of ethanol pollutes much more than gasoline, originated of fossil oil, because for this supposedly clean production, a great amount of diesel is used in tractors as fuel, and fertilisers/pesticides used to make it more efficient and profitable, polluting the environment even more than gasoline (Times of India 2009). When it comes to the livelihood of small producers, if there is a large area to produce biofuels, less available space for food production, probably generating a rise in food prices, which directly affects the poorest population, which spends more money than the usual due to supply and demand rule (Tomei and Helliwell 2016; Zimerman 2014).

According to data recorded by Land Matrix (2022), between the years 2000 and 2015 in Brazil alone, there were 274 large land foreign deals made for the area of 9,193,335 ha. It represents the total area of land deals carried out just below Argentina (with 310 deals and 10,502,167 ha) and Peru (with 147 deals and 18,898,302 ha) in the region. In the last decades there has been a change in the profile of investors in this sector, going through the financialisation of land, making it a commodity, and speculating in the market, through shares on the stock exchange, investment funds, pension funds and countless other ways to profit from the capital point of view. According to studies, Brazil receives about a third of global financial investments in land, with the rest shared with other countries (Clapp 2019; Reydon and Fernandes 2017a: 1163). This is a very large amount of capital that enters to speculate on the financial market and directly affects the lives of small producers and rural workers who are expelled from their original productive life.

In this way, State action in guaranteeing the reduction of inequalities in access to land becomes essential. Several countries impose restrictions on the acquisition of land by foreigners, which would leave more land for family farming and without being speculative, causing values to rise rapidly, preventing humbler people who live from agriculture from being able to buy land. A much-discussed solution is the implementation of agrarian reform, which would allow farmers to acquire land, in addition to technical knowledge and obtain financing under friendly conditions to produce and keep families away from poverty and misery. As already discussed above, Brazil has very high inequality in access to land and, despite the actions taken in recent decades to promote agrarian reform in the Brazilian countryside it's far from being achieved, which contributes to the fact that the numbers of conflicts over land disputes are still quite notorious. Below, we present a table with data on deaths resulting from these conflicts according to region, state and municipalities, allowed by the perception of inequality, including in the distribution of deaths due to land disputes in the country. We call them agrarian deaths, which are deaths resulted by land conflicts, registered by the Land

Pastoral Commission (CPT, in its acronym in Portuguese), which is part of the Catholic Church and takes care of issues related to land and land conflicts, helping the neediest communities in rural Brazil, and that have registered all kinds of rural conflict data since the 1970s.

The North region of the country stands out for having 48% of fatal victims in land disputes, while the Northeast has 27%, the Midwest 12%, the Southeast 8% and the South 5%, in the period of the research, from the establishment of the democratic regime in the country (1985). These numbers precisely demonstrate the existence of a heterogeneity that refers to conflicts and deaths in the country, requiring, in this order, focused actions of public policies for specific regions and places by the State, at federal, state and municipal levels. The interdisciplinarity of public policies, with security, education, health, employment, income supplement for the neediest and land policy to reduce inequalities also in the countryside, is essential to reduce violence generated by land conflicts as well (Table 6.5).

Furthermore, it is noted that regardless of which political entity is in the federal government, whether the PSDB or the PT (both that have alternated for no less than 24 years, even with 32 political parties represented in the Congress[3]) the number of deaths from land disputes remains unchanged and high. What is most striking is that the PT has a mobilising power in civil society and articulates directly with various social movements, including mass rural social movements, such as the MST (Zimerman 2017). And even so, there was no significant difference in the number of agrarian deaths, when compared to the PSDB, its direct electoral competitor at the federal level until 2018: in the Cardoso's administrations (1995–2002) there were 270 fatal victims, while in the Lula's governments (2003–2010) and Rousseff's (2011–2016), respectively, 304 and 185 deaths. That said, we present a graph with the number of agrarian deaths by land disputes per year, also covering the period from 1985 to 2020 (Chart 6.2):

In this context, we can analyse the small increase in deaths generated by land disputes in PT administrations from the frustration of those

[3] In 2022, there were 32 political parties registered with the Superior Electoral Court (TSE), the body responsible for registering political parties and for elections in Brazil. Data retrieved from https://www.tse.jus.br/partidos/partidos-registrados-no-tse.

Table 6.5 Agrarian deaths by Brazilian state, 1985–2020

Region	Federation unity	Deaths	Municipalities (deaths/municipality)	Deaths/100,000
North	RO (Rondônia)	150	52 (2.88)	**8.35**
	AC (Acre)	19	22 (0.86)	**2.10**
	AM (Amazonas)	70	62 (1.13)	**1.66**
	RR (Roraima)	31	15 (2.07)	**4.75**
	PA (Pará)	564	143 (3.94)	**6.43**
	AP (Amapá)	7	16 (0.44)	**0.81**
	TO (Tocantins)	46	139 (0.33)	**2.89**
North region total		887	449 (1.98)	**4,69**
Northeast	MA (Maranhão)	175	217 (0.81)	**2.45**
	PI (Piauí)	16	223 (0.07)	**0.49**
	CE (Ceará)	20	184 (0.11)	**0.22**
	RN (Rio Grande Norte)	8	167 (0.05)	**0.23**
	PB (Paraíba)	29	223 (0.13)	**0.72**
	PE (Pernambuco)	76	185 (0.41)	**0.79**
	AL (Alagoas)	25	102 (0.25)	**0.75**
	SE (Sergipe)	5	75 (0.07)	**0.22**
	BA (Bahia)	138	417 (0.33)	**0.92**
Northeast region total		492	1793 (0.27)	**0.93**
Southeast	MG (Minas Gerais)	86	853 (0.10)	**0.40**
	ES (Espírito Santo)	15	78 (0.19)	**0.37**
	RJ (Rio de Janeiro)	35	92 (0.38)	**0.20**
	SP (São Paulo)	16	645 (0.02)	**0.03**
Southeast region total		152	1668 (0.09)	**0.17**
South	PR (Paraná)	55	399 (0.14)	**0.47**
	SC (Santa Catarina)	13	293 (0.04)	**0.18**

(continued)

Table 6.5 (continued)

Region	Federation unity	Deaths	Municipalities (deaths/municipality)	Deaths/100,000
	RS (Rio Grande do Sul)	20	496 (0.04)	**0.18**
South region total		88	1188 (0.07)	**0.29**
Midwest	MS (Mato Grosso Sul)	47	78 (0.60)	**1.67**
	MT (Mato Grosso)	147	141 (1.04)	**4.17**
	GO (Goiás)	23	246 (0.09)	**0.32**
Midwest region total		217	465 (0.47)	**1.31**
Brazil total		**1836**	**5563 (0.33)**	**0.86**

Source Self-made, with extracted data from the Pastoral Land Commission reports, 1985–2020 and IBGE, while the population estimate dates to the year 2021

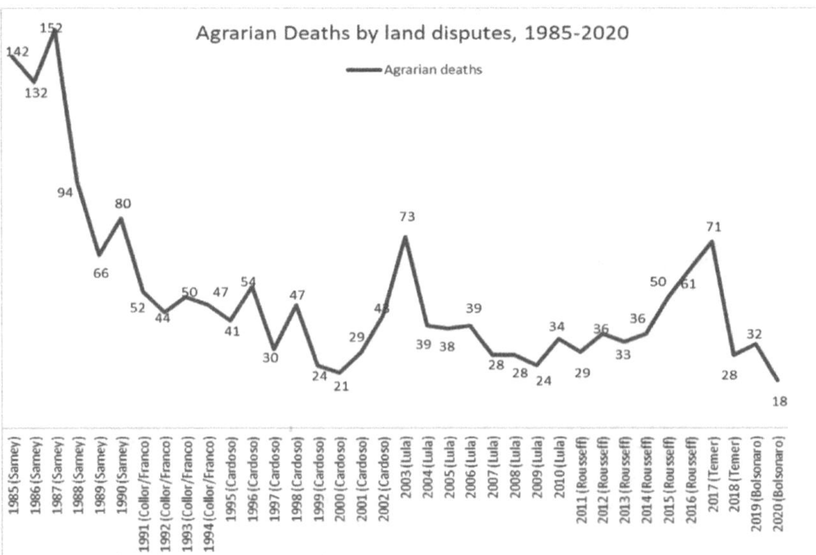

Chart 6.2 Agrarian Deaths in Brazil, 1985–2020 (*Source* Land Pastoral Commission, 1985–2020 and IBGE)

who understood that this government would defend more firmly the less favoured and would have greater action with regard to agrarian reform, which, in fact, in that matter, did not happen. The governance and alliances made by the PT, including the Rural Block, which seeks the status quo in land inequality, were probably responsible for maintaining this pattern of agrarian violence at its current level. This deals with the governability to pass laws, projects and amendments through the two parliamentary houses (Chamber and Senate), when there is a need for a broad coalition, whatever ideology it has. In this way, it is hard to fight for agrarian reform when a government that is ideologically in favour of this policy cannot implement it the way it would like due to the difficulty of keeping the government coalition intact.

In the short period of time in power by the Bolsonaro government (since 2019), although deaths resulting from land disputes in Brazil have greatly reduced, according to the Pastoral Land Commission reports, several other human rights violations of the most vulnerable rural population in the country occurred in different ways. It is important to elucidate the proliferation of various organisations and civil society movements in Brazil during the democratisation process, which deal with the issue of inequalities in access to land and, to a large extent, strive and guide their actions to achieve better distribution, through agrarian reform. In this sense, Zimerman (2016) points out the existence of 98 different active associations, between the years 2000 and 2011, acting on several fronts to demand a better redistribution of land to the less favoured in the countryside. Land occupations, as a way of claiming a more egalitarian redistribution and using the constitutional prerogative that the land must satisfy a social function, otherwise it can be subject to expropriation for social purposes, such as agrarian reform, has led to many social movements to perform this action. The MST and the CONTAG stand out with greater performance in occupations in the Brazilian case, which together account for 62.9% of occupation actions, or 2673 (52.5%) and 530 (10.4%), respectively, each of them in the period included in this same study (Zimerman 2016). Such a prerogative usually leads to confrontations between the parties involved and results in deaths due to land disputes, in addition to other kinds of violence.

Since the eighteenth century, riots have been the most common type of demonstration worldwide. Two-thirds of them are targeted at hunger and have become known as food riots (Griffin 2021). It is especially during food crises that food riots proliferate. For example, in the 2007–8 crisis food riots spread to about 30 countries (Patel and McMichael 2009). We still don't have significative studies on the serious food and unemployment crisis produced during the Covid-19 pandemic. Such demonstrations in the struggle for food subsistence in undeveloped and developing countries usually start out peacefully and are fought violently, generating a spiral of violence with security forces. Therefore, the victims add up. The struggle for access to land and the reduction of agrarian inequality by the more vulnerable who live in the countryside is a way of fighting hunger and fighting for human dignity. Democratic regimes should pay attention to their population and assist with emergency measures for those in need. However, it is important that the state be a mediator of conflicts, and a proposer of long-term solutions that go through the reduction of poverty and violence in the countryside and not just in emergencies.

Conclusions and Lessons Learned

We have seen in the course of this chapter that land inequality is present in all Brazilian regions in the countryside. This is one of the main factors responsible for the expansion of poverty in these areas, increasing the vulnerable rural population's dissatisfaction due to several governments' inaction or inability, spreading on spirals of violence and continuing absence of public policies to reduce the impact of land inequality in the livelihood of rural inhabitants. As there are more people than is officially announced living in rural areas, without public policies taking care of promoting the reduction of inequalities in this sector of society, for a mere matter of, in the official statistical data, a good part of this rural population is non-existent and not accounted for the promotion of socioeconomic well-being, there is no change in the attitude of the Brazilian State towards the issue of inequality in the countryside.

Currently, we can't treat issues related to the countryside as backwardness. Food production saves lives (especially in times of economic crisis), generates jobs and collects taxes. This sector must be seen and treated differently by the State than it is the case. Brazil is a reasonably well-placed country in the world ranking when only considering the situation of the average per capita income and land. However, the inequality present in Brazilian society is among the largest in the world and is persistent in its character, requiring political will and an effort on the part of the State to break this vicious cycle and reverse this trend that has plagued the country for a long time. Only in this way will large and small farmers be able to live side by side and help the country to grow. On the one hand, bringing in revenue and, on the other hand, reducing poverty and hunger. Likewise, we must see the two types of producers/farmers, not in a competitive way, but as supporters of a rich country, less unequal and with social justice.

Since the democratisation in Brazil (from 1985) several social movements and specifically rural social movements have emerged and make the represented masses within the countryside more aware of their own situation and the rights the full citizenship is about. These movements led in a struggle for agrarian reform where everyone could live with dignity and be able to support themselves with their own work in the countryside, reducing hunger and poverty. Nonetheless, the inherent violence of the confrontation between opposing interests of large landowners/agribusiness and the landless/small landowners, with the State not always knowing how to best match this conflict, caused victims, many of them fatal, to arise. Such struggles did not change the situation of inequality in the countryside and in the cities, and the agrarian reform was inexpressive, with the political system, through its elites, making it difficult to walk towards a more just and egalitarian society. The Brazilian elite has privileged the proliferation of agribusiness due to the lack of regulation and the creation of policies that encourage this model, making it difficult to continue and create new settlements for the most vulnerable population living in the countryside. Leaders need to position themselves in the face of present and future challenges, giving priority to caring for their vulnerable and poor so that society can be strengthened and move forward. Everything passes through the agricultural sector within developing countries, which is the key to poverty reduction and elimination of hunger.

The big obstacle in this scenario seems to be the political system, where a coalition of the president with other political parties is always necessary to compose a majority and have the governability as mentioned earlier. However, how is it possible to improve the situation of rural vulnerable people when a significant part of the government coalition is in favour of large extensions of land and agribusiness to the detriment of small landowners and the landless? Something needs to change and that involves political reform. Brazil is a rich country with many poor people. To modify this, it is necessary to fight rural poverty as a priority, not as governmental policy but as state policy, avoiding being hostage by different administrations. And what causes this extreme poverty in the countryside is the alarming land inequality, as seen. Almost four decades of continuous democratic governments in Brazil have shown that there are first- and second-class citizens. Democracy cannot persist in such a situation. Reducing violence over land disputes involves reducing inequality via the redistribution of land and income for a more just and prosperous society.

References

Abramovay, R. 2000. Funções e Medidas para a Ruralidade no Desenvolvimento Contemporâneo. *Texto para Discussão*, 702. Brasília: IPEA.

Ahmad, N.B. 2017. Blood Biofuels. *Duke Environmental Law and Policy Forum* 27 (2): 265–315.

Ansell, B., and D.Samuels. 2014. *Inequality and Democratization: An Elite-Competition Approach*. Cambridge Studies in Comparative Politics. Cambridge: Cambridge University Press.

Barbier, E.B. 2020. Is Green Rural Transformation Possible in Developing Countries? *World Development* 131: 104955.

Barbosa, L.P. 2020. Challenges Facing Latin American Peasant Movements Under Progressive Governments and New Right-Wing Parties—The Case of Brazil. *Latin American Perspectives* 234: 94–101.

Bojanic, A.H. 2017. The Rapid Agricultural Development of Brazil in the Last 20 Years. *Euro Choices* 16 (1): 5–10.

Borras, S.M., and J.C. Franco. 2012. Global Land Grabbing and Trajectories of Agrarian Change: A Preliminary Analysis. *Journal of Agrarian Change* 12 (1): 34–59.

Brown, D.S., C.J. Brown, and C. Brown. 2016. Land Occupation and Deforestation in the Brazilian Amazon. *Land Use Policy* 54: 331–338.

Calil, Y.C.D., and L. Ribera. 2019. Brazil's Agricultural Production and Its Potential as Global Food Supplier. *Choices* 34 (3): 1–12.

Carter, M. 2010. The Landless Rural Workers Movement and Democracy in Brazil. *Latin American Research Review* 45 (4): 186–217.

Chaherli, N., and J. Nash. 2013. *Agricultural Exports from Latin America and the Caribbean: Harnessing Trade to Feed the World and Promote Development*. Washington, DC: World Bank Press.

Clapp, J. 2019. The Rise of Financial Investment and Common Ownership in Global Agrifood Firms. *Review of International Political Economy (RIPE)* 26 (4): 604–629.

Dell'Angelo, J., G. Navas, M. Wittemana, G. D'Alisa, A. ArnimScheidel, and L. Temper. 2021. Commons Grabbing and Agribusiness: Violence, Resistance and Social Mobilization. *Ecological Economics* 184: 107004.

Desenvolvimento Humano. 2016. *O Que o Governo Dilma Fez (e não fez) pela Reforma Agrária?* Retrieved from https://www.socioambiental.org.

FAO. 2017. *Latin America and Caribbean Regions Present the Major Land Distribution Inequality*. Rome: Food and Agriculture Organization.

Ferranti, D., G.E. Perry, W. Foster, D. Lederman, and A. Valdés. 2005. *Beyond the City: The Rural Contribution to Development*. Washington, DC: World Bank Press.

Fitz, D. 2018. Evaluating the Impact of Market-assisted Land Reform in Brazil. *World Development* 103: 255–267.

Frankema, E. 2010. The Colonial Roots of Land Inequality: Geography, Factor Endowments, or Institutions? *Economic History Review* 63 (2): 418–451.

Frankema, E. 2009. The Colonial Origins of Inequality: Exploring the Causes and Consequences of Land Distribution. In *Poverty, Inequality and Policy in Latin America*, ed. S. Klasen and F. Novak-Lehman, 19–45. Cambridge, MA: MIT Press.

Garcia-Arias, J., A. Cibils, A. Costantino, V.B. Fernandes, and E. Fernández-Huerga. 2021. When Land Meets Finance in Latin America: Some Intersections between Financialization and Land Grabbing in Argentina and Brazil. *Sustainability* 13 (14): 8084.

Girardi, E.P. 2008. *O Rural e o Urbano: É Possível uma Tipologia?* Ph.D. thesis, Unesp, Presidente Prudente.

Gomes, I. 2013. O Que é Rural? Contribuições ao Debate. *Boletim de Geografia* 31 (3): 81–95.

Griffin, C.J. 2021. Rural Workers and the Role of the Rural in Eighteenth Century English Food Rioting. *The Historical Journal* 64 (5): 1230–1256.

Hammond, J.L. 2009. Land Occupations, Violence, and the Politics of Agrarian Reform in Brazil. *Latin American Perspectives* 36 (4): 156–177.

Hunsberger, C., E. Corbera, S.M. Borras, J.C. Franco, K. Woods, C. Work, R. de la Rosa, V. Eang, R. Herre, S.S. Kham, C. Park, and S. Sokheng. 2017. Climate Change Mitigation, Land Grabbing and Conflict: Towards a Landscape-based and Collaborative Action Research Agenda. *Revue Canadienne D'études Du Développement* 38 (3): 305–324.

Ioris, A.A.R. 2018. Place-making at the Frontier of Brazilian Agribusiness. *GeoJournal* 83 (1): 61–72.

Kay, C. 2019. Land Reform in Latin America: Past, Present and Future. *Latin American Research Review* 54 (3): 747–755.

Land Matrix. 2022. Land Portal. Retrieved from https://landportal.org/book/datasets/datlm0.

Long, J.C., Delamater, P.L. and Holmes, G.M. 2021. Which Definition of Rurality Should I Use? The Relative Performance of 8 Federal Rural Definitions in Identifying Rural-urban Disparities. *Medical Care* 59 (10), Suppl.5: S413–S419.

OECD. 2016. *A New Rural Development Paradigm for the 21st Century: A Toolkit for Developing Countries.* Paris: OECD Publishing.

Oliveira, R.C., and R.D.S. Silva. 2021. Increase of Agribusiness in the Brazilian Amazon: Development or Inequality? *Earth* 2 (4): 1077–1100.

Oyvat, C. 2016. Agrarian Structures, Urbanization, and Inequality. *World Development* 83: 207–230.

Patel, R., and P. McMichael. 2009. A Political Economy of the Food Riot. *Review* (Fernand Braudel Center) 32 (1): 9–35.

Pérez, E.C. 2001. Hacia una Nueva Visión de lo Rural. In *Una Nueva Ruralidad en América Latina?*, ed. N. Giarracca, 17–30. Buenos Aires: Asdi/Clacso.

Presidência da República. 2000. *Decreto 3338*, de 14/01/2000. Brasília.

Presidência da República. 2003. *Decreto 4723*, de 06/06/2003. Brasília.

Reydon, B.P., and V.B. Fernandes. 2017a. Financialization, Land Prices and Land Grab: A Study Based on the Brazilian Reality. *Economia e Sociedade* 26: 1149–1179.

Reydon, B.P. and V.B. Fernandes. 2017b. Financeirização, Preços de Terra e *Land Grab*: Um Estudo Baseado na Realidade Brasileira. *Economia e Sociedade* 26: 1149–1179.

Rizov, M. 2020. A Vision of the Farming Sector's Future: What Is in There for Farmers in the Time of the Second Machine Age? *Local Economy* 35 (8): 717–722.

Robles, W. 2018. Revisiting Agrarian Reform in Brazil, 1985–2016. *Journal of Developing Societies* 34 (1): 1–34.

Robles, W., and H. Veltmeyer. 2015. *The Politics of Agrarian Reform in Brazil: The Landless Rural Workers Movement*. New York, NY: Palgrave MacMillan.

Rodrigues, J.F. 2014. O Rural e o Urbano no Brasil: Uma Proposta de Metodologia de Classificação dos Municípios. *Análise Social* 211: 430–456.

Rulli, M.C., A. Saviori, and P. D'Odorico. 2013. Global Land and Water Grabbing. *Proceedings of the National Academy of Sciences of the United States of America (PNAS)* 110 (3): 892–897.

Sant'Anna, A.A. 2016. Land Inequality and Deforestation in the Brazilian Amazon. *Environment and Development Economics* 22: 1–25.

Sencebe, Y., F. Pinton, and A.A. Cazella. 2020. On the Unequal Coexistence of Agrifood Systems in Brazil. *Review of Agricultural, Food and Environmental Studies* 101: 191–212.

Silva, J.G. 1996. Por uma Reforma Agrária não Essencialmente Agrícola. *Agroanalysis*, March, 8–11.

Silva, J.M.M. 2015. *O Estatuto da Terra: A Construção da Reforma Agrária Limitada do Primeiro Governo da Ditadura Civil-militar*. MSc dissertation, UnB, Brasília.

Silva, M. 1946. Tentativa de Classificação das Cidades Brasileiras. *Revista Brasileira De Geografia* 3: 283–316.

Sondergaard, N. 2020. Food Regime Transformations and Structural Rebounding: Brazilian State–agribusiness Relations. *Territory, Politics, Governance*, forthcoming.

Sondergaard, N., and R.D. Silva. 2020. Reshaping the Policy Arena: How the Agro-export Policy Network Propelled Brazil within Global Agricultural Governance. *World Trade Review* 19 (4): 550–566.

Souza, H.G., F.J.S. Tabosa, and J.A. Araújo. 2020. *Income Elasticities and Inequality of Poverty in Urban and Rural Areas of the Brazilian States: A Spatial Approach*. New York: United Nations.

Souza, G.S. 2020. Technological Progress in the Brazilian Agriculture. *Socio-Economic Planning Sciences* 72: 1–10.

Times of India. 2009. Biofuels Could Pollute More, Warn Experts. *The Times of India (1861–2010)*.

Tomei, J., and R. Helliwell. 2016. Food Versus Fuel? Going Beyond Biofuels. *Land Use Policy* 56: 320–326.

Varrotti, P.S., and C. Gras. 2021. Network Companies, Land Grabbing, and Financialization in South America. *Globalizations* 18 (3): 482–497.

Veiga, J.E. 2004. A Dimensão Rural do Brasil. *Estudos Sociedade e Agricultura* 12 (1): 71–94.

Wineman, A., D.Y. Alia, and C.L. Anderson. 2020. Definitions of 'Rural' and 'Urban' and Understandings of Economic Transformation: Evidence from Tanzania. *Journal of Rural Studies* 79: 254–268.

Wittman, H., and J. Blesh. 2017. Food Sovereignty and Fome Zero: Connecting Public Food Procurement Programmes to Sustainable Rural Development in Brazil. *Journal of Agrarian Change* 17 (1): 81–105.

Wolford, W. 2010. *This Land Is Our Now: Social Mobilization and the Meanings of Land in Brazil*. Durham, NC: Duke University Press.

World Bank. 2002. World Development Indicators (rural population). Retrieved from http://data.worldbank.org/indicator/SP.RUR.TOTL.ZS.

World Bank. 2018. World Bank Staff Estimates Based on the United Nations Population Division's World Urbanization Prospects. Retrieved from https://data.worldbank.org/indicator/SP.RUR.TOTL.ZS?most_recent_value_desc=true.

Zimerman, A. 2014. Violence Against Peasants in Latin America: Land Deals and the Food/Fuel Crop. *International Journal of Contemporary Sociology* 51 (2): 117–142.

Zimerman, A. 2016. A Piece of Land or Peace on the Land: How Much Is a Peasant's Life Worth in Brazil? *Journal of Human Security* 12 (1): 37–51.

Zimerman, A. 2017. Os Números não Mentem e as Aparências Enganam: Polarização Artificial PSDB x PT na Política Agrária Brasileira e a Violência Rural. *Nineth Congress of the Latin American Political Science Association (ALACIP)*, Montevideo.

Zimerman, A. 2018. Land Conflicts and Violence in the Brazilian Amazon Region. *Global Journal of Human-Social Science: F Political Science* 18 (5): 38–43.

Zimerman, A., and F. Pinheiro. 2020. Appearances can be Deceptive: Political Polarization, Agrarian Policy, and Coalitional Presidentialism in Brazil. *Politics and Policy* 48 (2): 339–371.

7

The Agrarian Question and the Rural Development Paths in the Periphery of Argentina: Past and Present in the Territorialisation of Peasantry in Santiago Del Estero

Cristian Emanuel Jara and Raúl Gustavo Paz

Introduction

Studies on the *agrarian question* consist of a more than 130-year-old research agenda related to the expansion of capitalism in the rural world and its tensions with non-capitalist modes of production (Akram-Lodhi and Kay 2012; Azcuy Ameghino 2004; Byres 2009; Fernandes 2014; Hocsman 2014). As a paradigm, the agricultural question includes a set of theoretical and methodological elements to address rural transformation processes through the analysis of the formation of social classes, the conflicts and alliances established among or within them, and the role of

C. E. Jara (✉) · R. G. Paz
Universidad Nacional de Santiago del Estero, Santiago del Estero, Argentina
e-mail: cristianjara_cl@hotmail.com

R. G. Paz
e-mail: pazraul5@hotmail.com

A. A. R. Ioris and B. Mançano Fernandes (eds.), *Agriculture, Environment and Development*, https://doi.org/10.1007/978-3-031-10264-6_7

the state in these processes. The early stages of debate about the agricultural question were focussed on the paths of capitalism development in central countries, where capital gradually imposed conditions on peasantry to further transform this rural actor into proletarians or members of the bourgeoisie.

In Capital, originally published in 1867, Marx (2003) highlighted the violent origin of capitalism in England, where the *primitive accumulation* destroyed the modes of communal use of land.[1] Later, Kautsky (1998) and Lenin (1981) indicated agrarian transformation processes different from the ones occurred in the English countryside. In the case of Prussia, the *Junkers* (the political, military and landowner elite) transformed their feudal farms into businesses. In the USA, the development of capitalism was characterised by the formation of the farmer and the expansion of middle-sized capitalist properties (Azcuy Ameghino 2004). In Latin America, the colonial experience and the continuity of the dependent relationships left an imprint on the agrarian question. Bartra (1989) used the term "*non-classical paths*" to refer to the modes of penetration of capitalism that did not necessarily lead to the disappearance of peasantry, but that survived through a diversity of strategies, such as partial incorporations to the market.

Within the frame of these debates, the aim of this chapter is to present a *situated approach*[2] to the agrarian question to explain the particular characteristics of Santiago del Estero, a province in north-western Argentina. Analysing this case in the context of theoretical debates on the agrarian question is highly significant because Santiago del Estero is home to the greatest proportion of rural population in the country (31.3%, against 3% in Buenos Aires) and because of the presence of family farming run by indigenous peasants that live in *agricultural holdings with undefined boundaries.*[3] Those spaces have no legal titles or deeds.

---

[1] Thus, the English path entailed the appearance of three social classes in the rural sector: landlords, tenant farmers, who produced for the market, and hired labourers, who received a wage for the sale of their labour.

[2] Auat (2021) states that situated reflection does not refer just to the circumstances that leave a particular bias to each thought, but to *critically taking the point of view from which someone talks and thinks.*

[3] The 2018 National Agricultural Census (NAC) recorded a total of 4519 *agricultural holdings with undefined boundaries* of a total of 15,531 agricultural holdings in Santiago del Estero

Indeed, their inhabitants are vulnerable to evictions due to the advance of agribusinesses. Therefore, this is a very problematic agrarian structure (de Dios et al. 2020). On the other hand, we are interested not only in the particular characteristics of agrarian processes at the provincial level but also in identifying the characteristics in common with the rest of Latin America, such as land concentration, rural violence and the prevalence of primary activity. As Kay (2002) indicates, these factors implied serious obstacles to combat poverty and distinguishes the development pathways that have occurred in Latin America from those in other countries (like Taiwan, South Korea, Singapore and Hong-Kong), where the agrarian reform and the contribution of peasantry to the generation of surpluses were key factors in the industrialisation strategy.

The present chapter identifies processes of different temporalities that formed and conditioned the current agrarian question in Santiago del Estero. As Braudel (1970) stated, we bring into dialogue *events* (short-term), *situations* (mid-term) and *longer-term processes*, such as the patterns of land occupation that started to be conceived in colonial times along with inequities of a country that concentrated wealth in the central region and the port in Buenos Aires. Regarding methodological aspects, it should be noted that our procedures are different from those of agrarian historians, since we use secondary sources rather than archives or other methods of historical research. This differentiation is important because this is a social studies approach that attempts to have greater sensitivity to history in the analysis of current agrarian issues; however, it does not mean the adoption of methods used by historians (Wald 2016).

In this line, this analysis attempts to understand the path of capitalist development in an unequal, disorganised and, therefore, highly problematic agricultural structure, where peasantry has a strong presence,

---

province. de Dios et al. (2020) expressed serious doubts about the reliability of the results of the last NAC and considered that this type of productive holdings may be under-reported. This may be due to several reasons, such as deficient census coverage due to difficult access through impassable roads to areas with a strong presence of holdings with undefined boundaries. Moreover, no surveys were made in areas characterised as not suitable for agricultural and forestry use, where these holdings are also located, and several lands occupied by peasants were not reported as agricultural holdings because they were used for production for own consumption.

although it has not been given a significant role in rural development policies. To address this proposal, we present information in three sections. First, we briefly describe the penetration of capitalism in Santiago del Estero from a multi-scale approach (considering national and international processes). Second, we argue that penetration of capital in the province does not follow one but several paths that lead to resistance and recreation of peasantry at different moments. Finally, we address the narratives about rural development and the place assigned to peasantry within the frame of the new and old agrarian question.

The Agrarian Issue in Argentina and Its Several Periods

For the purposes of our examination, we identify five periods of penetration of capitalism in the rural sector and the respective tensions in the patterns of land control, the role of the state and the peasant dispossession/resistance processes in Argentina, especially in Santiago del Estero. It should be noted that this periodisation is merely indicative, since periods do not respond to drastic ruptures, but only serve to precisely show those relevant features that persist, disappear and/or reappear in the agrarian question.

The colonial period (1500–1816). The present-day Argentina, which was under the administrative control of the Viceroyalty of Peru, was incorporated into the Viceroyalty of the Río de la Plata in 1776. The Spanish court chose Buenos Aires as the capital city, due to its easy access from the mother country through the Atlantic Ocean, among other reasons. It should be noted that Santiago del Estero had been founded several years earlier (1553) and was scarcely linked to Buenos Aires, since it was strongly integrated into the mining complex in Potosí, mainly through the production and trade of textile crafts. It should be underlined that colonial reorganisation reinforced the large-scale land privatisation, either in law or in fact, through legal instruments of the empire, such as *mercedes reales*.[4] Thus, a system of large estates or *estancias* was created. Such land concentration, along with an improper regularisation of land titles, persists to this day, especially in provinces of northern Argentina. *The transition period (1816–1880)*. After the declaration of independence in 1816, a slow process of infighting started, which led to the organisation of the Nation State and which took about seven decades. During the period of absence of a central national government, the provincial *caudillos* (political-military strongmen) consolidated their power and demanded autonomy from the Buenos Aires port. These governors had local power based on charismatic leadership and personal army, while kept some distance from liberalism.

In the case of Santiago del Estero, the agrarian structure underwent significant changes during the government of the Taboada family (1851–1875),[5] since vast areas of fiscal lands were sold to obtain income for conducting public works. Many of these land plots were considered "desert" or "vacant", although they were inhabited by peasants and indigenous people. In general, a few families close to the government had access to these plots of land in exchange of political and military favours (Banzato and Rossi 2010). One of the consequences was the consolidation of the *latifundios* (large holdings inherited from colonial times) and the patronage system, since there were people living in the granted lands, who inhabited the land despite the lack of titles and who were also in a relationship of submission to the patron or landlord (Tasso 2004).[6] Thus, this new land enclosure was performed with active participation of the Argentine state, which was developing and recently liberated from the Spanish metropolis.

---

[4] *Merced Real* was a legal Iberian institution that arose between the fifteenth and sixteenth centuries. It generally referred to a royal "donation" (by the Crown) of real-estate in exchange of a service.

[5] By the mid-twentieth century, the Taboada, a traditional family, took power (in alliance with the liberal elites that governed the country) and ruled the province for two decades.

[6] The patronage system implied a type of economic-social supremacy, by which the rural poor people were subordinated to the major traditional landowners (the patrons). This occurred in different ways, for example: the patrons demanded fidelity and political support from rural people working for them in exchange of providing employment, granting land plots temporarily or facilitating access to irrigation water from the patron's channel. This was the only possibility for rural workers who had no or insufficient land (Tasso 2004).

Period of hegemony of the agro-export led model (1880–1930). The ruling elites, identified with the ideals of economic liberalism, but politically conservative, designed a model for the incorporation of the country into the capitalist world system as supplier of raw material. In this context, the capitalist relationships expanded in the Humid Pampas, which was strongly integrated to the international markets. Other actors in the agricultural sector occupied unfavourable positions in this model. The peasant and indigenous communities, despite the successive dispossessions, survived in different regions of the country, usually inhabiting lands considered marginal by the capitalist enterprises at that time. Similarly, it was difficult for European immigrants to have access to a fair distribution of the plots, since the agrarian colonisation was interrupted due to the high speculation on land.

Beattie (2009), who reflected on these processes, asks why capitalist development in Argentina was different from the process occurred in the United States, despite the great similarities between countries, especially the fertile prairies. One of the answers of this author is that while the United States adopted the model of parcelling out new lands to small family holdings, Argentina gave the lands to a small elite. Therefore, the persistence of *latifundios* and inequality in access to land were some of the subsequent obstacles to modernisation of agriculture and transference of resources from the agricultural sector to the industrial sector. We do not intend to refute Beattie's arguments, since other elements that contribute to the historical complexity of both national experiences should be also considered. However, we stress a key conceptual aspect of our approach to the agrarian question: the central role of distribution and regularisation of land ownership in the development of a country.

At the time of expansion of capitalist agriculture in the Argentine Pampas, the *quebracho colorado* (*Schinopsis balansae*) cycle started in Santiago del Estero (Paz and Jara 2020). The companies exploited the forest to meet the demand for sleepers necessary for the expansion of railroads, which connected the agricultural production sites with the port for export. By the mid-twentieth century, these companies started to abandon the province, with drastic social and environmental consequences (Dargoltz 2003). Several unemployed people that stayed in these deforested lands underwent a repeasantisation process, i.e., they started to live in and work the land on their own account, although the land property titles were not regularised. Later, this was another key element to understand the persistence of agricultural holdings with undefined boundaries and the consequent current land conflicts.

Import-substitution industrialisation period (1930–1976). The worldwide economic downturn of the 1930s revealed the collapse of the agro-export-led model. Consequently, the national government adopted a policy to promote the national industrialisation with the aim of stimulating the manufacture of previously imported products. The state adopted protectionist measures aimed at transferring resources from the rural sector to the industry. The model depended on the foreign exchange from the agro-export sector. The Argentine northern region, in general, was incorporated to this development model in a marginal way. Consequently, many inhabitants had to migrate to the provinces of the central region of the country, where the manufacturing sector established (especially Buenos Aires, Córdoba and Rosario) in search of employment (Paz and Jara 2020). In the case of Santiago del Estero, after the companies that exploited *quebracho colorado* abandoned the province, and the consequent repeasentisation of formerly salaried workers occurred, a limited agricultural modernisation process started in the 1960s through the incorporation of new technologies and infrastructure (like the construction of Río Hondo Dam) that allowed the use of lands for crop cultivation. This period was characterised by the booming of cotton as the main commercial crop. It should be noted that the land revaluation due to the new dynamics in the market led to a new conflict scenario, since agribusiness entrepreneurs, in collusion with judges and politicians, evicted rural families that lived in those lands and still had not regularised land tenure.

The Neoliberal period and reprimarisation (1976 to date). The last civic-military dictatorship (1976–1983) stressed the veering to a more open economy, which was then reinforced with the return to democracy in 1983 (especially during the presidential mandates of Menem), generating the conditions that facilitated the expansion of agribusiness and high land tenure concentrations in fewer hands.[7] In the extra-Pampas regions (including Santiago del Estero), the expansion of the agricultural frontier accelerated through the introduction of export crops. Indeed, traditional crops (cotton and maize) were replaced and more lands were incorporated to the agribusiness. Therefore, evictions of the peasant communities increased and MOCASE (Spanish acronym for Peasant Movement of Santiago del Estero) was founded in 1990 with the aim of taking collective actions in defence of land (de Dios 2010).

A new period started after the social, political and economic collapse occurred in Argentina in 2001 due to neoliberal policies, when Néstor Kirchner became president. In this period, the role of the state was redefined, and the debate about the development models was reopened. This first decade of the twenty-first century was characterised by the increase of primary production boosted by the high international prices. One of the reasons for the high prices was the demand for agricultural products by China, a country that had been experiencing sustainable economic growth. Although the Humid Pampas was still the most dynamic region due to its integration to the agro-industrial complex, the price of land increased in other peripheral regions. Therefore, the level of social conflict also increased due to different territorial conflicts between agribusiness and peasant and indigenous communities.[8] During those years, the successive national governments played a dual role in the strong tensions of the agrarian question. On the one hand, the expansion of capital was facilitated, since the agro-export activity had historically been a generator of foreign exchange and incomes to the state coffers. On the other hand, there was an attempt to keep some levels of legitimacy through redistributive policies, despite the strong tensions with

[7] Between 2002 and 2018, 25.5% of the Agricultural Holdings (AHs) disappeared. One per cent of the most concentrated AHs occupy 36% of the total surface area of the sector, whereas 55% of the smaller AHs has 2% of the land (Manzanal and González 2021).

[8] Some of the tragic events occurred in Santiago del Estero related to land conflicts include: the death of Eli Juárez, who fainted and then and died when trying to stop eviction; Cristian Ferreyra was killed in the department of Copo in 2011; he bled to death after being shot by one of the private security guards of the soybean producers of the region.

the export sectors. Likewise, policy measures aimed at family agriculture were recognised and institutionalised.[9]

The mandate of President Mauricio Macri (2015–2019) was supported by economic groups linked to agribusiness, international finances and concentration of media ownership. In line with these sectors' interests, some of the actions taken were the reduction of export taxes on soybean and the closure of the National Registry of Rural Workers and Employers (RENATEA), an agency in charge of inspecting employment in the rural sector; then, massive redundancies of workers from the Family Agriculture Secretariat took place (Jara et al. 2019). In a scenario of macroeconomic failure of the conservative policies, Alberto Fernández won the presidential elections in 2019. After a few months of the start of his mandate, the COVID-19 pandemic was declared. Tensions between the advance of agribusiness and the resistance of rural populations were not interrupted during the pandemic (which had different phases). Monitoring conducted by Greenpeace (2020) revealed that over March–September 2020, 16,805 ha were deforested in Santiago del Estero, the province with the highest deforestation rate in the country, followed by Salta, Formosa and Chaco.

Overall, this section highlighted that one of the long-term dimensions of the agrarian question in Argentina is related to the continuity of its mode of insertion to the global capitalism as supplier of raw materials and an interrupted industrialisation process. That situation entailed a system of political, commercial and financial subordination to the different powers (mainly Spain, England, United States and China). In turn, this dependence had particular characteristics related to the different accumulation-dispossession phases. We especially highlight the

[9] The creation of FONAF (Spanish acronym for National Forum of Family Farming Organisations), in August 2006, with the participation of representatives of several rural organisations, generated the guidelines for public policies aimed at Family Agriculture. In this line, important regulations were issued, such as the Emergency Law of Indigenous Community Property No. 26160 (which suspended eviction proceedings from indigenous common lands and established a national-wide survey), Law N° 26331 (Native Forest act) and Law N° 26737 (which restricts foreign investment in land). In addition, Law N° 27118, on the "Historical Reparation of Family Agriculture for the Construction of a New Rurality in Argentina", was passed, but its implementation was not regulated.

capacity of peasants to persist, despite the continuous dispossession and violence.

The Agrarian Question and Its Heterogeneities

In this section, we explain why the classical paths (which were identified in European countries) are not applicable to analyse the development of capitalism in Santiago del Estero. Firstly, the English path is not applicable. Unlike what happened in the Humid Pampas, when by the end of the nineteenth century, landowners leased out lands to the immigrants for commercial production (Ansaldi 1993), in Santiago del Estero, the expansion of the capitalist relations in the agricultural sector followed another trajectory. Tasso (2004) analysed the capitalist expansion between 1870 and 1940. The author examined the impact of the expansion of irrigation and railroads on the formation of commercial agriculture in the central region of the province with the arrival of European immigrants. Those farmers (a family farmer with some degree of capitalisation), locally known as *finqueros*, made profits and were able to have access to land. However, during the 1930s, in the context of the great economic crisis, a *refeudalisation* phenomenon occurred when these immigrants that had achieved some accumulation decided not to take the risk, abandoned agriculture and went to the city to live off the interest on their own capital. Thus, the formation of an agricultural business class was a process that came to a standstill. Indeed, the traditional *estancia*, the indigenous peasant holding, the medium and small capitalisation farms in the risk zone, the *latifundios* in the dry areas and the salaried rural workers continued to coexist.

Secondly, the Prussian path does not apply to Santiago del Estero because the local large landlords did not show a complete process of modernisation from above. The elites, who inherited the land from colonial times or obtained the land through dubious business with the state, preferred to accumulate plots of land with the purpose of speculation and social prestige rather than production (Banzato and Rossi 2010). Many of these traditional families have maintained the titles without holding possession. With the increase in land value in the last decades of the twentieth century, the holders of the titles made transactions that generated long-term rural conflicts with populations that have inhabited those lands and worked de facto there (de Dios 2010).

Thirdly, the American path (capitalism from below) does not apply to Santiago del Estero, since farmers' colonisation projects in the province were very scarce and isolated. In addition, the situation of small-scale farmers that held possession (de facto) of land but without property titles and who have lived in and worked that land for decades, was not regularised; therefore, accumulation of their capital was not guaranteed. Tenti (2002) investigated the first agricultural colonisation in Santiago del Estero, which took place early in the twentieth century in the locality of Colonia Dora (southeast of the province). It was promoted by a Portuguese that organised the land parcelling out and encouraged the arrival of Italian and Jewish immigrants. This colony developed by the railway and near the Salado River. However, the experience had serious limitations due to water deficit and lack of government support (Tenti 2002). Likewise, during the 1970s, another attempt was made, Colonia El Simbolar, a locality where the programme efforts were focussed on achieving the formation of a group of family farmers through the modernisation of peasants (Forni et al. 1991). Nevertheless, that project faced great difficulties due to the successive economic crises that affected the country, among other factors.

Therefore, this particular complexity of the trajectory of the agrarian development in the province can be metaphorically characterised by resorting to Tasso's image of the *baetón* (Tasso 2004), a type of blanket woven in a loom by traditional craftspeople. In this sense, the agrarian question in Santiago del Estero is like this fabric woven with multiple strands: there they overlap, mix, divide, compete and resist different lines and colours.

The Agrarian Question and the Narratives of Rural Development

In this segment, we ask the following question: how has rural development been conceived in Santiago del Estero? Although the development concept was disseminated after the Second World War, it is possible to explore how the local elites gradually constructed different narratives based on the notion of "progress" since the mid-nineteenth century. This allows us to define some traits of the symbolic disputes over the agrarian question at the local level. From a retrospective view, we can identify at least three great competing narratives about rural development: (a) agricultural modernisation, (b) extractivism and (c) the model of indigenous peasant family farming. Based on the arguments put forward in this article, we can infer that the former two narratives have been hegemonic, but they were not the only ones; instead, they have coexisted with other alternative conceptions. The discourse about agriculture modernisation considers that rural development would be achieved through production increase. For that purpose, some of the following conditions should be generated: technological innovation, farmers' capitalisation, guaranteed property ownership and access to markets. Rural development, conceived as peasantry modernisation, reappeared in several periods of the history of the province, as in the end of the nineteenth century, in the context of the railroad expansion and irrigation agriculture (Banzato and Rossi 2010).

During the government of Absalón Rojas (1884–1892), agro-industrial projects were strongly supported, particularly sugar mills, which soon faced a crisis due to different reasons, especially overproduction at the national level.[10] Similarly, the successive droughts and the lack of infrastructure for irrigation were big obstacles to this attempt of

[10] Large-scale sugar cane cultivation triggered the agro-industry expansion. A French businessman, Pedro Saint Germes, established the Contreras sugar mill in 1879, with the initial sugarcane cultivation of 50 hectares for its own supply. Ten years later, there were seven sugar mills, located in the Capital city, Banda and Robles (in the centre of the province), which produced 1200 tonnes of sugar and 400,000 litres of sugar cane spirits. Sugarcane cultivation became a paradigm of the economy that politicians and businesspeople of that time wanted to encourage (Tasso 2004). However, overproduction at the national level reduced the sugar mill incomes, and those mills that were too indebted went bankrupt.

agricultural modernisation. By the mid-twentieth century, the strength of the narrative of agriculture modernisation regained importance. In the context of the green evolution, the governments of Santiago del Estero built water-engineering projects for land reclamation and applied new technological packages.[11] As previously mentioned, this period was marked by the expansion of the agricultural frontier, mainly for cotton production, which increased. However, towards the end of the 1960s, the international price of cotton dropped due to the substitution of this textile product with synthetic fibre, among other reasons (Paz and Jara 2020).

The second narrative understood rural progress, and later rural development, from a logic of extractivism of natural resources. Svampa (2017) stated that extractivism is an accumulation pattern based on the overexploitation of increasingly scarce natural goods, most of them non-renewable, as well as on the expansion of the frontiers to territories formerly considered unproductive. As explained above, one of the emblematic expressions was the exploitation of *quebracho colorado* in Santiago del Estero. Between 1890 and 1940, native forests were unlimitedly exploited for railroad construction and tannin extraction, leading to large, devastated areas, poverty and companies that then abandoned the region (Dargoltz 2003). That extractivist logic (with short-term returns and no environmental or social sustainability) is present in the soybean expansion process occurred in the last decades of the twentieth century. Thus, the increase in grain production for export occurred due not only to the new technological packages but also to the incorporation of other lands to agricultural activity by means of deforestation, evictions of peasants and soil depletion (Svampa 2017).

With subtle differences, modernisation and extractivism narratives had a strong impact on the ruling elites in the twenty-first century; therefore, overall, peasant family agriculture has never been regarded as an active subject of progress. Let us take an example that is closer to us in time. In 2005, The Strategic Territorial Plan (*Plan Estratégico Territorial*, PET) was implemented; it was focussed on an agro-industrial model and

[11] The construction of Los Quiroga Dam ended in 1949 and Río Hondo Dam in 1968 (Tasso 2004). The "Cero" Dam construction started in 1944 and was complemented with the Figueroa reservoir upstream, with greater capacity (Jara et al. 2019).

was aimed at complementing and improving the hydric infrastructure with the road network, and with energy works within the frame of *development corridors* (water, roads, energy) to put in production vast areas that had still not been fully incorporated to the modern formal economy (PET 2005). Reading of PET encourages the analysis of the following aspects. On the one hand, the government agencies perceive the peasantry as a social problem, as a group that lacks conditions to be included as an active player of development. Therefore, the proposal for transforming peasantry into family agriculture with a commercial logic is put forward once again. On the other hand, while there is a concern about the sustainable use of natural resources, little is mentioned about equal access to land and regularisation of possession.

Indeed, the rural social movements were the ones that claimed that indigenous peasant family agriculture was an active subject of other narratives about rural development. It should be noted that this alternative narrative arises in the context of the fight over land. Thus, besides conducting protest actions (against the detrimental effects of agribusiness) and requesting guarantees of possession rights from the state, the movements gradually created specific proposals to challenge the dominant ideas of development. This undoubtedly implies a dispute over material resources allocated to rural territories, as well as a symbolic struggle to make family agriculture sectors visible in official statistics and for their recognition of rights and capacities by government agencies (Paz et al. 2018).[12]

In that respect, it is relevant to consider the reflection of a leader of MOCASE in a dialogue with the authors of this work (our translation): "*In our opinion, development is not about increasing capital; instead, it is about reaching a balance with nature*". Thus, the equivalence between progress and economic growth is perceived as a problem by the rural leader. The interviewee links the notion of equilibrium to the multidimension nature of development. In turn, that leader comments that the concept of rural development promoted by the social movement has the challenge of being more inclusive, of going beyond the interests of

[12] A detailed analysis of experiences of alternative development whose main actor is the peasant in Santiago del Estero can be found in the book *Sistemas comunales, estructura agraria y explotaciones sin límites definidos* (Paz et al. 2018).

a sector, "the problem of land is not a problem of peasants, but of all". Social movements face the difficult task of framing their struggle beyond the trade union or corporation level and take on the challenge of questioning different sectors of society about the multiple interests at play in the struggles of peasant and indigenous agriculture at present.

Three Contributions of a Situated Approach to the Agrarian Question

The aim of this chapter was to recover the classical and contemporary debates about the paradigm of the agrarian question but from a situated and multi-scale approach. Along those lines, we consider that the analytical perspective that we adopted in this work contributes in three directions: (a) the *criticism* of the multiple inequalities in the rural world; (b) the understanding of the dispute over immaterial territories in the agrarian transformation process, and (c) a dialectical view of the accumulation/dispossession processes along with the active presence of indigenous peasant family agriculture. The processes analysed in this work suggest that the modes of accumulation/dispossession of the means of production and the class relations are still key to understand the agrarian question. However, they are not enough to understand the complex matrix of inequalities that intersect and reinforce one another.

In that sense, the approach to the agrarian question allows us to analyse the complex network of external dependence and asymmetries within the states. Being integrated to the world as a supplier of raw material has been a long-term aspect of the difficulties to advance into a development model that is more economically, socially and environmentally sustainable. At the same time, the formation of the Argentine state stressed the dominant role of Buenos Aires and wealth concentration in the bourgeoisie of the Pampas region. In turn, provinces like Santiago del Estero became the *periphery of the periphery.*

The agrarian question cannot be separated from the question of ethnicity, since class and ethnicity inequalities intercept. At present, the presence of indigenous populations continues to be the main characteristic of rurality in Santiago del Estero. According to the survey of

Proyungas Foundation, there are at least 78 indigenous communities in Santiago del Estero, which belong to six big ethnic groups: Diaguita-Cacano, Guayurú Lules Vilela, Sanavirón, Tonokoté and Vilela (García Moritán 2017). The colonial history, as already stated, simultaneously forged the formation of the *latifundios* and subjection of indigenous populations. The image of a white, European and urban Argentina imposed by the elites of the Pampas outlined a country with no black or indigenous people or peasants: that "other" rural sector was barbarism (Jara 2016). There are certainly many other inequalities of the agrarian question in Argentina that should be critically examined and that are beyond the scope of this work. A still unresolved question is the overlap between the agrarian question and the gender(s) question. Moving in this direction will allow us to better understand criteria of division of labour in the rural sector and the roles of unpaid work and of peasant women in the defence of land.

On the other hand, it should be noted that the situated debate of the agrarian question proposed in this chapter is part of a dispute over immaterial territories. In this sense, the empirical evidence and the arguments here presented a break with the hegemonic view built about the Argentine rural sector from the hegemonic academic spheres of the Pampas region. This can be observed when processes that are specific of a given region are generalised for the whole country, or when de-contextualised categories are extrapolated. A clear example is the concept of pampeanisation to understand the agricultural expansion process in Santiago del Estero, which entails the risk of making a simplification. From the perspective of the processes occurred in the Pampas region, capital expansion would inexorably lead to the disappearance of family agriculture holdings. Unlike expected in the central region, very different processes can be observed in Santiago del Estero, depending on the historical times, actors involved, soil quality, types of land tenure, and production profiles. Indeed, family farmers are still present, resisting and displaying their potential towards new alternative development designs.

References

Akram-Lodhi, H., and C. Kay. 2012. *Peasants and Globalization*. London: Routledge.

Ansaldi, W. 1993. La Pampa es Ancha y Ajena: La Lucha por las Libertades Capitalistas y la Construcción de los Chacareros como Clase. In *La Problemática Agraria: Nuevas Aproximaciones*, ed. M. Bonaudo and A.R. Pucciarelli, 71–101. Buenos Aires, Tomo II: Centro Editor de América Latina.

Auat, A. 2021. ¿Qué Hacer con el Neoliberalismo? *Erasmus* 23: 1–19.

Azcuy Ameghino, E. 2004. *Trincheras en la Historia: Historiografía, Marxismo y Debates*. Buenos Aires: Imago Mundi.

Banzato, G., and M.C. Rossi. 2010. El Mercado de Tierras en las Fronteras Interiores Argentinas: La Expansión Territorial de Buenos Aires y Santiago del Estero en la Segunda Mitad del Siglo XIX. *América Latina En La Historia Económica* 34: 7–34.

Bartra, A. 1989. Campesinado: Base Económica y Carácter de Clase. *Cuadernos De Antropología Social* 3: 3–8.

Byres, T.J. 2009. The Landlord Class, Peasant Differentiation, Class Struggle and the Transition to Capitalism: England, France and Prussia Compared. *The Journal of Peasant Studies* 36 (1): 33–54.

Dargoltz, R. 2003. *Hacha y Quebracho. Historia Ecológica y Social de Santiago del Estero*, 5th ed. Marco Vizoso: Santiago del Estero.

de Dios, R. 2010. Los Campesinos Santiagueños y su Lucha por una Sociedad Diferente. In *Movimientos Sociales y Derechos Humanos en Argentina*, ed. B. Pereyra and P. Vommaro, 25–46. Buenos Aires: CICCUS.

de Dios, R., R. Paz, and C. Rossi. 2020. Censos Nacionales Agropecuarios, Coberturas y Procesos de Transformación Agraria en Santiago del Estero. *Realidad Económica* 334: 123–181.

Fernandes, B.M. 2014. Questão Agrária e Capitalismo Agrário: O Debate Paradigmático de Modelos de Desenvolvimento para o Campo. *Reforma Agrária* 35 (2): 41–53.

Forni, F., R. Benencia, and G. Neiman. 1991. *Empleo, Estrategias de vida y Reproducción*. Buenos Aires: Centro Editor de América Latina.

García Moritán, M. 2017. *Mapa Relevamiento Comunidades Indígenas*. Tucumán: Fundación ProYungas.

Greenpeace. 2020. *A Pesar de la Cuarentena, se Deforestó Dos Veces el Tamaño de la Ciudad de Buenos Aires*. Retrieved from https://www.greenpeace.org/argentina/story/issues/bosques/a-pesar-de-la-cuarentena-se-deforesto-dos-veces-el-tamano-de-la-ciudad-de-buenos-aires. Recuperado el 18/01/2021.

Hocsman, L.D. 2014. Cuestión Agraria Actual en Perspectiva Regional/Global: Bloque de Poder Agrario Modernizado y Complejo Corporativo. *Contemporânea - Revista De Sociologia Da UFSCar* 4 (1): 43–60.

Jara, C.E. 2016. ¿Qué Es un Campesino? La Construcción de un Sujeto Político Ambiguo en Santiago del Estero (Argentina). *Astrolabio* 16: 340–361.

Jara, C.E., R.R. Sperat, L.F.R. Manrique, and A.G. Herrera. 2019. Desarrollo Rural y Agricultura Familiar en Argentina: Una Aproximación a la Coyuntura desde las Políticas Estatales. *Revista De Economia e Sociologia Rural* 57: 339–352.

Kautsky, K. 1998. *A Questão Agrária*. Brasília: Instituto Teotônio Vilela.

Lenin, V. 1981. *Obras Completas*. Moscú: Editorial Progreso.

Manzanal, M., and F. González. 2021. *Desigualdad, Territorio y Agricultura Familiar: Discusiones Teórico Metodológicas*. Buenos Aires: Estudios Rurales. Publicación del Centro de Estudios de la Argentina Rural, Universidad Nacional de Quilmes.

Marx, K. 2003. *El Capital: El Proceso de Producción del Capital*. México: Siglo XXI.

Paz, R., and C.E. Jara. 2020. Danzando en el Tiempo: Transformaciones Agrarias y Persistencia del Campesinado en Santiago del Estero (Argentina). *European Review of Latin American and Caribbean Studies* 110: 21–38.

Paz, R., R. Rodríguez, and C.E. Jara. 2018. *Sistemas Comunales, Estructura Agraria y Explotaciones sin Límites Definidos*. Santiago del Estero: Editorial EDUNSE.

PET. 2005. *Plan Estratégico Territorial para la Instalación del Modelo Agroindustrial Descentralizado de Desarrollo Humano Sustentable 2007–2016*. Santiago del Estero: Gobierno de Santiago del Estero.

Svampa, M. 2017. Continuidad y Radicalización del Neoextractivismo en Argentina. *Perfiles Económicos* 3: 87–97.

Tasso, A. 2004. Un Caso de Expansión Agraria Capitalista Seguido por Depresión. *Población y Sociedad* 10 (11): 109–136.

Tenti, M.M. 2002. Los Orígenes De La Colonización Agrícola in Santiago Del Estero: El Caso de Colonia Dora. *Nuevas Propuestas* 31: 79–92.

Wald, N. 2016. Historical Paths to Current Unrest: Extending the Temporal Lens in Analysing Geographies of Agrarian Change and Conflict. *Geoforum* 76: 38–47.

8

The Empty Food Bowl: Discourse Disconnection of Australian Agriculture

Alana Mann

Introduction

In 1997 John Wilson wrote that "while the hunger of Aboriginal Australians continues to be both a national and international scandal, hunger experienced by many other Australians is best described as hidden or silent hunger, not readily acknowledged publicly" (Wilson 1997: 14). Sadly, this is still the case. Against the backdrop of high rates of local food insecurity, worsened by the Covid-19 pandemic and the failure of the just-in-time logic of global supply chains, Australian farmers and other food producers are under increasing pressure to meet the needs of both distant and local markets under volatile trade and environmental conditions. Yet the nation lacks a long-term vision that balances agricultural

A. Mann (✉)
Sydney Environment Institute, University of Sydney, Sydney, NSW, Australia
e-mail: alana.mann@sydney.edu.au

A. A. R. Ioris and B. Mançano Fernandes (eds.), *Agriculture, Environment and Development*, https://doi.org/10.1007/978-3-031-10264-6_8

exports with domestic food security and, for Aboriginal people dispossessed of their homelands and access to productive resources, the impacts of white settler colonialism persist.

This chapter discusses the contradictions of dominant discourses that promote Australia as "food bowl to Asia" in the context of local food insecurity, environmental mismanagement, and climate change. The pandemic has exposed the fragility of a food system built to rely on interconnected and complex global food supply chains that facilitate trade between nations. Rising food prices can be directly attributed to the impacts of movement restrictions and illness on local markets and the availability of agricultural labour, but the roots of the problem extend much deeper to a colonial imaginary that continues to marginalise those on the wrong side of the socio-economic divide, including rural populations and especially Indigenous Australians.

Of the multiple worldviews, perspectives, and policy lock-ins that construct Australia's food history, and its future, the colonial imaginary of "food bowl to the world" persists, along with the notion of the pioneering farmer. Yet today's imaginary is coupled with emerging alternatives such as the "eco-innovator" embracing regenerative agriculture and agroecology (Cross and Ampt 2017) and there is growing appreciation for Indigenous knowledges, cultures, and practices regarding land management. Many farmers are focussing on land regeneration and increasing on-farm biodiversity through practices that include strategic grazing, pasture cropping, and native perennial grass regeneration. There is also a growing Australian Native Foods or "First Foods" industry, capitalising on the climate-readiness of species and the cultural knowledge embedded in Aboriginal custodianship of the land, seas, and waterways over thousands of years. Fundamental is the socio-economic participation and leadership of Traditional Owners in the growth of this industry which, while not sufficient to restoring Indigenous food sovereignty, can secure the economic participation of local communities.

Pandemic Foodways and the View from Australia

Globally and disproportionately, food supply is extremely vulnerable to global warming. At the same time, the way we grow and consume food generates significant emissions. Yet despite our extensive knowledge of how and why the conventional food system is unsustainable the actions and processes we need to achieve sustainability transformations remain poorly understood (Béné et al. 2019). The role that local food systems may play in sustainability transformations is increasingly drawing the attention of the general public, policy-makers and researchers, especially in high-income countries (De Schutter 2017). During the Global Financial Crisis (GFC) and the COVID-19 pandemic, re-localisation has emerged in many national jurisdictions as key to the development of more resilient food systems (Stephens et al. 2020). In 2009, the USDA established the "Know Your Farmer, Know Your Food Initiative", which invested over one billion US$ in local and regional food projects across a five-year period. During the pandemic in Southern Europe local producers have demonstrated their value as suppliers of fresh produce, able to respond quickly and flexibly to market changes to feed people healthy food (van der Ploeg 2020). British horticultural producers report changing crop plans and increasing production as well as turning over new land to grow staple vegetables for local markets; co-benefits include a move to healthier diets and a reduction in the ecological impacts of agriculture (Willett et al. 2019). The breakdown of conventional distribution channels has driven the upscaling of innovative models linking producers with consumers such as the Open Food Network (OFN), a digital platform cooperative that supports food enterprises to sell online via food hubs, experienced an incredible spike here and overseas in demand as new food enterprises joined and turnover increased (Schultz et al. 2020). Formal policy instruments like Canada's "Local Food Infrastructure Fund" (CAN$50 million, 2021–2022) initiative aim to support "community-based organisations with a mission to reduce food insecurity by establishing and/or strengthening their local food system". In 2020, the European Commission (EC) developed the "Farm to Fork Strategy" designed "to enhance resilience of regional and local food systems in order to create shorter supply chains … reducing dependence on long-haul transportation" (cited in Enthoven and Van den Broeck 2021).

Australia has no such re-localisation initiatives nor a long-term policy strategy that deals with the tension between producing crops for export ($120 billion in 2021) and ensuring a strong domestic supply of fresh, affordable food. A diversity of growers producing a diversity of foods is essential to building capacity for resilience, especially as we face longer droughts and extreme weather events. Further, the nations' food production systems are still heavily reliant on fossil fuels. Despite the formation of the Western Australian AgZero2030 Group and National Farmers Federation (NFF) policy to position agriculture to take advantage of the "social, environmental and economic opportunities" of a low-emissions future (NFF 2020), the unsustainability of the food system is not acknowledged by our political leaders.

This includes high levels of food insecurity—18% of the national population experienced food insecurity in the previous 12 months *prior* to the COVID-19 pandemic (Foodbank 2018). National health surveys estimate the prevalence of food insecurity in Australia to be 5% of the population, meaning that approximately 1 million Australians are experiencing some form of food insecurity (Yii et al. 2020). Australians who are unemployed, residing in remote or rural areas, Indigenous and recent migrants are more likely to face food insecurity (Lambie-Mumford and Dowler 2015). The increase in food insecurity has been followed by the growth of the charitable food sector in Australia, a reflection of declining government investment in welfare across many sectors (Lindberg et al. 2015). National programmes to address these outcomes have been critiqued for their technocratic, top-down approach. Interventions such as the Food and Nutrition Policy (1992) and Eat Well Australia (2000–2010) have included recommendations and actions to address food and nutritional insecurity, but these programmes have long been regarded as being inadequately resourced, limited and peripheral in nature (Bastian 2011).

Historically, these interventions have been disconnected from rigorous evaluation efforts and neglect inclusion of the perspectives of people at the frontlines of hunger and malnutrition in assessing the key determinants of their own food insecurity (Elkharouf et al. 2021). A key example is the series of high-level federal strategies, plans, and inquiries related to addressing the disproportionate rates of food insecurity among Aboriginal and Torres Strait Islander (ATSI) populations. Such policies have limited ATSI people's participation in relevant decision-making processes, which has limited the success of the programmes (Davy 2016; Spurway and Soldatic 2016). The establishment of the National Aboriginal and Torres Strait Islander Peoples in Australia Nutrition Strategy and Action Plan (NATSINSAP) in 2000 failed to take into account the agency and interrelated rights of ATISI people. A major reason for NATSINSAP's failure was the policy's embedded assumption that low-socioeconomic households make poor choices simply because they lack education. In doing so, the programme inadequately addressed the underlying factors contributing to food insecurity based on the paternalistic assumption that those targeted simply needed to be taught about nutrition, resulting in justified resentment from the ATSI community (Davy 2016; Spurway and Soldatic 2016). Meanwhile, Australian food policies have been geared towards increasing productivity and profits in export agriculture.

Historic imaginaries include Australia as "food-bowl to the world" or SE Asia. Two-thirds of the food and fibre produced are exported (ABARES 2021). Investment is focussed heavily on innovation to promote competition in global markets. For example, in the Upper Spencer Gulf in South Australia, embracing the cities of Port Augusta, Port Pirie, and Whyalla with a combined population of 53,000 including 22,000 workers and 2000 small businesses, is now described as the "home of arid innovation". The Gulf is scaling-up to enter the globally competitive intensive horticulture and aquaculture industries through controlled environment agriculture (CEA). One example is leading innovator Sundrop Farms whose commercial greenhouse facilities integrate solar power, electricity generation, freshwater production, and hydroponics. These industries are providing social and economic benefits including local jobs and high-volume food production for domestic and international markets. However, the shift from traditional industries including mining, shipbuilding, steel manufacture, and railways is having profound but poorly understood and under-researched impacts on local communities, many of which experience food insecurity.

Regional food systems generate 70% of fresh produce in Australia. They enable food to flow from source to plate through inter-linked practices that cover the growing, processing, distribution, acquisition and consumption of food, and the management of food waste. In Australia, these holistic systems are threatened by a set of rapidly evolving forces and risks that residents and policy-makers alike are grappling to understand. Climate change means rising temperatures, more "Black Summer" bushfire events, unpredictable regional rainfall, and the movement of fish stocks south as oceans warm. Ruptures in the global trade regime, such as the decline in relations with China and the imposition of carbon tariffs by the European Union, are impacting specific industries and products. Given over half of the Australian regions have an "emission intensive" employment profile jobs and incomes are likely to be adversely affected. Importantly, the pandemic has revealed the vulnerability of Australian agriculture to crises which affect international mobility of migrant labourers. It has highlighted how workers in the food system are among the lowest paid and endure poor work conditions (CSIRO 2021). These evolving risks are acute for rural communities that already experience higher rates of poverty, food insecurity, and poor health than urban Australians.

The Myth of the Food Bowl

Hunger has been a constant in Australian history since colonisation—the immediate concern upon the arrival of the First Fleet was how to grow enough food to survive. Accordingly, the search for arable land drove exploration inland. Aboriginal populations were driven from their traditional lands and food sources, facing starvation or barely surviving on rations, an insidious and flexible instrument of control for the colonial government (Rowse 1998). A huge pastoral industry developed in the north and northwest of Australia, built with the forced labour of Aboriginal people. Today, the region is the site of a sustained resource boom benefiting from the global extractive economy in mineral ore. Almost half of Western Australia's Aboriginal population live in the Kimberley, where remoteness, socio-economic disadvantage, and the high price of fresh, nutritious food (and the limited places to purchase it) mean diets are poor. Notably, this region is one of the few places where Aboriginal communities still practice traditional food gathering and cooking to alleviate their poverty and hunger.

Federal Government policy does little to remedy this, based on the stated principle that "most Australians can afford to buy the food they need and can access safe and nutritious food" (Australian Government 2013: 94). This thinking is embedded in proposals for the development of the region, which encompasses the Ord and Fitzroy River Valley. The Pivot North framework proposes to develop the region through intensive agriculture and irrigation based around the Fitzroy Valley, the La Grange Aquifer, and Kununurra in the east (Australian Government 2014). In terms of intensive agriculture, the Ord River Irrigation Scheme has been described as a "failed experiment" for rice, horticulture, sugar, and cotton. High salinity, pests, diseases, transport challenges, and volatile world agricultural commodity prices have been blamed for several ill-fated political projects to create a food bowl in this region since the 1960s, leading agricultural scientist Bruce Davidson (1967) to coin the phrase "the Northern Myth". Yet emerging European markets for quinoa and plantago (source of psyllium husk) are encouraging new efforts to farm the Ord Valley.

The Northern Myth is an example of the dominant imaginary of Australia as a colonial outpost supplying meat, wool, and grain to England. Conjuring romantic visions of the nation's pioneering history, the arid north is the last frontier for successive governments from both sides of politics intent to double Australia's food production through new irrigation schemes and novel new crops. Fifty years, and more than a billion dollars, since Davidson did the math more than 60 crops have been trialled and failed. Those that have had some success, like sandalwood, you can't eat. In the remote areas of the West Kimberley, the high cost of living, including food expenses and the generally low incomes of residents mean that food insecurity is common, particularly among Aboriginal communities.

Many studies have detailed the formidable barriers to large-scale, profitable, and sustainable irrigation in the North. These barriers are not only climatic, physical, and economic but also social and cultural—this is not terra nullius as traditional colonial histories would have us believe but the land of several Aboriginal language groups, including the Ngarrawanji, Malarngowem, and Yurriyangem Taam. A land area the size of Belgium, 34,000 square kilometres or 93.5% of the Kimberley region, is now recognized as native title lands but the claims process, often decades long, is onerous and has taken its toll on Traditional Owners. Many of the original claimants have passed before the title is granted. In the meantime, forcible removal of communities from land and the imposition of mining and large-scale agricultural schemes violates indigenous relationships with "country" and the law of the land. In his address to the Federal Court Kalkara man Matt Dawson explains, "one of the main responsibilities for country [is] to look after it and protect it. This means protecting it from harm from outsiders, messing it up. It gets messed up if people don't follow the law. If someone is on country that does not belong to it or know about that place, the country can also hurt that person because they are doing the wrong thing…If someone gets hurt on my country then I am responsible for them doing the wrong thing. This is why being responsible for country is not small thing; it is a very serious responsibility" (cited in Allam 2019).

As Dawson's comment reflects, "country" in Aboriginal English is both a common and proper noun. Country is addressed as a person; it "knows, hears, feels, smell, takes notice, takes care, is sorry or happy … Country is a living entity with a yesterday, today and tomorrow…a consciousness, a will toward life … nourishment for mind, body and spirit" (Rose 1996: 7). Indigenous cultures' deep ecological knowledges and connections to Country and other species have nourished them physically and spiritually for thousands of years. Customary norms and practices, notions of custodianship, and stewardship obligations protect ecosystem health and manage species distribution (Williams and Hardison 2013). Totems, described as *borrungur* by the Noongar peoples of Southwestern Australia, are plants and species adopted by Indigenous people in a process whereby "they become you and you become them" in a deep kinship that engenders care and protection (Honeybone and McAllisoter 2019). First Nations peoples' relationships with Country shares a more powerful sense of entitlement than "ownership through purchase or appropriation, because with entitlement goes the "bond" of responsibility and a culture of land care that is passed down through generations in order to ensure the livelihood of future generations" (Goodall 2019: 171). Custodianship of land remains central to Aboriginal cosmologies and cultural practices in Australia.

The Case for a New Paradigm

It is now internationally recognised that "agricultural practices that include indigenous and local knowledge can contribute to overcoming the combined challenges of climate change, food security, biodiversity conservation, and combating desertification and land degradation". Australia is one of the countries with the greatest potential to reduce diet-related greenhouse gases if it seriously embraced its native foods. Aboriginal peoples' careful stewardship of the Australian continent is an example of sophisticated land management, an advanced "proto-agriculture" that included cultivation of edible food species, food storing and harvesting, and the strategic selection and organisation of land according to its suitability for specific uses (Gammage 2012; Pascoe 2014). Upstream of Bourke on the Barwon River at Brewarrina are rock fish-traps that are not only among the first examples of aquaculture but possibly the oldest human-made structures in the world (Simons 2020).

The mosaic burning regimes, or fire-stick farming, described in Bruce Pascoe's *Dark Emu Black Seeds: Agriculture or Accident?* (Pascoe 2014) are now being reintroduced in Aboriginal led carbon farming projects as pathways to inter-generational learning, connection to country, and wealth generation for communities throughout Australia. Pascoe notes that in countering global warming "new ideas and methods will arise out of the very oldest land use practices" (Pascoe 2014: 148). Among these are experimenting with grazing indigenous animals and growing indigenous crops: Aboriginal people knew that by very deliberately and systematically caring for the land, they influenced what food was available and how much. As Pascoe notes, native grains and tubers have been domesticated by Aboriginal people and proven "good to eat, good for us after 100,000 years of testing, and good for the country. Most are perennial so there is less tilling involved, so less carbon release, plus we don't have to devote scarce water resources and fertilisers. And we reduce our dependence on using weedicides and pesticides."

By growing Aboriginal plants and grains like warrigal greens, cumbungi (bulrush), and murnong (yam daisy), Australia "could reach our carbon emission targets easily". So, Pascoe asks, "Why don't we? Are we that paranoid about Aboriginal claims to the land that we can't acknowledge the plants tested and domesticated over a longer period than anywhere else on earth?" (Pascoe 2018). He notes that where there is substantial interest in commercialising traditional foods Indigenous people must have ownership "so large food companies don't put a brand on it and dispossess us one again…we don't want to be dispossessed twice" (Pascoe cited in Vernon 2019). Given colonial histories of exploitation, lack of recognition and respect for values and rights, and appropriation of resources, Indigenous people are understandably wary of "partnership arrangements" and particularly those that do not incorporate a "spirit of mutualism and reciprocal accommodation" (Williams and Hardison 2013: 542). Western understandings of traditional ecological knowledge (TEK) are frequently framed in a material, transactional sense that does not fit the spiritual beliefs and kinship relations traditional knowledge-holders share with their environments (Little Bear 2000).

Failure to embrace traditional Aboriginal knowledge leaves Australia's food-producing regions at extreme risk in a rapidly warming climate. A stand-out case is the Riverina in the eastern state of New South Wales, which extends from the Snowy Mountains northwest though the Murrumbidgee River catchment area to the plains of Hay and Carrathool. It comprises several regional cities and has a population of 170,000 people, primarily employed in the agriculture and service industries. The Riverina produces one-fourth of the agricultural production of NSW, a diverse mix of wheat, rice, oilseeds, citrus, nuts, vegetables, grapes, and livestock. Competition for water in the region has been described as "irrigation hunger games" as prolonged drought has reduced harvests of staple crops including rice and contributed to the contraction of the dairy industry in the depleted Murray-Darling Basin, where milk production has declined by 25% since 2014. The future of the Murray Darling basin as Australia's most vital inland water system is of deep concern. Bruce Pascoe is among those critical of water management characterised by corporate deal-making that favours extractive industries, including coal seam gas mining or fracking, over water for communities. He says "Capitalism provides a platform for decisions among fellow capitalists but shudders under the load of persuading communities over vast areas of the country. If that were not so we would not have reached such impasse ... we would never consider leaving a state in our Federation without drinking water, we would not have laws which allow coal seam gas miners to ruin a farmer's land and threaten the very groundwater of the continent" (Pascoe 2014: 155).

The decline of the river system is a tragedy of epic proportions for the millions who rely on it and particularly for Indigenous people for whom it represents a "cultural highway, a massive Appian way" (Pieri 2019). It embraces four states including Queensland, where a large inland geological basin, the Galilee, spans 247,000 square kilometres encompassing 12 Indigenous language groups. Relying on the Northern section of Australia's Great Artesian Basin, the hydrology of which is extremely complex, the Galilee includes the headwaters of seven major river basins. It is also the site of the hugely controversial Adani coal mine. Central to the concerns of the traditional owners of the mine site, the Wangan and Jagalingou people, is the impact of the mine on the local Doongmabulla Springs, fed by 60 tributaries and covering more than 10 ha. This unique ecosystem is one of the few permanent sources of water in the area and vital to many species found nowhere else on the planet. For a region in almost perpetual drought this is an essential resource for farms and communities. Yet the state government exempted Adani from new restrictions regulating water use in 2017. It takes 250 L of water to produce a tonne of coal. Only 36% of over 300 spring complexes surveyed in the Galilee Basin at the turn of last century are still active after 120 years of drawdown for agriculture (Bradley 2019).

Agriculture accounts for approximately 70% of water consumption in Australia, with cotton and irrigated pastures historically accounting for the higher use. Water markets are highly localised, with trade occurring between users within a single river catchment with the exception of the southern Murray-Darling where water trading between systems and across state boundaries is enabled through hydrological connectivity (ABARES 2019). Water prices have reached their peak in the southern Murray-Darling Basin, leading to an inquiry into the water market. Foreign investors, corporate speculations, poor water-sharing policy, and the introduction of new almond crops have been variously blamed for water shortages, exacerbated by reduced winter rainfall and higher temperatures directly related to climate change. In January 2019 temperatures exceeded 35 °C over 14 days, driving demand for irrigation from 4.5 to 7 gigalitres a day (Hughes, 2019).

One of the most visible and shocking impacts of the Murray-Darling's ill health has been a series of mass death events among native fish populations. In early 2020, 20 different locations in the Macquarie, Namoi, Gyydir, Border, Barwon-Darling, Lachlan, Upper Murray, Murrumbidgee, and Lower Darling rivers experienced blue algae breakouts and bushfire ash runoff (Readfearn 2020). These episodes followed a catastrophic event across December 2018—January 2019 when over a million fish died on a forty kilometre stretch of the Darling River, downstream from Menindee Lakes, a vital waterbird habitat and important cultural site for the local Barkandji people. Their voices are conspicuously absent in water management. Under the NSW Water Management Act, traditional owners possess Native Title rights, known as Basic Landholder Rights. Despite the recognition of Native Title rights for 128,000 square kilometres of Barkandji land in 2015—after an eighteen-year legal battle—the Barwon-Darling Water Sharing Plan provides zero allocation for Native Title (Thompson 2019).

This is just one example of government failure to recognise and accommodate Aboriginal rights in Australia's mismanaged water governance regime. There is plentiful evidence that Indigenous-driven co-governance, distributed appropriately through families and communities, offers the best prospects for successful integration of TEK and Western science in managing social-ecological systems (Hill et al. 2012). Yet engagements in Australia are characterised by "top-down consultation meetings which are then ignored or … 'fake consultation'" (Simons 2020: 97). Its consequence is "a river system run to the edge of its ability to survive", jeopardising the future of a country that relies on it for food production (ibid.: 68). As an investigation into water theft, bureaucratic incompetence, and political corruption proceeds, journalist Margaret Simons reflects that "blaming individuals, individual communities, or the growers of certain crops obscures the large failure—of our politics when faced with a complex challenge. Underlying that is the failure of our ability, as Australians, to recognise common interests".

An Emerging Research Agenda

Food and environmental change are two Australian National Research Priority areas, supported by an investment of $156 million in Commonwealth Scientific and Industrial Research Organisation (CSIRO) missions tackling drought resilience, agri-food exports, and future protein sources (Marshall 2021). The knowledge produced needs to be complemented by sociological insights regarding the potential of food systems change for sustainable economic growth, job creation, and community well-being across different industries and supply chains and within specific locales where the production, growing, and processing of food takes place. There is a clear need for a research agenda that develops these understandings if Australia is to grow diverse and resilient local food economies in which all citizens, and stateless-people including asylum seekers, participate. Research engaging in questions of agency, justice, power, and resistance against the conventional food system can make a significant contribution to debates concerning the efficacy of various transitions and transformations as theories of change. This distinction is crucial in the case of food producers debating the merits of transitions to "sustainable intensification" versus transformation to agro-ecological farming or Indigenous land management methods like savannah enrichment (Courtenay and Walbidi 2020).

Advocates of localisation argue that connecting growers, entrepreneurs, and eaters in strong relationships that increase transparency and meet environmental, health, and labour sustainability objectives drive the most successful transitions to sustainable local food supply networks. These networks underpin a food system that "achieves and maintains food security under uncertain and dynamic social-ecological conditions, through respecting and supporting the context specific cultural values and decision-processes that give food social meaning, and the integrity of the social-ecological processes necessary for food provisioning today and for future generations" (Eakin et al. 2017: 759). A people-centred, systems approach aims to democratise and redistribute power within food systems by enabling inclusive access to participation in the food system (including the growing, harvesting, gathering, processing, and distribution of food), as well as universal access to sustainable, healthy, culturally appropriate, and affordable food.

The goal of people-centred transformation is challenging unjust power relations at all levels and for all actors. This relies upon democratic communication processes which engage all citizens in participatory food planning with the support of local authorities and town councils. Creating stronger urban–rural linkages, foreground the role of communication, and building relationships on mutual recognition of voice invites participation (Mann 2019: 121). Truly participatory models differ from purely market-driven alternative food networks (AFNs) that support, rather than transform, the conventional food economy. For example, AFNs that provide organic food those with the capacity to pay more encourage apolitical consumption or individual behavioural rather than community resilience and remain largely silent on matters of social justice and disadvantage. A coordinated approach to supporting local food ecologies based on these values maximises their potential to contribute to social wellbeing, sustainability, and local economic development, supported by a diverse range of sustainability-oriented actors including growers' cooperatives, benefit corporations, and social enterprises working with traditional businesses and municipal authorities to create resilient local food networks.

The sustainable expansion of local and regional food markets can contribute to economic development and food security for all actors along the food value chain. For example, local food sales provide localised economic benefits if farmers can source intermediate inputs, labour and capital locally. When this occurs, local food sales can result in regional economic activity that is a greater "multiple" of the initial level of sales than would otherwise have occurred. Local food markets also may provide market access and business opportunities for food producers who otherwise would not be farming (O'Hara and Pirog 2013). Inclusiveness in food value chains is recognized as an important element to improve the economic viability or even the long-term sustainability of business (Helmsing and Vellema 2010). Inclusive value chains operate through broad-based engagement across sectors to enable the co-construction of relevant integrated policy frames to build resilient local food networks that increase access to healthy, affordable, culturally appropriate food; generate decent jobs and income; increase regional resilience against shocks and lessen the dependence on distant supply sources; foster rural–urban linkages and promote ecosystem and natural resources management and support participatory governance.

This approach foregrounds the engagement of community members, so they feel empowered to make decisions to improve their situation. It is potentially transformative in that it recognised that the spaces and structures that invite and support open and equitable participation do not occur naturally; they require deliberative design and maintenance with attention to principles including voice, diversity, agency, and the navigation of difference. Fundamental is a "strengths-based approach", based on respect for local knowledge and context, strong relational ties between participants, and a commitment to communication in culturally appropriate ways (AbCF 2021). This approach recognises that communities possess the assets they need to respond not only to new, climate-related threats but also to longstanding inequities driven by poverty, poor policy, and other structural impediments that limit both people's access to healthy, nutritious, culturally appropriate food, and their social and economic participation (Elkharouf et al. 2021). Strengths-based approaches recognise food systems as intensely local and deeply embedded in culture, a point often missed in the sustainability literature. Research design grounded in the belief that socially just sustainability transitions support the capabilities of diverse food system actors to participate in change on their own terms include fair procedures in decision-making and participation in the design of transitions. It assumes that the economic, institutional, and social relationships that exist between the different actors within food systems makes them intimately dependent on each other and that better access to information, stronger cooperation, more inclusion, and higher levels of aspiration and self-efficacy contribute to the building of resilience across the community (Béné 2020). Beyond communicating that "our ability to live on this planet is what is at stake" (Black 2011: 177) this research agenda aims to provide recommendations that inform policy development and offers a practical framework for addressing climate threats to our food systems in an equitable and inclusive way.

Conclusion

Views and analyses of food systems begin from different levels and scales. National policy and global markets look different when viewed from the bottom up while local food cultures look different when viewed from the level of global trade systems or national policy narratives. There is a vital need for more research integrates these disparate levels and scales of analysis. Combining cultural approaches with political-economic ones, and viewing policy through the lens of local studies promotes methodological convergence between locally grounded, bottom-up approaches and systemic thinking, research, and practice. Knowledge about how regional food communities is innovating in response to current and emerging risks to food production and distribution can inform the development of guidelines for policy, planning, and design of the city and regional food systems that unlock equitable, inclusive, and sustainable economic and employment opportunities for all Australians.

Strong regional food supply networks assure access to food when longer supply chains are impacted by climate shocks, financial crises, and pandemics. These rely on strong stakeholder and community networks that are complex and heterogeneous, comprising multiple socio-ecological actors, activities, processes, and factors which interact in often unpredictable ways. The diversity of Australian regions is a strength in a changing climate, but regions are also more exposed to disruption and a failure to plan. While on- and off-farm socio-tech innovations and entrepreneurship can generate healthy and sustainable food products along with broader social and economic co-benefits, more coordinated, participatory approaches are needed to unlock these assets for Australian communities. These approaches will vary according to different geographic, socio-demographic, and climatic conditions, and according to import and export flows within regions.

Globally, the disruption created by COVID-19 has opened opportunities for innovation and for less-powerful actors to shift the regime towards resilient food systems that are sustainable, just, and inclusive. Community-led decision-making clearly empowers people to share responsibility for their food resilience in times of need. Australian communities are already challenging conventional agri-food practices by co-creating emergent sustainability innovations. Commercial interest in Indigenous or First Foods is growing with recognition that "Indigenous, local and traditional knowledge systems and practices, including Indigenous peoples' holistic view of community and environment, are a major resource for adapting to climate change" (IPCC 2014: 79–80). Traditional Owners and their communities are at the forefront in field sites include West Kimberley, WA, site of the Nyamba Buru Yawuru irrigation project, and the Bidyadanga Community's large-scale harvesting of Kakadu Plum.

We need to understand how these place-based efforts contribute to systemic societal change and how they can be supported and scaled up. An assets-based approach is grounded in the belief that learning in the context of food systems change is open-ended and transformative but also rooted in the life-worlds of people and the encounters they have with each other. Sharing this knowledge requires us to hail people from their epistemology rather than performing consultation and assimilation, which disempowers knowledge holders such as Traditional Owners (Black 2011: 170). In contributing to our understanding of how people learn, how they recognise and transcend normative thinking that prevents change, and how they negotiate the dissonance experienced in encountering alternative ways of knowing, this project will help us "walk together", in the words of Yulparija elder Merridoo Walbidi, as we must to solve the unique challenges that climate change poses to Australia and our food future.

References

ABARES. 2019. *Snapshot of Australian Water Markets*. CC BY 4.0. Canberra: Australian Bureau of Agricultural and Resource Economics and Sciences. https://doi.org/10.25814/5cabfa02a6538.

ABARES. 2021. *Snapshot of Australian Agriculture*. Research Report 20.1. CC BY 4.0. Canberra: Australian Bureau of Agricultural and Resource Economics and Sciences. https://doi.org/10.25814/5d9165cf4241d.

AbCF. 2021. *Core Benefits Verification Framework*. Cairns: Aboriginal Carbon Foundation.

Allam, L. 2019. The Kimberley is Now 93.5 per cent Title: But Indigenous Victory is Bittersweet. *The Guardian Online,* May 24. Retrieved from https://www.theguardian.com/australia-news/2019/may/24/the-kimberley-is-now-935-native-title-but-indigenous-victory-is-bittersweet.

Australian Government. 2013. *National Food Plan: Our Food Future*. Canberra: Australian Government.

Australian Government. 2014. *Pivot North: Inquiry into the Development of Northern Australia: Final Report*. Canberra: Joint Select Committee on Northern Australia.

Bastian, A. 2011. The Future of Public Health Nutrition: A Critical Policy Analysis of Eat Well Australia. *Australian and New Zealand Journal of Public Health* 35 (2): 111–116.

Béné, C. 2020. Resilience of Local Food Systems and Links to Food Security: A Review of Some Important Concepts in the Context of COVID-19 and Other Shocks. *Food Security* 12: 805–822.

Béné, C., P. Oosterveer, L. Lamotte, I.D. Brouwer, S. de Haan, S.D. Prager, E.F. Talsma, and C.K. Khoury. 2019. When Food Systems Meet Sustainability: Current Narratives and Implications for Actions. *World Development* 113: 116–130.

Black, C.F. 2011. *The Land is the Source of the Law: A Dialogic Encounter with Indigenous Jurisprudence*. Abingdon, Oxon: Routledge.

Bradley, J. 2019. How Australia's Coal Madness Led to Adani. *The Monthly*, April. Retrieved from https://www.themonthly.com.au/issue/2019/april/1554037200/james-bradley/how-australia-s-coal-madness-led-adani#mtr.

Courtenay, K., and M. Wilbidi. 2020. *"We Must Walk Together": How Ancient Cultural Practices and Modern Agronomy are Cultivating New Hope and Industry in the Australian Bush*. Melbourne: Australian Book Connection.

Cross, R., and P. Ampt. 2017. Exploring Agroecological Sustainability: Unearthing Innovators and Documenting a Community of Practice in Southeast Australia. *Society and Natural Resources* 30 (5): 585–600.

CSIRO. 2021. *Labour Impacts on Agriculture*. Canberra. Retrieved from https://www.csiro.au/en/research/environmental-impacts/sustainability/labour-impacts-on-agriculture.

Davidson, B. 1967. *Northern Myth: Study of the Physical and Economic Limits to Agricultural and Pastoral Development in Tropical Australia*. Melbourne: Melbourne University Press.

Davy, D. 2016. Australia's Efforts to Improve Food Security for Aboriginal and Torres Strait Islander Peoples. *Health and Human Rights* 18 (2): 209.

De Schutter, O. 2017. The Political Economy of Food Systems Reform. *European Review of Agricultural Economics* 44: 705–731.

Eakin, H., J.P. Connors, C. Wharton, F. Bertmann, A. Xiong, and J. Stoltzfus. 2017. Identifying Attributes of Food System Sustainability: Emerging Themes and Consensus. *Agriculture and Human Values* 34: 757–773.

Elkharouf, O., K. Cox, D. Schlosberg, A. Mann, and E. Perroni. 2021. In the Land of the "Fair Go": Global Food Policy Lessons beyond the Charity Model. *Local Environment* 26 (10): 1192–1204.

Enthoven, L., and G. Van den Broeck. 2021. Local Food Systems: Reviewing Two Decades of Research. *Agricultural Systems* 193: 103226.

Foodbank. 2018. *Foodbank Hunger Report 2018*. Retrieved from https://www.foodbank.org.au/wp-content/uploads/2018/12/2018-Foodbank-Hunger-Report.pdf?state=wa.

Gammage, B. 2012. *The Biggest Estate on Earth: How Aborigines Made Australia*. Sydney: Allen and Unwin.

Goodall, J.R. 2019. *The Politics of the Common Good*. Sydney: NewSouth Publishing.

Helmsing, A.H.J., and S. Vellema. 2010. Governance, Inclusion and Embedding. In *Value Chains, Social Inclusion and Economic Development:*

Contrasting Theories and Realities, ed. A.H.J. Helmsing and S. Vellema, 1–19. London: Routledge.

Hill, R., C. Grant, M. George, C.J. Robiinson, S. Jackson, and N. Abel. 2012. Typology of Indigenous Engagement in Australian Environmental Management: Implications for Knowledge Integration and Social-ecological System Sustainability. *Ecology and Society* 17 (1): 23.

Honeybone, E., and K. McAllisoter. 2019. Adopting an Indigenous Totem Could be a Simple Way to Care for Country. *ABC News*. Retrieved from https://www.abc.net.au/news/2019-06-01/every-aussie-should-adopt-a-totem/11116978.

Hughes, N. 2019. Drought and Climate Change are Driving High Water Prices in the Murray-Darling Basin. *ABC News*. Retrieved from https://www.abc.net.au/news/2019-07-22/drought-and-climate-change-are-driving-high-water-prices-in-the/11329502.

IPCC. 2014. *Climate Change 2014: Synthesis Report*. Contribution of Working Groups I, II and III to the Fifth Assessment Report of the Intergovernmental Panel on Climate Change, Geneva, Switzerland.

Lambie-Mumford, H., and E. Dowler. 2015. Hunger, Food Charity and Social Policy – Challenges Faced by the Emerging Evidence Base. *Social Policy and Society* 14 (3): 497–506.

Lindberg, R., J. Whelan, M. Lawrence, L. Gold, and S. Friel. 2015. Still Serving Hot Soup? Two Hundred Years of a Charitable Food Sector in Australia: A Narrative Review. *Australian and New Zealand Journal of Public Health* 39 (4): 358–365.

Little Bear, L. 2000. Jagged Worldviews Colliding. In *Reclaiming Indigenous Voice and Vision*, ed. M. Battiste, 77–85. Vancouver: University of British Columbia Press.

Mann, A. 2019. *Voice and Participation in Global Food Politics*. Abingdon, Oxon: Routledge.

Marshall, L. 2021. *A Vision for Australian Agriculture*. Retrieved from https://blog.csiro.au/future-australian-agriculture.

NFF. 2020. *Climate Change Policy. National Farmers Federation*. Retrieved from https://nff.org.au/wp-content/uploads/2020/08/2020.08.06_Policy_NRM_Climate_Change.pdf.

O'Hara, J.K., and R. Pirog. 2013. Economic Impacts of Local Food Systems: Future Research Priorities. *Journal of Agriculture, Food Systems, and Community Development* 3 (4): 35–42.

Pascoe, B. 2014. *Dark Emu: Black Seeds: Agriculture or Accident?* Broome: Magabala Books Aboriginal Corporation.

Pascoe, B. 2018. *Let's Talk about Real Australian Food*. Retrieved from https://www.sbs.com.au/food/article/2018/05/29/comment-lets-talk-about-real-australian-food.

Pieri, S. 2019. The Basin and the Kill. *Arena*. No. 159, February. Retrieved from https://arena.org.au/the-basin-and-the-kill-by-stefano-de-pieri.

Readfearn, G. 2020. Murray-Darling: Thousand of Fish Have Died in NSW in the Past Two Weeks. *The Guardian*. Retrieved from https://www.theguardian.com/australia-news/2020/feb/02/murray-darling-thousands-of-fish-have-died-in-nsw-in-past-two-weeks.

Rose, D.B. 1996. *Nourishing Terrains: Australian Aboriginal Views of Landscape and Wilderness*. Canberra: Australian Heritage Commission.

Rowse, T. 1998. *White Flour, White Power*. Cambridge: Cambridge University Press.

Schultz, I., C. Freeman, L. Kelly, and J. Sheridan. 2020. *Recipes for Resilience: The Impacts of COVID-19 on Values-based Food Supply Networks in Australia and the Sector's Response*. https://doi.org/10.6084/m9.figshare.13377401.

Simons, M. 2020. Cry Me A River: The Tragedy of the Murray-Darling Basin. *Quarterly Essay*. Retrieved from https://www.quarterlyessay.com.au/essay/2020/03/cry-me-a-river.

Spurway, K., and K. Soldatic. 2016. "Life Just Keeps Throwing Lemons": The Lived Experience of Food Insecurity among Aboriginal People with Disabilities in the West Kimberley. *Local Environment* 21 (9): 1118–1131.

Stephens, E.C., G. Martin, M. van Wijk, J. Timsina, and V. Snow. 2020. Editorial: Impacts of COVID-19 on Agricultural and Food Systems Worldwide and on Progress to the Sustainable Development Goals. *Agricultural Systems* 183. https://doi.org/10.1016/j.agsy.2020.102873.

Thompson, B.M.R. 2019. Aboriginal Voices are Missing from the Murray-Darling Basin Crisis. *The Conversation*. Retrieved from https://theconversation.com/aboriginal-voices-are-missing-from-the-murray-darling-basin-crisis-110769.

Williams, T., and P. Hardison. 2013. Culture, Law, Risk and Governance: Contexts of Traditional Knowledge in Climate Change Adaptation. In *Climate Change and Indigenous Peoples in the United States*, ed. J.K. Maldonado, B. Colombi, and R. Pandya, 23–36. Cham, Switzerland: Springer.

van der Ploeg, J.D. 2020. From Biomedical to Politico-economic Crisis: The Food System in Times of Covid-19. *The Journal of Peasant Studies* 47 (5): 944–972.

Vernon, C. 2019. In Conversation with Bruce Pascoe. *Greater Sydney Landcare Network Inc.* Retrieved from https://greatersydneylandcare.org/in-conversation-with-bruce-pascoe.

Willett, W., J. Rockström, B. Loken, M. Springmann, T. Lang, S. Vermeulen, T. Garnett, D. Tilman, et al. 2019. Food in the Anthropocene: The EAT–Lancet Commission on healthy diets from sustainable food systems. *The Lancet* 393: 447–492.

Wilson, J. 1997. Australia: Lucky Country/Hungry Silence. In *First World Hunger: Food Security and Welfare Politics*, ed. G. Riches, 14–45. Toronto: University of Toronto Press.

Yii, V., C. Palermo, and S. Kleve. 2020. Population-Based Interventions Addressing Food Insecurity in Australia: A Systematic Scoping Review. *Nutrition and Dietetics* 77 (1): 6–18.

9

Say Agribusiness but Mean Genocide: Grabbing the Guarani-Kaiowa World

Antonio Augusto Rossotto Ioris

The Genocidal Trail of Agribusiness

The genocide of indigenous peoples is one of the most evident attributes of the ever-expanding agribusiness, as it releases land, labour and resources then used for the production of commodities and the circulation of capital. Contemporary genocides actualise and maintain processes characterised by intense violence, given that the underlying logic of agribusiness seems to be 'let's end everything before everything ends'. The genocidal basis of colonisation continues, therefore, to evolve in order to meet the present-day needs of an agri-food sector dominated by corporations, land speculators and banks (rentier capital). One of the main indigenous groups affected by enslavement, exploitation and displacement during the long history of colonisation and nation-building in South America (Maybury-Lewis 1997) was the Guarani, who used to

A. A. R. Ioris (✉)
Cardiff University, Cardiff, UK
e-mail: IorisA@cardiff.ac.uk

A. A. R. Ioris and B. Mançano Fernandes (eds.), *Agriculture, Environment and Development*, https://doi.org/10.1007/978-3-031-10264-6_9

occupy large parts of the Plata basin and was accordingly assaulted and enslaved from the early decades of Portuguese and Spanish conquest. Among the sub-groups of the large Guarani nation subjected to this invasion, there is the Guarani-Kaiowa (Mura 2019), who especially in the last century were severely impacted by the invasion of their land and their confinement in small, utterly inadequate reservations (Ioris 2020a). Because of the prime agricultural value of their ancestral land, the strategic importance of the region for national development and the hostile attitudes of farmers, the practice of violence was the main channel of communication between the increasing number of settlers and the indigenous population. In addition to more regular aggressions in the form of massacres, cases of genocide typically happened when the Guarani-Kaiowa demonstrated their opposition to conquest and attempted to survive as a cohesive ethnic group. If brutal pressures were not sufficient to reduce their determination to recover the lost areas and restore key elements of traditional community life, genocide was the answer.

The Guarani-Kaiowa are the second largest indigenous people in Brazil today (with around 50,000 individuals, the largest outside the Amazon), and maintain close connections with a population of the same ethnic group on the other side of the Paraguayan border, as well as with other indigenous nations in the State of Mato Grosso do Sul (located on the border with Paraguay and Bolivia), particularly the Guarani-Ñandeva, who also belong to the large Guarani population and speak almost the same dialect (Pereira 2016). Numerous other confrontations have taken place in the region and all over the region, attracting negative media attention and bad publicity for the farmers, although this does not seem to concern them particularly. The situation certainly became easier to manage with the election of a neo-fascist president in 2018, who intensified the anti-indigenous and anti-life tendencies in national and local politics. The authors of violent, criminal attacks are typically abusive landowners who share discriminatory attitudes against 'the sub-human Indians' and operate in alliance with politicians (most of whom are landowners themselves) and through their private militias, known as *pistoleiros*. Because of the proliferation of farms and aggressive regional development policies, the Guarani-Kaiowa have lost around 99% of their ancestral land and been confined to the fringes of the hegemonic agribusiness-centred economy. Their existence has been significantly redefined by the struggle for land, and their world has been dramatically undermined and compressed.

The Guarani-Kaiowa have certainly paid a heavy price for who they are and where they live, amounting to a challenging geography that is complicated by the fact that their existence and intense socio-spatial interactions are deeply interconnected with the economic transformation of the region and the expansion of agribusiness production units (Ioris et al. 2019). Most observers believe that the situation is nothing other than genocide and that those responsible for the genocidal fate of the Guarani-Kaiowa, including farmers, political leaders and members of agribusiness support organisations, bear criminal responsibility. As in the case of North America, Guarani-Kaiowa leaders understood and denounced the fact that nation-building and frontier-making involved not only land grabbing but also the intention to totally annihilate their communities, people and nation; in other words, the hostile intent behind development was serious and tangible. Between 2000 and 2019, the Guarani-Kaiowa was the indigenous group most severely assaulted in the country, with an annual average of 45 new cases and the assassination of 14 political leaders (CIMI 2020). In the years 2015 and 2016 alone, 33 attacks were perpetrated by paramilitary groups against Guarani-Kaiowa communities. Moreover, the ongoing genocide in Mato Grosso do Sul has meant much more than just the loss of land and assassination of community members, but is rather a brutal mechanism of spiritual, social, economic and environmental destruction.

A genocide is essentially predicated upon, and starts with, the subtraction of key socio-spatial relationships that define particular ethnic groups (called here geocide), as has happened in processes of intense spatial and social unravelling in the Gaza Strip, Chechnya, Kashmir and Somalia. As destructive as the grabbing of land, the killing of leaders and immiseration of Guarani-Kaiowa families is the denial of their humanity and the imposition of institutional rules centred on the market value of land and the short-term profitability of agribusiness commodities. Although genocide is a word commonly used by journalists and activists in relation to the dramatic challenges faced by indigenous peoples in Brazil, the significance of the Guarani-Kaiowa genocidal experience is not casual or merely sloganised. The indigenous genocide unfolding in Mato Grosso do Sul is not just a case of hyperbolic violence or widespread murder, but something qualitatively different from other serious crimes. The phenomenon combines strategies and procedures based on direct opposition between groups of people who have been turned into irreconcilable enemies by the pattern of regional development and the balance of political power. The situation in Mato Grosso do Sul is even more painful because the Guarani-Kaiowa is fully aware of being at the centre of an unstoppable genocide that is only the most recent stage in a long genocidal cycle. In this brutal context, resistance has been crucial for the Guarani-Kaiowa to have any chance of surviving, and resist they do.

Several authors working in this field, have demonstrated the continuity between colonial and contemporary processes of genocide. The lived, often tragic, trajectory of the Guarani-Kaiowa was central for the sustenance of the exploitative, rentist and wasteful politico-economy of Brazilian extractivist and agrarian capitalism. Yet it is still to be demonstrated that, whereas the subjugation of the Guarani-Kaiowa represented an important chapter of the colonisation of South American countries, the present-day genocide continues to be crucial for the maintenance of the regional economy and for the consolidation of export-oriented agribusiness in Mato Grosso do Sul. The Guarani-Kaiowa are both survivors and victims of a genocidal cycle that continues because very little has changed in economic or moral terms over the centuries. The Guarani-Kaiowa have endured various genocides over several generations, and their current existence remains a perennial struggle to contain and reverse these processes. This calls for a more careful consideration of the causes and ramifications of a genocidal tragedy that is constantly being denounced by the victims and their closest allies (to no avail).

There was a real chance of compromise in 2007 when the federal government signed an agreement ordering the return of a minimal amount of land to the Guarani-Kaiowa. However, the land was evidently never returned. In 1988 a similar solution had been agreed and simply ignored. No laws or agreements aiming to redress even a small part of the damage caused by land grabbing have been acceptable to those 'masters of the universe' in charge of (indigenous) life and death. Once again, national politics forced marginalised groups, living below the threshold of whiteness, status and property, into a socio-spatial position outside the hegemonic economy, politics and the oppressive rule of law. Just as Germany today is less than what it could have become if not for Nazism, and the United States is dwarfed by its own indigenous Holocaust, Brazil is haunted by the failure to rectify, at least partially, this major socio-spatial liability. Life through genocide is the perpetuation of centuries of socio-ecological devastation and Western intellectual, economic and religious arrogance. Genocidal crimes were not only committed against the Guarani-Kaiowa during colonisation, they happened yesterday, are being committed today, and most likely will happen again tomorrow and next year. This large-scale waste of human lives seems unstoppable and is even accelerating.

It is perhaps odd to interrogate the extent of today's genocide, taking into account that for the indigenous peoples in the Americas—also described as native, ancestral, first nations, Fourth World or aboriginal peoples—the world, by and large, ended after the arrival of the European invaders several centuries ago. They know, better than anyone else, the meaning and the consequences of genocide. The indigenous genocide was just part of the massive effort to deal with mounting scarcities in Europe. While abundance was promised at the new frontiers, new rounds of scarcity emerged in both areas due to the internal dynamics of capitalism, notably the exploitation of society and of the rest of nature (Ioris 2018). As an important chapter of that long geography of conquest and annihilation, the genocidal pressure on the Guarani-Kaiowa basically reproduces, and 'modernises', forms of prejudice and oppression employed during colonisation and the early history of Brazil, when indigenous peoples were basically treated as exotic relics of an ignoble past that had to be overcome. More than three centuries of intermittent clashes with the colonial enterprise, the commercial demands associated with the Second Industrial Revolution and the geopolitical stability that followed the end of the Paraguayan War (1864–1870) and triggered increasing exploitation of indigenous labour. From the last two decades of the nineteenth century, large-scale harvesting of *erva-mate* leaves was carried out by transnational corporations, making use of a semi-enslaved indigenous labour force. The process of land grabbing and commodification, which began in the early years of the last century and was augmented from the 1960s onwards with the expansion of export-based agribusiness, and led to the removal of the most remaining vegetation, the aggravation of land disputes and, eventually, Kaiowcide (Ioris 2021).

The relationship of the Guarani-Kaiowa with genocide is more complex than the passive victimisation of human rights discourses. A key message from Guarani-Kaiowa theology is that genocide is not unprecedented. However, that does not make it any less awful and despicable. The eschatological perspective of the Guarani-Kaiowa adds some very special features to their life through genocide since colonisation. For instance, Guarani people have a particularly troubled relationship with death and are always extremely concerned about losing relatives and the possibility of dying alone. It is unacceptable for them to show pictures of dead bodies and they carefully avoid images of deceased people, because these may attract bad spirits which will try to take them to the next world. According to Guarani religious beliefs, death is not the end of the story but brings additional troubles to all involved. The Kaiowa feel particularly demoralised when, as happens quite often in attacks organised by hostile farmers, a relative is murdered and the body simply disappears (Morais 2017). Another lesson from their tragic experience is that those at risk of suffering total destruction should mobilise the accumulated knowledge of the world, combined with past memories and spiritual support, and persevere in the pursuit of justice and shared goals. The Guarani-Kaiowa seem to have been doing all that for many years. They rapidly understood the methods and direction of colonisation and land grabbing, and the values and attitudes of those coming to their territory in ever greater numbers, and had to develop adaptive responses to somehow mitigate the losses and coexist with these aggressive enemies. Guarani-Kaiowa spatial controversies demonstrate that very few groups, if any, are more attuned to contemporary trends, or have a more active socio-spatial protagonism (Ioris 2022).

This chapter is basically a reflection on indigenous genocide as a form of acute ethnopolitics, which is based years of engagement with Guarani-Kaiowa communities and tries to interpret their socio-spatial condition from the perspective of Kaiowcide. It comprised community visits, meetings, interviews and attendance at religious ceremonies and public events, during field trips between 2017 and 2020. It first entailed the construction of an equitable and productive dialogue with indigenous communities that required, first of all, an ethical and political commitment to avoid patronising stereotypes and utilitarian oversimplifications. This created an opportunity to bring together the emotional insights of personal narratives and instances of the cruel geography of exclusion and deprivation that underpins Kaiowcide. The study also mobilised other empirical evidence, newspaper articles, government and NGO reports, statistics, historical data and multiple theoretical points to build an argument on the community-based political economy of reciprocity realised by the Guarani-Kaiowa—which has been under attack from wider socio-economic forces and also subject to internal community tensions—without going down the route of neo-populism or schematic localisms. Rather than a naïve attempt to 'give voice to indigenous people' which normally produces a simulacrum of their opinions and perspectives, the intention was to engage with real individuals and try to capture some of the complexity of their lived space. The next section will relate long genocidal trends with the anti-life management of the Covid-19 pandemic in 2020 which was very emblematic of long-lasting trends.

The Geocide-Genocide-Massacre Nexus

It was the invasion of the Americas that marked the beginning of European expansion and it was the Indians of the Americas who have borne the brunt of their indigenous status for the longest time. The main tool used by the invaders was outright genocide, which paved the way for their territorial and extractive plans and for the consolidation of a mercantilist economy based on slavery and racially driven killing. The American landscapes, which had been transformed over millennia by socio-ecological interaction between human and more-than-human agents, were simply considered *terra nullius* by the European invaders, that is, it was nobody's property, free to be grabbed by those moving from the East hungry for profit. In effect, the aggressors themselves nullified and ruined everything blocking their way to personal enrichment, equipped with firearms and royal and papal decrees guaranteeing them access to the world of peoples never heard of before. Social destruction clearly happened not only through the direct assassination of individuals and groups but also through the spread of diseases and the imposition of the European religion. The primary motive for the elimination of the autochthone nations has been territoriality (settler colonisation), but not always, as in the case of the fur trade in Canada or the exploitation of labour in the Peruvian mines. This rationale of displacement and conquest continued to underpin national development and state building when the management of the former colonies was transferred to newly formed 'independent' governments. Evans and Thorpe (2001) aptly propose the concept of 'indigenocide' to describe the theoretical and practical procedures that made indigenous peoples less valued than the land they inhabited and which was wanted by the invaders. 'Indigenocide' contrasts with state-driven, industrial and bureaucratic genocides, such as the Jewish Holocaust, and happens when land is intentionally invaded, for as long as it is necessary or possible, leading to the killing of the original inhabitants, classified as the lowest form of humanity and deserving of extermination.

The idea of 'indigenocide' is certainly sound and helpful in terms of understanding the specificities of indigenous genocides, but it is necessary to scrutinise it in relation to the various criteria discussed above. First, the question of intentionality, which in the definition of 'indigenocide' seems to echo the provisions of the 1948 Convention, has already been extensively considered in the previous sections, where it was argued that indirect acts of violence and incidental measures, such as the spread of contagious diseases, can also amount to genocide. Second, it appears restricted to the classical model of land invasion followed by social and religious abuses and concentrated mass assassination, which are certainly relevant for the characterisation of the crime, but fall short of exhausting other possible genocidal strategies. 'A discussion of genocide as practiced against indigenous peoples should not (…) focus solely or even principally on deliberate attempts to massacre entire societies. Often the widespread dying resulted not so much from deliberate killing but from the fatal circumstances imposed by the imperialists on the conquest' (Maybury-Lewis 2014: 82). In effect, the historical experience shows that in almost all indigenous genocides, including the decimation of entire nations in the sixteenth century in the Americas, most victims perished in the course of short-term clashes interspersed with long phases of low belligerence but sustained antagonism and brutal treatment by the state and by companies or groups of individuals in search of resources, land or labour. The genocide of indigenous peoples is normally lived and endured over several years or decades, during which the victims are not passive but try to react and resist whenever the conditions are ripe.

Third, it is not the scale or rate of killing that determines whether an indigenous genocide is taking place but the systematic and brutal imposition of oppressive actions and norms that make it extremely difficult for indigenous people to survive and reproduce. An indigenous genocide, as a period of intense social destruction and the loss of a significant proportion of the population, is not an isolated phenomenon but is preceded and supervened by a broader process of world grabbing, that is, the subtraction and invalidation of the indigenous world. This longer and deeper process can be described as *geocide*, which is not merely the recognition of a major antagonism between indigenous and non-indigenous peoples but encapsulates the intolerance and destructive impetus of more powerful invaders. Genocide is dialectically connected with, and predicated upon, the subtraction of indigenous worlds that is geocide. Geocide is the language and technology of colonisation, which gradually eliminates any remaining opportunities for indigenous people to maintain their collective and individual lives. It entails micro- and macro-dynamics of violence that consolidate prejudices, difference and rivalries and, depending on the nature of the disputes and the balance of power, can spark a genocide. There are also isolated or more circumscribed cases of lethal violence in the form of massacres, which also have geocide as their deep-rooted motivation. Indigenous genocides are predicated upon geocide, which is the more permanent driving-force underpinning the socio-spatial relationship between indigenous people and the potential perpetrators of genocide.

Geocide is an expression of what the indigenous scholar Taiaiake Alfred (2004: 90) describes as situations in which native peoples have inherited 'relationships founded on hatred and violence and a culture founded on lies to assuage the guilt or shame of it all', and where victims and perpetrators alike continue to deny their shared past and the corresponding moral implications. World subtraction, which is implicit in geocide, is more than just the grabbing of land and resources, but entails the erosion of existing or potential social relations. Before an actual genocidal experience, a series of 'preconditions' must be fulfilled, first of all, the reduction of the victims to something less than human, worthless and 'outside a web of mutual obligations', as well as the degradation of the perpetrators to criminals or pathological individuals (Chalk and Jonassohn 1990: 28). It is perhaps ironic that the UN Declaration on the Rights of Indigenous Peoples itself incorporated a sense of geocide in its definition of such social groups. The text of this declaration states that indigenous peoples are those with a historical and geographical continuity with pre-invasion and pre-colonial societies, and that still consider themselves distinct from other sectors of the societies now prevailing in their territories. In this sense, they form minority, non-dominant sectors of national societies and struggle to preserve, develop and transmit to future generations their ancestral lands and ethnic identity which are the basis of their continued existence according to their own social institutions and traditional practices. The Declaration suggests that for a group to become indigenous (in general rather than in terms of any particular ethnic denomination) their land must have been invaded and there must be some geocidal antagonism from other social groups. As such, the definition has an inadvertent backward-looking connotation that perpetuates the idea that geocide is essential for a group to be considered indigenous, which is certainly important as a political gesture but is also ontological restrictive.

Fourth, and also very important, significant controversy has arisen over whether the definition of genocide should be limited to the elimination of a *genos*, an ethnicity or nationality, evidently following the perpetrators' definition of an ethnic minority (Bauer 1999), or whether its scope should be broadened to include atrocities based on class identity and politico-economic disputes (Mann 2005) amplified by references to ethnicity and belonging. Nonetheless, this is a false dichotomy and questions related to ethnicity of indigenous people, as demonstrated by the incidence of racism, discrimination and segregation, cannot be separated from the exploitation of the indigenous labour force by the owners of the means of production. This apparent separation between ethnic or class-based oppression, which could independently lead to genocides, is reconciled through the control of state action. There exists a politico-economy of genocide that is not restricted to past events and continues to shape political interventions in favour of or against new genocides. As observed by Bauman (2000), genocide is a rare event but not without major socio-political repercussions, and it reveals a great deal about the shortcomings of Modernity (particularly in the twentieth century) and helps us to interrogate the present condition of the world. The same global society that made the Holocaust possible still exists, and there was nothing in that society that could stop it from happening during the war or that could stop it from happening again today. Some take genocide to be an aberrant act, but its roots are really in the 'process of historical development out of which our entire, global, political-economic system has emerged' (Levene 2005: 9).

Genocide is the contingent and concentrated manifestation of more perennial and dispersed forms of socio-spatial violence involved in the subtraction of the indigenous world (geocide). The long and gradual unfolding of geocide—according to a politico-economic and ideological regime that connects local circumstances with national trends—guarantees the necessary conditions for the occurrence of regular genocidal episodes during conquest and colonisation. The notion of geocide refers to a long-term process of brutal disqualification, exploitation and appropriation of land from the ancestral inhabitants of a territory who find themselves in the way of the economic, spatial or political gains expected by invaders and colonisers. Genocide, in turn, is the more intense destruction of social groups, carried out over a relatively shorter time, through direct or indirect measures (direct measures could include the use of firearms and the capture and execution of target populations, while indirect measures could include spreading disease or provoking regular road accidents). Genocide is nonetheless predicated upon the progress of geocide, and therefore the connection between geocide and genocide is not just temporal, but deeply dialectical and associated with acute politico-economic disputes. The above conceptualisation of the socio-spatial association between *geocide, massacres* and *genocide* will now facilitate the comprehension of indigenous genocides as quantitatively and qualitatively different from comparable non-indigenous phenomena.

Kaiowcide: Consolidating the Power of Agribusiness

The long struggle of the Guarani-Kaiowa for the recognition of their most basic rights has important parallels with the class-based struggle of landless peasants and marginalised urban groups in Brazil. Each indigenous group is unique, and defining features of the Guarani-Kaiowa include precisely their ability to preserve their language (a semi-dialect of Guarani) and maintain a relatively large and unified social identity amidst a series of interrelated genocides. It has been reported in several documentaries, movies and UN reports, and the images of protest, police repression, dead bodies, miserable living conditions and dirty children have circulated around the word. Still, the Guarani-Kaiowa remains the most threatened indigenous population in Brazil, denied recognition of their original lands and subject to systematic abuses and exploitation. The indigenous groups and extended families that are now described as Guarani-Kaiowa (and Paĩ-Taviterã in Paraguay) have been living through a series of genocides for more than four centuries. The most recent and ongoing process of genocide since the 1970s—described here as Kaiowcide—corresponds to the solidification of the agribusiness-based economy, the growing neoliberalisation of production, rapid urbanisation, the serious deterioration of living conditions inside and outside the reservations and the introduction of formal democratic legislation (Ioris 2017). From the perspective of the agribusiness sector, the presence of a contemporary indigenous population is no more than a leftover from violent skirmishes that happened decades ago during the conquest of the territory, and indigenous people constitute a horde of desolate, strange people who 'only have themselves to blame' for their fate.

While new Brazilian legislation has recognised the rights of ancestral peoples to maintain their indigenous identities indefinitely (rejecting assimilation and tutelage), geocide continues through neoliberal economic and ideological constructs that guarantee high levels of alienation and homogenisation through market consumerism, evangelical protestantism and the financialisation of all aspects of life. Despite legal and constitutional improvements, most public authorities prefer to look the other way and hope that the indigenous population will renounce their ethnic claims and become indistinguishable from other poor Brazilians. Consequently, Kaiowcide has entailed killing both through the imposition of market-based interpersonal relations (e.g. hyper-exploitation of indigenous workers and the renting out of indigenous land to agribusiness) and, as in the past, 'conventional' murder by state police or paramilitary militias (and increasingly by drug dealers too). The main claim here is that Kaiowcide has dramatically impacted the Guarani-Kaiowa in recent decades—which remains clearly connected to geocide and is the continuation of previous genocidal phases—because it is a *counterreaction of land grabbers, reactionary judges and politicians and the repressive agencies of the state apparatus against a legitimate and determined indigenous mobilisation for the restoration of land-based relationships and for better social, political and economic opportunities.* Kaiowcide is a renewed, bespoke and ongoing phenomenon of brutal socio-spatial elimination taking place in a context dominated by agribusiness farmers and the prevalence of globalised, urban values.

It is precisely because the Guarani-Kaiowa decided to react to the geocidal violence associated with agribusiness production, making good use of novel politico-institutional spaces, that they have been targeted for further rounds of genocide, now in the form of Kaiowcide. In other words, Kaiowcide is not happening because of a lack of political resistance, but precisely as a result of the ability and determination of the Guarani-Kaiowa to fight for what they consider legitimate. The fundamental demand is, obviously, the return to their ancestral areas grabbed by farmers in the course of agrarian development, what happens in the form of reoccupation and retaking of land (called *retomada*). The main argument here is that Kaiowcide is a form of genocide that has occurred because of the political reaction of the Guarani-Kaiowa, since the late 1970s, against a long genocidal process that escalated with the advance of an agribusiness-based economy. The key analytical challenge involved in making sense of Kaiowcide is to connect the widespread hardships faced by the communities with the collective mobilisation of groups dispersed in the territory and capable of coordinating effective political initiatives (such as the *retomadas*). In historical terms, Kaiowcide corresponds to the violence and deception of neoliberalised agribusiness, which is both explicit and embedded in aggressive mechanisms of mass production and elitist property rights. The ambiguity of Kaiowcide, combining both innovative and apparently archaic forms of cruelty, is also an emblematic hallmark of neoliberalised agribusiness, which seems to offer a solution to food insecurity but in effect maintains and aggravates malnutrition, risks and socio-ecological degradation (Ioris 2016).

The more recent genocidal phase combines elements of state abandonment and political persecution with a range of violent measures stimulated and facilitated by the exploitative pattern of regional development. Kaiowcide has certainly incorporated additional unique features, such as the need to respond to international public opinion and give the impression that the actions of agribusiness organisations are legal and legitimate, but it also dialectically preserves elements of the most primitive brutality employed by the Jesuits, kings and *conquistadores* in the past. Even so, there is a subtle but important difference between previous genocides associated with space invasion and ethnic cleansing and the systematic attempts to contain and undermine the Guarani-Kaiowa socio-political revival since the 1970s through Kaiowcide. In the previous phases, the Guarani-Kaiowa was subjugated by Catholic missionaries and attacked by *bandeirantes* and *encomienderos*, were converted into semi-enslaved labourers working in the production of *erva-mate* and occasional farm labourers [*changueiros*] recruited (ironically) for the removal of the original vegetation, while also being expected to remain in small, inappropriate reservations and having their identity rapidly diluted as a consequence of individualising policies (as in the case of the division of the reservations into family plots of land instead of collective areas). These past experiences form the basic etiology of Kaiowcide, considering that the long process of colonisation, territorial conquest and settler migration paved the way for the subordinate insertion of Brazil into globalised agribusiness markets and the consolidation of agrarian capitalism in Mato Grosso do Sul.

The more diffuse and less evident basis of indigenous genocides, which the literature often treats as politicide, gendercide and culturicide, was certainly present in the previous two phases of the long Guarani-Kaiowa genocidal experience, but the important difference is that in the past the aim was to assimilate and proletarianise the indigenous population, while under Kaiowcide the goal is to contain the possibility of political revolt through mitigatory measures, alienating religiosity and encouraging consumerist behaviours, as well as intimidation and the suppression of legitimate land claims through lengthy court disputes complemented by the operation of paramilitaries and private farm militias. Note that the deadly features of Kaiowcide go beyond the boundaries of politicide, as the victims have been targeted because of the perpetrators' prejudices against Guarani heritage and ethnicity. Moreover, it is also more than ethnocide (as conceptualised by Mann 2005) because there has been a clear intention to kill the leaders of the indigenous mobilisation. It also has elements of culturicide, but goes beyond that because Kaiowcide entails forced movement and murder. In practice, all these processes converge and reinforce each other. The genocidal practices of Kaiowcide have been greatly facilitated by the fabricated invisibility and neglect of indigenous communities by the vast majority of the regional population, who prefer to remain ignorant of the crude realities of life for the Guarani-Kaiowa.

As a result, Kaiowcide has not only lasted for several years now but has created a self-reinforcing mechanism in the stimulation of novel forms of reaction and counteraction. There is a vicious circle that constantly reinvigorates Kaiowcide: the Guarani-Kaiowa react to previous genocides and then have to be crushed through new cycles of genocide, which inevitably, because of their active political agency, spark fresh reactions and enflame additional genocidal measures. The agribusiness community and state armed forces have the necessary resources to buy equipment, guns and bullets, recruit mercenaries, lawyers and judges, and sustain an aggressive media campaign in defence of the highly perverse status quo. Their strength is entirely dependent on the brutality of the state apparatus, the inscrutability of the legal system and the profitability of export-oriented agriculture production. The Guarani-Kaiowa, by contrast, forge ahead through a horizontal power network based on a family support system, reliant on the wisdom and prestige of respected community elders and religious notables (often the same individuals). The brutal elimination of any one person is bitterly felt and will be always remembered by the communities, but the non-Western mentality of indigenous peoples is much less individualistic, and they are therefore able to more effectively respond to the losses of chiefs, elders, shamans and community members. Cross-community political alliances and mutual support are centred around personal reputation, family bonds and interactions with the common ancestral land rather than money, material resources and external lobbying influence. Highly prestigious leaders live in apparently destitute conditions, whilst retaining voice and influence because of their knowledge, religious status and ability to provide effective guidance.

Kaiowcide is, therefore, the coordinated attempt to annihilate, repress and contain the Guarani-Kaiowa precisely when they have been able to challenge the accumulated consequences of agrarian capitalism and frontier-making in the service of conservative national development. Distinct from the graphic narratives of other contemporary genocidal processes, the victims of Kaiowcide are not calculated in thousands of deaths. The total number of fatalities in the last four decades, including the murders of leaders and community members using firearms and other weapons, amounts to several hundred victims. But this is only a small fraction of the total losses due to road accidents, suicides of adults and teenagers, child mortality and deaths related to the miserable conditions in the reservations, encampments and shanty towns. In other words, if violent murders of indigenous persons are recurrent news in Mato Grosso do Sul, a much larger number of deaths have been caused by a lack of proper housing, malnutrition, contaminated water, mental illness and inadequate medical assistance (often related to the lack of transport connections and infrastructure needed to transport patients to distant hospitals). As Levene (2005) theorises, genocides presume a reified treatment of the victims, who are considered generic strawmen and no longer seen as real people. Indigenous identities and social organisation have been devalued, and aspects of the Guarani-Kaiowa way of life are persistently scorned and even criminalised (as in the case of collecting medicinal plants on private farms).

Similar to other regional and global experiences, Brazilian society has become increasingly individualistic and dysfunctional because of mass consumption pressures and the growing role played by fundamentalist groups and evangelical churches with a message of individualism and a focus on personal success (as well as condemnation of central elements of Guarani-Kaiowa identity). The sense of a shared social purpose has been eroded by the combined forces of conservatism, individualism and privatisation that have challenged the role of the state as protector and ultimate provider. Very few people have been directly involved in the physical elimination of indigenous persons, but large sections of the population have been indirectly complicit in perpetuating anti-indigenous trends and have indirectly contributed to indigenous genocides including Kaiowcide. Regional development and agribusiness in Mato Grosso do Sul have been established through productivist, short-lived economic goals, while most of local society has tended to be intolerant towards ethnic diversity and the conservation of ecosystems (both considered superfluous and unimportant), all of which has been instrumental for the advance of Kaiowcide. All these embarrassing abhorrent trends became even more pertinent after the election of Bolsonaro in 2018, who at last provided a 'legitimate' reason for the violence and elitism of the agribusiness economy (and also facilitated the spreading of the Covid-19 outbreak). These broad tendencies that underpin the process of Kaiowcide and its evolution will be examined in more detail below. The three main phases of Kaiowcide are connected to wider geocide pressures, as well as to national politics and related developments on the ground.

Living to Overcome Kaiowcide

Kaiowcide is, effectively, the most recent phase of a long genocidal process that has, since the seventeenth century, attempted to destroy the Guarani-Kaiowa people and significantly destabilised their sociospatiality through invasions, enslavement and persecution. Kaiowcide is the reincarnation and revival of an old genocidal practice. While the focus in recent years may have shifted from assimilation and confinement to abandonment and confrontation, there remains the same intention to destabilise and eliminate the original inhabitants of the land through the asphyxiation of their religion, identity and, ultimately, geography. As a background to the actual genocide, the trend of aggression and world robbery—defined here as geocide—intensified in the second half of the last century and produced multiple consequences at individual and community levels, including severe mental health issues, alcoholism, domestic violence and high levels of suicide. When it became evident that the government would continue to prevaricate, the collective decision was made to start a coordinated reoccupation of ancestral areas lost to development (the retomadas), which triggered a corresponding reaction from farmers and the authorities in the format and language of Kaiowcide. In practice, this means that in addition to the obstacles faced by any subaltern class or social group in the highly unequal, racist and conservative society of Brazil, the Guarani-Kaiowa also face the monumental challenge of continuing the fight to recover their land in order to rebuild basic socio-spatial relationships in the midst of a genocide.

Because of the multiple difficulties within communities and beyond the small spaces where they live, where their ethnicity is at least respected and cherished, the Guarani-Kaiowa are relentlessly propelled into a daily anti-genocidal struggle for social and physical survival. In the words of Quijano (2000: 226), they constantly have to be 'what they are not', that is, there are major barriers to acceptance for their ethnic specificities and their most fundamental needs as a distinctive social group. Regular murders of Guarani-Kaiowa, both during the retomadas and in isolated hostilities, have become so common that many incidents now do not even make the headlines. Between 2003 and 2017, around 45% of the homicides involving indigenous victims in Brazil were committed in Mato Grosso do Sul (461 in total), and 95% of these were Guarani people (CIMI 2018). In the same period, 813 indigenous suicides were registered in the State. This means that through suicides and murders alone, around 3% of the Guarani-Kaiowa population was eliminated in less than 15 years. When other causes of death are factored in, such as loss of life due to hunger, malnutrition and food insecurity, poor sanitation, lack of safe water, drug use and acute mental health problems, among others, it is not difficult to perceive the widespread impact of genocide in Guarani communities and settlements.

Although from the perspective of Guarani-Kaiowa geography itself the boundaries of their land are not absolute but associated with the long and dynamic presence of extended families in the terrain, non-indigenous institutions have imposed borders and fences in the name of national sovereignty and the sanctity of private rural properties. Because of the need to present their claims before the apparatus of an antagonistic state, their idiosyncratic understanding of space had to be translated into objectivity-seeking maps, anthropological surveys and legally valid proof of socio-spatial connections. All these legal and bureaucratic requisites mean that only the areas with the most compelling evidence of recent indigenous presence have even a minimal chance of being returned to the indigenous claimants. These are the most significant areas under dispute, or which already have some level of regularisation. Note that, despite the violence and the genocide, the indigenous demands are relatively small in relation to the total size of the region. Note also that, even if all those areas are one day restored to the original inhabitants (as stipulated in the legislation), it will remain a true archipelago of isolated indigenous 'islands' in a sea of hostile agribusiness activity. In any case, most areas in this image, especially the largest ones, are merely aspirational, as they are still controlled by the farmers and their return blocked by the courts.

Such a dialectic of forced invisibility and immanent protagonism has ethnicity as a central but highly contested category. Rather than separating indigenous people into an entirely distinct politico-economic condition, ethnicity influences land and labour relations (i.e. facilitating land grabbing and the over-exploitation of labour-power) and also the mechanism of adaptation and political reaction. Thus, there exists a crucial tension between an identity that is tolerated by the stronger groups only inasmuch as it increases economic gains and a disruptive alterity that rejects exploitation and is constantly revitalised by the ethnospatial practices of the Guarani-Kaiowa. This lived reality defies any simplistic politico-economic categorisation. The prejudices of the non-indigenous sectors give rise to concrete forms of exploitation and, not infrequently, hyper-exploitation in the form of modern slavery. In July 2020, right in the middle of the Covid-19 pandemic, a group of 24

Guarani labourers, four of whom were aged only 15, and their families, including six young children, were freed by the authorities after secretly contacting an indigenous community leader. The group had been forced to stay on the farm because of unpaid debts and had to live in miserable conditions, sleeping on thin and dirty mattresses in minuscule, cold rooms, with appalling kitchen and toilet facilities. Criminal cases involving modern slavery have been common since the 1990s, when enslaved people were frequently rescued from sugar cane plants; on one occasion, around 900 people were liberated in a single day. In a context of sustained transgressions committed by public authorities and businesses, Kaiowcide continues to unfold through an accumulation of anti-indigenous pressures that go beyond land-related controversies to include a whole range of ethnic-related aggressions.

The most relevant form of resistance and reaction to such a genocidal state of affairs is, clearly, the mobilisation for the retaking of indigenous areas—retomadas—which involves not only the material dimension of the land but is also a source of collective hope and reinforces a sense of joint political purpose. If the reoccupation of farmland became even more dangerous after the election of the openly fascist and pro-indigenous genocide government in 2018, this has not curbed the determination to demand that the state resolve the dispute and allow the indigenous families to return to the land of their ancestors. The main pillar of the land recovery action by the Guarani-Kaiowa is their awareness that politics must be a shared endeavour that presupposes interpersonal reciprocity. This turns individual land recovery actions into a collective territorial strategy because of the common will to be recognised as a distinctive and valued social group. In that regard, the Guarani-Kaiowa are in a position of strength, because their life is intensely based on social interaction, particularly among members of the same extended family. The long road back to their ancestral areas typically ends with an intense and mixed feeling of achievement, loss and realisation of what comes next.

Indigenous genocide is the forename, surname and address of agrarian capitalism and rural development in Mato Grosso do Sul. In recent decades and since the reorganisation of the State as a self-governing member of the Brazilian federation in 1978, the last round of genocidal action came in the form of Kaiowcide, that is, a coordinated, cross-scale reaction against the well-organised mobilisation for land rights and material compensation for past aggression. Previous phases of genocide resulted in the tragic disintegration of lived, ethnic-referenced spaces [*tekoha*] and the confinement of the Guarani-Kaiowa in overcrowded sites with the worst social indicators in Brazil and unimaginable levels of human misery. When the Guarani-Kaiowa sensed that their total annihilation was the shared plan of farmers, businesspeople and the government, they started to organise, together with the Guarani-Ñandeva, large assemblies [*Aty Guasu*] to better connect with other indigenous peoples, campaign for political recognition, send their children and young people to school and university and take back areas from where the elders and deceased ancestors had been expelled. Because the powerful sectors in Brazilian society only use genocidal language to communicate with indigenous peoples, once the Guarani-Kaiowa began confronting and denouncing the illegitimate order, the authorities put into practice what they were already experts at: a new genocide in the form of Kaiowcide. If the Guarani-Kaiowa believed from the 1970s that they could recover from the tragic legacy of previous genocides, they only received what the powerful decided was right for agribusiness-based development: more destruction, persecution and death.

The above discussion has presented the genocidal tragedy but also the defiant political struggle of the Guarani-Kaiowa, a group that is desperately trying to resist and overcome genocidal pressures associated with agribusiness-based development, using this experience to provide a heuristic account of the importance of political ontology as a tool for interrogating the impacts of Western modernity and its socio-spatial legacy (Ioris 2020b). The Guarani-Kaiowa had to be partially assimilated and their social institutions severely undermined so that they could be exploited through depersonalised market-based relations. Socio-spatial differences were manipulated to render them invisible from a development perspective and to justify the appropriation of indigenous land and other illegal and racist practices by the state and business sector. At the same time, the Guarani-Kaiowa's own singularisation is their best hope of resistance and the main force that allows them to continue hoping for a better life under a different world order. The refutation of the reductionism of a single, given reality of the world represents an ontological political practice based on the political dimension of ontology and on the ontological dimension of politics (Escobar 2015). The reconstruction of their socio-spatial settings is also a form of resistance against labour exploitation, the alienating influence of evangelical churches and the homogenising pressures of urban pop culture.

An indigenous genocide such as Kaiowcide cannot be judged in terms of the number of people, the extension of reservations, seats in the parliament or media coverage, but must first and foremost be judged by the monstrousness of past and present relations between 'Indians and non-Indians', which continue to be based on violence, neglect and racism. If the indigenous problem is important and disturbing, more important still is the prospect of the 'Indian-political' widening their role in local and national politics. This threat or hope, depending on how one perceives it, is like a shadow hanging over the Brazilian national government and sectors of civil society, because the indigenous list of demands and their higher moral ground is clear, as much as their ability to forge alliances and subvert the orderly flow of public affairs. But the identity of the 'Indian-political' is constituted through the struggle and becomes more powerful the more they know about their condition and are able to reclaim (affirm) their cultural life. There is a profound politics of identity based on the understanding that ethnic and cultural identification is neither immutable nor essentialist, but subject to various influences and fraught with internal tensions. The 'Indian-political' is not necessarily coherent and does not have a fixed, unchangeable goal, but is able to creatively learn from acts of mobilisation, confrontation and negotiation. The Guarani-Kaiowa are already doing this extremely well, even at the cost of devastating sacrifices made by many for a few, but tangible, accomplishments in their fight to overcome Kaiowcide.

References

Alfred, T. 2004. Warrior Scholarship: Seeing the University as a Ground of Contention. In *Indigenizing the Academy: Transforming Scholarship and Empowering Communities*, ed. D.A. Mihesuah and A.C. Wilson, 88–99. Lincoln: University of Nebraska Press.

Bauer, Y. 1999. Comparison of Genocides. In *Studies in Comparative Genocide*, ed. L. Chorbajian and G. Shirinian, 31–43. New York: St. Martin's Press.

Bauman, Z. 2000. *Modernity and the Holocaust*. Cambridge: Polity.

Chalk, F.R., and K. Jonassohn. 1990. *The History and Sociology of Genocide: Analyses and Case Studies*. New Haven and London: Yale University Press.

CIMI. 2018. *Violência contra os Povos Indígenas no Brasil: Dados de 2017*. Brasília: Conselho Indigenista Missionário.

CIMI. 2020. *Violência contra os Povos Indígenas no Brasil: Dados de 2019*. Brasília: Conselho Indigenista Missionário.

Escobar, A. 2015. Territorios de Diferencia: La Ontología Política de los 'Derechos al Territorio.' *Cuadernos De Antropología Social* 41: 25–38.

Evans, R., and B. Thorpe. 2001. Indigenocide and the Massacre of Aboriginal History. *Overland* 163: 21–40.

Ioris, A.A.R., ed. 2016. *Agriculture, Environment and Development: International Perspectives on Water, Land and Politics*. Basingstoke: Palgrave Macmillan.

Ioris, A.A.R. 2017. *Agribusiness and the Neoliberal Food System in Brazil: Frontiers and Fissures of Agro-neoliberalism*. London: Routledge.

Ioris, A.A.R. 2018. Amazon's Dead Ends: Frontier-making the Centre. *Political Geography* 65: 98–106.

Ioris, A.A.R. 2020a. Indigenous Labor and Land Resources: Guarani-Kaiowa's Politico- Economic and Ethnic Challenges. *Resources* 9 (7): 84.

Ioris, A.A.R. 2020b. Ontological Politics and the Struggle for the Guarani-Kaiowa World. *Space and Polity* 24 (3): 382–400.

Ioris, A.A.R. 2021. *Kaiowcide: Living through the Guarani-Kaiowa Genocide*. Lanham, Maryland: Lexington Books.

Ioris, A.A.R. 2022. Indigenous Peoples, Land-based Disputes and Strategies of Socio-spatial Resistance at Agricultural Frontiers. *Ethnopolitics* 21 (3): 278–298.

Ioris, A.A.R., T. Benites, and J.D. Goettert. 2019. Challenges and Contribution of Indigenous Geography: Learning with and for the Kaiowa-Guarani of South America. *Geoforum* 102: 137–141.

Levene, M. 2005. *The Meaning of Genocide: Genocide in the Age of Nation-State*, vol. 1. London: Tauris.
Mann, M. 2005. *The Dark Side of Democracy: Explaining Ethnic Cleansing*. Cambridge: Cambridge University Press.
Maybury-Lewis, D. 1997. *Indigenous Peoples, Ethnic Groups, and the State*. Boston: Allyn and Bacon.
Maybury-Lewis, D. 2014. Colonial Genocide. In *Genocide: A Reader*, ed. , J. Meierhenrich, 82–84. Oxford: Oxford University.
Morais, B. 2017. *Do Corpo ao Pó: Crônicas da Territorialidade Kaiowá e Guarani nas Adjacências da Morte*. São Paulo: Elefante.
Mura, F. 2019. *À Procura do "Bom Viver": Território, Tradição de Conhecimento e Ecologia Doméstica entre os Kaiowa*. Rio de Janeiro: Associação Brasileira de Antropologia.
Pereira, L. 2016. *Os Kaiowá em Mato Grosso do Sul: Módulos Organizacionais e Humanização do Espaço Habitado*. Dourados: UFGD.
Quijano, A. 2000. Colonialidad del Poder, Eurocentrismo y América Latina. In *La Colonialidad del Saber: Eurocentrismo y Ciencias Sociales—Perspectivas Latinoamericanas*, ed. E. Lander, 201–246. Buenos Aires: CLACSO.

10

Land and Food Access in the Context of Climate Change: Implications to Rural Development in Mozambique

Natacha Bruna and Máriam Abbas

Introduction

Developing countries are considered the most vulnerable to climate change and concentrate about 98% of the undernourished people in the world, with Africa being the region with the highest prevalence of hunger (19% of the population—more than twice the world average of 8.9%) (Ford et al. 2015; FAO et al. 2020). Most of these people live in rural areas, having agriculture as their main source of income and subsistence, with high dependence on natural resources (such as access to land, water, forestry resources, among others), therefore are the ones that will be most disproportionately impacted by climate change (FAO et al. 2020). The current environmental crisis represents a threat to many underdeveloped countries as it considerably increases their socioeconomic vulnerability. The impacts range from diminished food production and access to food

N. Bruna (✉) · M. Abbas
Observatório do Meio Rural (OMR), Maputo, Mozambique
e-mail: natachabruna89@gmail.com

A. A. R. Ioris and B. Mançano Fernandes (eds.), *Agriculture, Environment and Development*, https://doi.org/10.1007/978-3-031-10264-6_10

and land, internally displaced people as a result of extreme weather events in high-risk areas, among many more.

Mozambique is the 44th most vulnerable country in the world regarding climate change, and the 25th least ready country to face its impacts (ND-GAIN 2022). In the last decade, the country has been systematically hit by extreme weather events with negative implications to rural population and development. Besides extreme weather events, many regions have experienced droughts and lack of rain, which directly impacts subsistence agriculture (food and cash crop production) and rural livelihoods. Climate-related disasters are among the main drivers of food insecurity, being droughts the main source of vulnerability in most farming systems in Sub-Saharan Africa and in Mozambique (Dixon et al. 2001; Marques et al. 2009). Most farming systems in Mozambique are dependent on rainfall, considering that agriculture is mainly practiced in a rainfed system.

Lack of rain has been considered the main cause of food shortage by most households in different farming systems in Mozambique (Abbas 2022b). Given the importance of climate drivers in farming system choice, the fact that these are subject to changes in time will affect farmers' decisions on what and how to produce (Abbas 2022a). According to Abbas (2022a), farming systems whose choice is dependent on climate drivers are also the ones that are already vulnerable to climate change, either due to their location or due to their low adaptive capacity to face the impacts of climate change. This is why the adaptation policies are considered urgent in rural Mozambique. While mitigation is considered urgent globally. On the other side, Mozambique's biodiversity potential is internationally acknowledged, as around 25% of the national territory has conservation potential (ANAC 2015). The crucial combination of the 'need to adapt' and the biodiversity potential of the country makes it a great receiver of climate funds and a strategic destination for climate change adaptation and mitigation projects. Due to the combination of existing 'marginal lands,' big biodiversity potential and high levels of vulnerability to climate change, Mozambique has currently become the third largest recipient of climate funds in Sub-Saharan Africa, with approximately USD 147.3 million approved in 2016 from multilateral recipients for adaptation and mitigation projects until 2017 (CIAT and The World Bank 2017).

Consequently, a national public entity aiming at administration of conservation areas (*Administração Nacional das Áreas de Conservação*—ANAC) was established with the direct support of the World Bank along with many other projects aiming at the support and administration of climate change-related projects in the country. Most of the projects implemented through these funds are under the umbrella of the REDD+ national strategy. The REDD+ and CSA became among the main priorities of the two ministries in charge of agricultural, land and environmental issues in the country (MTA—Ministry of Land and Environment and MADER—Ministry of Agriculture and Rural Development). Despite the fact that Mozambique's CO_2 emissions per capita is one of the lowest in the world and in Sub-Saharan Africa, around 0.1 metric tonnes against the average of 0.7 of Sub-Saharan Africa, 0.8 from Low-Income countries and 3.5 of Middle-Income countries (The World Bank 2007), environmentally friendly policies are prioritised, promoted and enforced by multiple international organisations that have high influence over Mozambican policy (whether it is by their finance or aid relations). However, regarding the implementation of REDD+, many scholars argue that the project and related policy programmes have the potential to reinforce existing inequities and social exclusions (Corbera 2012; Corbera et al. 2017; Phelps et al. 2010). For instance, Bruna (2021) puts forward four study cases that portray how climate change adaptation and mitigation policies (REDD+ and CSA) have negative implications to rural subsistence, including land expropriation without adequate and just compensation, restricted access to forest resources, expropriation of ecological resources, among others.

Therefore, both direct impacts of climate change and the implications of the policies to address it (indirect impacts of climate change) are shaping rural livelihoods, access to land and food in countries such as Mozambique. This chapter aims to further understand these trends and their role in shaping rural development. Thus, the research question is the following: how does the environmental crisis and the policies to address it (mitigation and adaptation) shape access to food and land rights? By answering that question, one is able to grasp how rural development might be shaped by both trends of impacts (direct and indirect) regarding climate change. To answer the research question, a documental analysis was conducted including the review of the legal framework related to climate, agriculture and land topics. This chapter was also based on the doctoral research of both authors (Bruna 2021; Abbas 2022a) that basically relied on primary and secondary data about rural Mozambique, rural livelihoods and agricultural data sets.

Climate, Land Use, Farming System Change and Food Security

Climate projections indicate expected increases in temperature in Africa at a faster rate than the global mean increase (Tadross et al. 2017). For Mozambique, climate projections indicate that the annual average temperature is expected to increase by 1.2 °C (optimistic scenario—the sustainability narrative) to 5.4 °C (severe scenario—fossil-fuel development narrative [Hausfather 2018]) by 2100 (Abbas 2022a). Projections for average rainfall for Mozambique do not indicate substantial changes at a national level in the optimistic scenario, however, the situation is expected to be exacerbated in the most severe scenario, with decreases in overall rainfall, on average between 50 and 100 mm, along the coast and in central and northern regions (Abbas 2022a); see Fig. 10.1. The current most humid areas are the ones expected to experience the highest reduction in rainfall levels (up to 165 mm) (Abbas 2022a).

Besides the effect of changes in temperature and rainfall, crop yields could decrease substantially across Mozambique due to increased frequency and intensity of natural disasters, in particular droughts affecting the semi-arid regions of the country, floods affecting the rich valleys of the rivers where population density and economic activities are concentrated, and cyclones affecting the coastal zones of Mozambique where the majority of people lives (Brito and Holman 2012: 33). In the last four years (2019–2022), Mozambique has been hit by at least

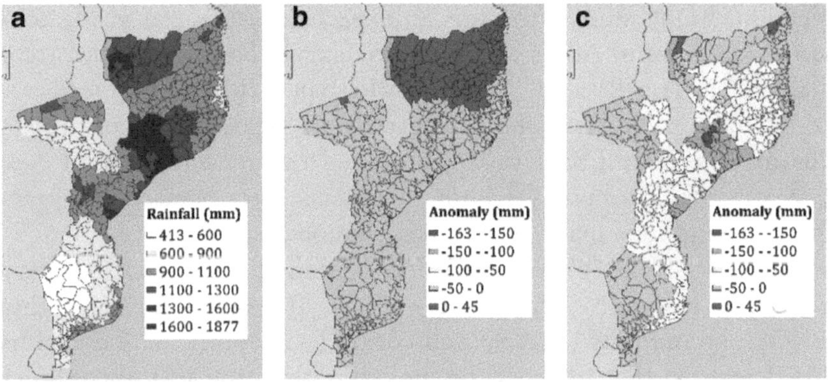

Fig. 10.1 Observed (1970–2000) (**a**) and expected changes (2081–2100) for two climate scenarios—optimistic scenario (**b**) and severe scenario (**c**)—for annual precipitation (mm), based on the median of eight Global Climate Models (GCMs). *Future changes are expressed as anomalies, representing the difference from the baseline scenario compared to the future climate scenario* (*Source* Abbas [2022a])

8 tropical cyclones/storms—compared to 13 in 52 years, from 1956 to 2008 (Tadross et al. 2017)—such as tropical cyclones Idai and Kenneth in 2019 (tropical storm), Chalane in 2020, Eloise and Guambe in 2021 and more recently in 2022 tropical cyclone Gombe and tropical storm Ana.

As mentioned before, climate-related disasters are among the main drivers of food insecurity. Increases in temperature and reductions in water availability will have a direct effect on agriculture through changes in soil characteristics, water availability and crop productivity as well as indirect effects through changes in farm management systems, pest and disease life cycles and competition for land and water resources (Abbas 2022a; Lötter 2017). Even though projections show that average annual rainfall has not changed much, there is an increasing inter-annual variability of rainfall (in particular in the central and southern regions of the country) which negatively affects the agricultural activity, considering that it is practiced mostly in a rainfed system. Lötter (2017) highlights the fact that by 2030 Southern Africa could face a decline of up to 50% in agricultural productivity due to water scarcity and insufficient irrigation. Many studies confirm an expected decrease in yields across many crop types in Africa and, in particular, in Southern Africa (Knox et al. 2012; Lobell et al. 2008; World Bank 2010; Thornton et al. 2011). Most projections show a decrease in yields for major crops in Mozambique. World Bank (2010) projected a decrease in yields of 2–4% of major crops (cassava, sorghum, soybeans, sweet potatoes and yams, wheat, groundnuts, maize, millet and potatoes).

Warner et al. (2016) projected, for the period 2010–2055, decreases in yields for cassava (−2%), maize (−24%), soybeans (−25%), sorghum (−17%), groundnuts (−20%) and cotton (−24%). Brito and Holman (2012) referred that maize is the most sensitive crop to future climate, showing the largest decrease (−11.1%), compared to other crops: soybeans (−6.4%), groundnuts (−4.6%), cassava (−4.2%), sorghum (−3.5%) and cotton (−2.9%). Overall, projected future changes in climate will affect land suitability and crop productivity significantly, having a substantial impact on food production (Marques et al. 2009; Brito and Holman 2012; Abbas 2022a). Currently, Southern Africa is not producing enough food to satisfy the needs of the population (Lötter 2017); and the same is verified for Mozambique (Abbas 2017), which will be exacerbated under climate change. Abbas (2022a) refers that in the context of climate change, it is expected an expansion of the semi-arid and arid areas throughout the country (Fig. 10.2).

These changes in climate will considerably affect farmers' productive decisions, with a significant impact on farming systems choice, moving farmers from one farming system to another (Abbas 2022a), influencing farming system spatial distribution in Mozambique, with impacts on land use and food security. As the country becomes drier farmers will no longer be able to produce certain crops that require a more humid climate, unless the proper socioeconomic conditions, such as irrigation, are provided. In this sense, most farming systems in Mozambique are expected to be replaced by others due to climate change, being expected an intensification of the current most food insecure systems, producing mainly staple food crops (Abbas 2022a). These farming system transitions may imply large changes in terms of land use, knowledge and

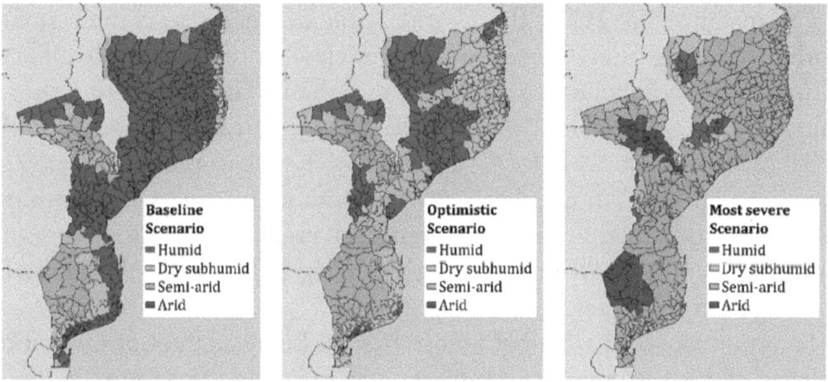

Fig. 10.2 Observed (1970–2000) and expected future (2081–2100) climate type based on aridity index for two scenarios (optimistic and severe), based on the median of 8 GCMs (*Source* Abbas [2022a])

investment needed to manage new systems, putting farmers under pressure from changing farming systems, with a direct impact on rural livelihoods and food security.

Climate Change Mitigation and Adaptation Policies in Mozambique

Concerns about the global warming and the current global environmental crisis emerged decades ago and the impacts are increasingly visible. The last section showed some of the most concerning ones for the case of Mozambique, however, experienced by many other developing countries. This is why, throughout the years, institutions such as the United Nations Framework Convention on Climate Change (UNFCCC) and the Intergovernmental Panel for Climate Change (IPCC) have been seeking global solutions to such crisis in order to mitigate and adapt to it. Thus, climate change mitigation and adaptation policies became mainstream and global. The set of policies put forward constitutes the main guidelines to be followed to reduce Greenhouse Gas (GHG) emissions, including sectors such as agriculture, extractive industry and others.

For instance, in 2010, the World Bank called for a 'climate-smart' world, meaning that climate-smart policies should be implemented. Climate-smart policies are those that enhance development, reduce vulnerability and finance the transition to low-carbon growth paths (The World Bank 2010). As a result, the global economy may be becoming 'greener' in all phases of the production processes, including packaging and distribution. As a result, a range of so-called 'environmentally friendly' investments arose claiming to be under the umbrella of mitigation strategies.

For countries in the Global South, besides being a target of mitigation policies and investments, because of their social and economic vulnerability, they are in higher risk of acute impacts of climate variability and extreme weather events. Thus, at the same time, they become hosts of adaptation policies, especially regarding the rural and agricultural sector, on which the poorest and rural population depend on. Mozambique is considered one of Africa's most vulnerable countries to climate change (Irish Aid 2018; Warner et al. 2016; World Bank 2010). In the past years, Mozambique has been considerable affected by extreme climate events such as floods, droughts and tropical cyclones and storms. From 1956 to 2008, Mozambique has suffered 67 natural disasters: droughts (10 events), floods (20), tropical cyclones (13), epidemics (18), windstorms (5) and earthquake (1) (Queface 2009).

Given the context and the global perception of the need to adapt and mitigate, the Government of Mozambique received around USD 8.8 million just to elaborate the national strategy of REDD+ and legal/administrative instruments to operationalise it (MITADER 2017). The Mozambican REDD+ national strategy aims 'to reduce emissions from deforestation and forest degradation, forest conservation, sustainable management and increase of carbon reserves through planted forests' (MITADER 2016) by focussing in three main sectors: agriculture, forest and energy. The REDD+ strategy also integrates mechanisms of climate change adaptation such as the promotion of sustainable agricultural practices in line with 'climate-smart' techniques (MITADER 2016). Most of these policies are land-based.

Among other goals, it aims to enforce sustainable agricultural practices among small-scale producers, promote alternative energy use, strengthen conservation and enforce sustainable forest management including creating a favourable operational environment for forest plantation companies (MITADER 2016, 2017). It was within this joint framework established by the Mozambican Government within the guidelines of the UNFCCC that climate change mitigation and adaptation projects started to be implemented.

The REDD+ strategy in Mozambique aims to reduce carbon emissions based on multisectoral integrated landscape interventions. MITADER (2016) claims that REDD+ National Strategy aims to achieve multiple benefits that go beyond reducing emissions as it integrates the promotion of rural development as a main pillar while attracting green investments in agriculture, forest, energy and infrastructure sectors. It reinforces and appoints conservation areas and claims that the community should use the land in a way that is compatible with conservation and tourism based on nature. As an integral part of REDD+ in Mozambique, CSA is directly concerned with adaptation as well as mitigation of climate change in the agriculture sector. FAO (2013) and The World Bank (2011) consider CSA a way to strengthen food security and still provide environmental benefits by increasing productivity in a sustainable way, strengthening farmers' resilience and reducing agriculture's GHG emissions.

In the context of land grabbing debates related to the environmental projects, specifically regarding conservation-driven projects, Fairhead et al. (2012) put forward the discussion of 'green grabbing' and its implications that range from prevention of livelihood practice and resource uses and restructuring of labour relations (Fairhead et al. 2012; Seagle 2012). But another major concern regarding environmental crisis is related to climate change fuelling the creation of new markets and new mechanisms of accumulation (Foster et al. 2010). For instance, climate-smart agriculture (CSA) is considered, by both the World Bank and FAO, one of the essential tools for achieving sustainable development, claiming that 'CSA meets these expectations by improving productivity, enhancing resilience and reducing greenhouse gas (GHG) emissions' (FAO 2013). The World Bank (2011) considers CSA a way to strengthen food security and still provide environmental benefits because CSA 'seeks to increase sustainable productivity, strengthen farmers' resilience, reduce agriculture's greenhouse gas emissions and increase carbon sequestration.' Despite being considered an 'effective' response to climate change by mainstream institutions, it is also seen as the solution for developing countries' agricultural sector as these institutions claim that agricultural practices of small-scale farmers are among the major causes of emissions.

Apart from the set of policies that are directly answering to climate change and within the framework of the National Strategy for Climate Change Mitigation and Adaptation, other national public policies also incorporate environmental goals and climate change discourses, turning into the transversal intersectoral goal. Table 10.1 shows how the main rural and agricultural policies of the country, which are also land-based, prioritise environmental goals and/or private economic interests.

Public policies directed at the agricultural sector in Mozambique have been neglecting small farmers and food production for the domestic market, focussing on the production of food and cash crops for the external market. These policies, with impacts on land distribution and food security, have been addressing the climate and/or food security problem in its discourses, however satisfying the international and private economic interests (Abbas et al. 2021). The national strategy of climate change in Mozambique (including all adaptation and mitigation policies) is directly related to agricultural public policies, highly reliant on land and highly dependent on market mechanisms (especially international markets, particularly carbon markets). So, besides the direct impacts of climate change—decreases on yield for food and cash crops, changing farming systems, decrease land suitability for current uses, among others—the indirect impacts of climate change (through the policies to address it) are also going to affect land distribution and food security—this will be further discussed in the next section.

Table 10.1 National rural and agricultural policies

Policy	Objective	Climate and agricultural problem
Strategic Plan for Development of the Agriculture Sector (PEDSA, 2011–2020)	Aims to contribute to food security and income for farmers in a competitive and sustainable way, guaranteeing social and gender equity	Adopts a value chain approach, focussed essentially on food and cash crops production for the external market. PEDSA has not implemented its vision of increasing food production and ensuring food security as most of the support has gone to the private sector (Abbas et al. 2021)
National Investment Plan for the Agriculture Sector (PNISA, 2013–2017)	Aims to accelerate staple food crop production; to guarantee income for farmers; to ensure secure access to and ownership of the necessary natural resources; to boost the development of areas with high agro-ecological and commercial potential; and others	PNISA emphasises the allocation of funds to the private sector, focussing in commodities to export rather than food to supply the domestic market (Abbas et al. 2021)
Strategic Plan for Development of the Agriculture Sector II (PEDSA II, 2022–2031) (in elaboration)	Aims to promote accelerated transformation of the agricultural sector through competitive, inclusive and sustainable growth, ensuring greater inclusive engagement of the private sector, and contributing to improving food and nutrition security, employment, strengthening resilience and adapting to shocks and disasters and poverty reduction	Maintain the focus on the external market and economic private interests, from the previous PEDSA Considers forest plantations (such as pine and eucalyptus) as reforestation measures Support the integration of new agricultural practices (e.g., conservation agriculture); however, it does not refer that these should be adapted to the local context and culturally accepted according to local and traditional (sustainable) knowledge

(continued)

Table 10.1 (continued)

Policy	Objective	Climate and agricultural problem
SUSTENTA (Nationwide rural development policy)	Aims to stimulate the rural economy, through the integration of rural families in the development of sustainable value chains, based on agriculture and forestry, in order to improve their income and quality and life, with respect for environmental conservation	Aiming environmental conservation, this programme promotes agricultural techniques that are considered to be 'environmentally friendly', claiming that smallholder's traditional farming techniques should be changed (MADER 2021)
Ongoing revision of the 1995 National Land Policy Revision	To implicitly commodify land and its natural resources: one of the main goals is to turn land titles (the right to use the land—as the land is still State property) transmissible. Meaning that, indirectly, land will be able to be sold and bought	Another secondary goal of this revision is to promote payment of environmental services schemes to respond to global carbon markets
National Reforestation Strategy (2010–2030)	Aims to increase the contribution of the forestry sector to poverty alleviation, economic, social and environmental development through the establishment of a forest plantation industry	It does not consider the negative social and environmental impacts of monocrop plantations. It assumes reforestation as the establishment of exotic species on a large scale and the replacement of native forest. It prioritises forest plantations over food production (Abbas et al. 2021)
National Climate Change Adaptation and Mitigation Strategy (2013–2025)	Aims to identify adaptation and climate risk reduction as a national priority and presents eight strategic actions aimed at building resilience and reducing climate risk in communities, ecosystems and the national economy	It does not address critically the care taken in the use of new production technology and its impact on pollution and the emission of toxic substances into the environment and ecosystems (Abbas et al. 2021)

Source Based on government instruments and Abbas et al. (2021)

Implications of Climate Change Policies to Rural Subsistence and Land Use

Socio-Economic Implications to Rural Livelihoods

As mentioned before, besides the direct impact of climate variability on rural population, they are also being affected by the multiple policies directed to rural areas that aim to mitigate and adapt to climate change. REDD+, CSA and others may also be linked in parallel with debates around 'appropriation of land and resources for environmental ends,' the so-called, green grabbing (Fairhead et al. 2012). Critiques and questionable agendas are approached regarding specificities around CSA and its 'triple-win objectives.' Clapp et al. (2018) claim that the emerging forms of policies to confront climate change are reshaping the markets and technology where inequality is not adequately addressed. For instance, Taylor (2018) points out the significant backing from the private sector and ways to strongly link farmers to markets and oriented production and distribution of food rather than where it is needed.

Critical agrarian scholars, such as Giraldo and Rosset (2017) underline the idea of the existing risk that agroecology being co-opted, institutionalised, colonised and stripped of its political content and potentially ending up offering more tools for industrial agriculture. In other words, the authors claim that rather than picking up the transformational potential of agroecology, mainstream actors use it as a technical tool, creating a real threat of co-optation and a new frontier of accumulation as agroecology might be seen as an area to generate rents for agribusiness if linked to international markets. The guidelines of CSA (FAO 2013; The World Bank 2011) consider changes in production techniques looking for a more environmentally friendly set of decisions such as prioritising climate resilient seeds and/or crops which are not directly answering local needs. These policies may not be expropriating land from people, but they are definitely expropriating the control of these people's land and striping them from their right to emissions; controlling what to produce, how to produce and how much to produce in a given area in order to protect the environment and reduce emissions, but with no regards to local needs and priorities (Bruna 2021).

Debates regarding the effectiveness of these policies are also increasingly arising, particularly questioning the effectiveness and studying the implications of REDD+. These include reinforcing existing inequities and social exclusions (Corbera 2012; Corbera et al. 2017; Phelps et al. 2010). Hunsberger et al. (2017) summarise the risks in REDD+ design and implementation: (1) disregard of rural communities' views while failing to address causes of deforestation; (2) local communities losing access and use of forest resources; deepening of existing inequalities if elites capture the policy's benefits; (3) reducing the forest to a single commodity value by assigning a price to it and (4) uncoincidental layers of interest among actors: international, national and local institutions. Corbera et al. (2010) argue that policies promoting conservation areas haven't been effective as sustainable forest management programmes have performed poorly because it is generally less profitable than alternative uses to individual actors or groups, and there is often a lack of secure land tenure or effective rights to forests, which may result in conflicts over land allocation. These conditions are worsened by the existence of illegal logging and trade networks. Additionally, Corbera et al. (2007) point out the need to incorporate equity and legitimacy in environmental decisions and the discourses around land use and its role in GHG emissions rather than just focussing on environmental effectiveness and economic efficiency.

Besides these critiques, it is already possible to observe operational land-based mitigation and adaptation projects on the ground scattered around biodiversity-rich areas of the country, with already negative implications for rural livelihoods. One of the first to be implemented was in the Gorongosa National Park Buffering Zone, one of the first targets of such policies. Many studies were conducted to grasp the impacts of REDD+ projects in poverty alleviation and livelihoods in Mozambique in a carbon project in this region. This project consists of carbon sequestration by reforestation (planting trees) and 'community development' (such as the implementation of microenterprises and promoting sustainable cash crops) through the Payment for Environmental Services scheme (Groom and Palmer 2012; Jindal et al. 2012). The results indicated that development activities unrelated to carbon sequestration, such as getting employed in those microenterprises, had a much bigger impact on poverty alleviation, but just for a minor share of households. Our previous research showed that the experience of implementing REDD+ and CSA in Mozambique, showed a poorly and ineffectively implemented policy that was ultimately being used as a tool to facilitate processes of land grabbing by a multinational forest plantation corporation in the name of development and under environmental discourses with negative implications to rural livelihoods (Bruna 2021).

Land Use in the Context of Green Policies: Towards Climate Smart Land Use and Rural Development

Mozambique is already a stage for land rush and resulting conflicts due to the intensification of foreign direct investment (FDI) either by extractive industry multinationals or by large- to medium-scale agribusinesses (Capaina 2019; Mandamule 2016; Mandamule and Bruna 2017; Mosca et al. 2016). In the last two decades, land use in rural Mozambique has been directly shaped by FDI mostly in the agricultural and mining sector (Bruna 2021; Mandamule and Bruna 2017). Now, Mozambique is at the same time one of the top targeted countries in the world in terms of transnational land acquisitions and in parallel the Government appointed 25% of national territory to conservation areas (MITADER 2017).

Some studies have revealed that between 2007 and 2008 more than two million hectares were allocated to biofuel projects and 3.4 million hectares were conceded to mining companies in Tete (Capaina 2019; Human Rights Watch 2013; Manuel and César 2014). Some data from 2014, shows that from the total area of Mozambican territory, around 12 million hectares were conceded to economic enterprises (tourism, agriculture and others) and mining concessions with more than 500,000 people affected by them (Manuel and César 2014). Although being a work in process, according to the statistical data collected by Salimo (2017), the government has mapped 933 communities that occupy an area of 17,256,260 hectares within the ten provinces of Mozambique, excluding the Capital City. However, with such approach from the climate change national policy and overall rural and agricultural policy,

Mozambique has become a host of a wave of 'green' investments based on land and natural resources (see Table 10.2). Additionally, the conservation areas in Mozambique occupy approximately 18.6 million hectares (25% of the national territory—some are still not well established) which include seven National Parks, nine National Reserves, twenty hunting areas, three community conservation areas and fifty farms (ANAC 2015).

As explained previously, most of the mitigation and adaptation policies implemented are land-based, whether it is to lower emissions (CSA) or aiming carbon sequestration (REDD+) or both. In the same line,

Table 10.2 Implications of REDD+ and CSA to rural livelihoods and subsistence

Mitigation/Adaptation policy	Region	Implications to rural livelihoods
REDD+ for conservation and CSA (Bruna 2021)	Gilé National Reserve—Zambézia province	Restricted access to forest resources determinant to rural livelihoods and subsistence Inefficient community development projects (did not compensate for the loss of resources) Threatened food security as they were forbidden to hunt Social exclusion and increased localised poverty
REDD+ for reforestation (Bruna et al. 2021)	Nhambita—Gorongosa National Park—Sofala province	Loss control of land and indirect land grabbing Rupture of rural livelihoods Restriction of access to land Threatened food security as the food production was substituted by reforestation
Tree plantation under REDD+ and CSA (Portucel Moçambique company and Eucalyptus plantation) (Bruna 2021)	Several districts in Zambézia and Manica provinces	Land loss (Land expropriation without appropriate compensation—land grabbing under environmental discourses) Labour exploitation and rural unemployment Inefficient community development projects (did not compensate for the loss of resources) Social exclusion and increased localised poverty Threatened food security as Eucalyptus plantation predominates in the area

(continued)

Table 10.2 (continued)

Mitigation/Adaptation policy	Region	Implications to rural livelihoods
Biofuel production under REDD+ (Procana and other subsequent companies) Bruna (2021)	Massingir District—Gaza Province	Land loss (Land expropriation without appropriate compensation—land grabbing under environmental discourses) Social exclusion and increased localised poverty Rupture of rural livelihoods Restriction of access to land Intercommunity conflicts due to resettlement in areas with resident communities Food insecurity due to restricted access to land by the resettled communities

Source Bruna (2021) and Bruna et al. (2021)

the IPCC (2019) report argues specifically for land-based solutions to climate change. This implies an increase in demand for land justified by the fight against climate change. And, as shown in Table 10.3, rural subsistence and, ultimately, rural development is increasingly being shaped by the implementation of mitigation and adaptation policy.

Currently, land use and distribution in Mozambique are expected to meet climate-related goals, whether it is by allocating and reinforcing conservation areas (REDD+), implementing 'greener' investments (forest plantations) or promoting alternative energy sources (biofuel production), all following the scheme of climate-smart land politics referred to by Borras and Franco (2018). Borras and Franco (2018) focus on the implications for land politics and observe the emergence of what they call 'climate-smart land politics' which consists of the combined processes of incorporating the twin objectives of purportedly combating the inefficient and destructive use of natural resources. An emerging trend of resource grabbing backed up by the fight against climate change is depicted in rural Mozambique.

With a total of over 80 million hectares of Mozambican territory, around 7% is cultivated by the subsistence sector, nevertheless, it is indeed the same areas that green investments are allocated to as well as climate change mitigation policies such as REDD+ for conservation and reforestation projects; promoting land expropriation, endangering

Table 10.3 Land use in Mozambique

Category	Hectares	Objective
Land-based investments under REDD+		
Tree plantations (actually planted—in 2016)	65,000	
Portucel Moçambique	350,000	Eucalyptus forest plantation and CSA
Chikweti forest	63,040	Establishment of exotic and commercial plantation
Green Resources	133,000	Establishment of exotic and commercial plantation
Procana	30,000	Sugarcane/ethanol plantation
Largest conservation areas		
Limpopo National Park	1,123,316	Reestablishment of a conservation area
Gilé National Reserve	286,000	Reestablishment of the area as national reserve
Niassa Special Reserve	4,200,000	Reestablishment of the area as a national reserve
Quirimbas national park	750,000	Reestablishment of the area as a national reserve
Gorongosa National Park	408,600	Enlargement of the conservation area to other districts
Licensed areas		
Forest concessions	10,500,000	
General land use (national)		
Cultivated areas[a]	5,579,116	
Re-established Conservation areas (ANAC/BIOFUND)	11,900,000	
Forest area (native)[b]	34,000,000	

[a]Cultivated areas include only subsistence sector (small and medium farms) (MADER 2021). On average, 70% of these farms have less than 2 hectares, and 90% have less than 3.5 hectares (Abbas and Mosca 2021)
[b]Inventário Nacional Florestal (2017)

land access and threatening rural livelihoods, in particular, food security as the next section will show. Overall, there is a competitive approach to land when it comes to environmental goals and social goals, as rural livelihoods are being adversely affected to accommodate internationally desired climate outcomes.

This section showed how gradually Mozambique is leaning towards climate-smart land use with implications such as land expropriation justified by environmental discourses, rupture of rural livelihoods and risks of food insecurity, among other adverse implications to rural subsistence. Most of the cases, land use is changing from subsistence food production to biodiversity (conservation areas, reforestation) or subsistence food production to green investments (tree plantations and biofuel production). The remaining question is how the rural population is going to guarantee their subsistence (food security and sovereignty) if the land used is being shaped primarily by environmental (and economic) goals.

Climate Smart Land Use and Options for Sustainable Farming Systems and Rural Development

The combination of the direct and indirect impacts of climate change brings along new waves of expropriation that further cuts into the necessary consumption of rural livelihoods. Rural livelihoods are thus threatened by both the direct impacts of climate variability and the policies seeking mitigation and adaptation of climate change. From now on, the future of smallholders and rural livelihoods could be highly reliant on the carbon/profit nexus of capitalism as the 'Future land use depends, in part, on the desired climate outcome and the portfolio of response options deployed' (IPCC 2019: 4).

Insights from both empirical cases explored in this chapter show how this new wave of green investments and projects, directly linked to climate change concerns, are further expropriating livelihoods and fuelling capital accumulation in the name of the fight against climate change. However, it is important to clarify that environmental policies and investments that go beyond the climate change agenda, for example, circular economy, green economy and so on, could also be a trigger or constituent, intersecting with the green frontier of accumulation (see Fig. 10.3).

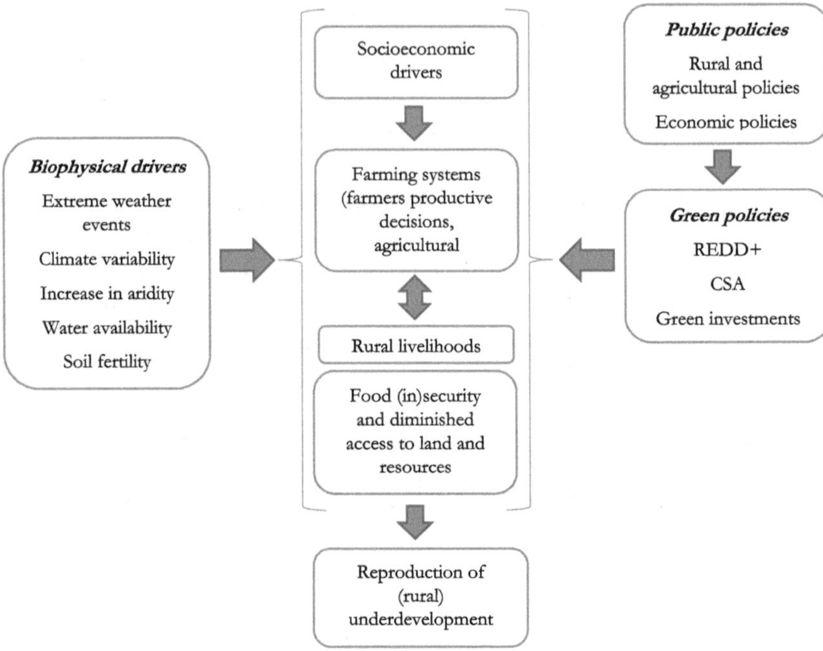

Fig. 10.3 The implications of climate change and green policies to rural development trends

This chapter underlined the need to rethink the set of solutions to the current climate crisis and deepen the understanding of the direct implications to rural livelihoods, land access and food production. It sheds light on the fact that both direct and indirect impacts of climate change are adversely affecting livelihoods and causing land expropriation in rural areas. Both of these impacts will shape rural development in the next decades in countries such as Mozambique. Thus, there is the need to think about more adequate and socially sustainable solutions and climate actions to mitigate and adapt to climate change. Regarding subsistence food production, there is general agreement on the important role of farming systems to improve food security under a changing climate (Dixon et al. 2001; Abbas 2022a). Abbas (2022a) refers that diverse mixed crop-livestock systems are expected to better tolerate the changing climate. The resilience of farming systems to climate change depends on the combination of activities that initially exist on the farm, offering adaptation possibilities. For the case of Mozambique, the higher resilience found in mixed crop-livestock systems to climate change is consistent with the fact that these systems already grow some drought-tolerant crops (e.g., sorghum) and (small) livestock, which is claimed to be less exposed to the impacts of climate change and can be sold in times of food shortages (Abbas 2022a).

Considering the (direct and indirect) impacts of climate change on land use change and rural livelihoods, an integrated natural resource management is needed, which includes diversifying farming systems, enhancing natural capital and building on local and traditional knowledge (Markwei et al. 2011). Sustainable farming systems, as well as their supporting ecosystems, are key to ensure food and nutrition security; and local and traditional foods can help improve the livelihoods of households and also food and nutrition security (Markwei et al. 2011; Mohamed-Katerere and Smith 2013; Vinceti et al. 2013). Agricultural and food security policies should consider a farming system multisectoral approach, a participatory decision-making, integrate ecosystem factors, commit to climate change action and recognise wild resources; and for its effectiveness policies should focus on food resilience through diversity, natural infrastructure and social justice (Mohamed-Katerere and Smith 2013; Abbas 2022a).

Others have argued for agroforestry systems in order to improve food security and nutrition, as it provides food, cash income and other subsistence benefits for rural population; and can increase resilience of poor households (Dewees 2013; Jamnadass et al. 2013; Mohamed-Katerere and Smith 2013). Agroforestry can contribute to food production by providing indigenous foods that are rich in micronutrients, fibre and protein; contributes to increase yields of some crops as well as stabilise crop production in drought periods and improves water-use by crops; it also contributes to increase income; allows the production of wood that is less harmful when burnt and has higher energy content; provide important ecosystem services, and, act as safety nets in time of crisis (Dewees 2013; Jamnadass et al. 2013; Vinceti et al. 2013). An ecosystem-aware approach to food security aims not only to alleviate hunger in the short term but also to build long-term food resilience (Mohamed-Katerere and Smith 2013).

Nevertheless, this approach is only valuable if policies (including rural, agricultural projects, economic and particularly mitigation and adaptation policies) are designed in a participatory way in order to ensure that local needs and priorities are tackled. So far, most of the policies have been designed in a top-down manner, mainly responding to international goals regarding the climate crisis, mainly by market-led mechanisms which ultimately undermines sustainable rural development. Additionally, there is the need to incorporate debates about climate justice and the historical causes of the environmental crisis so that countries such as Mozambique, which did not contribute considerably to the current situation, don't carry out its impacts unfairly.

References

Abbas, M. 2017. Segurança Alimentar. Auto-suficiência Alimentar: Mito ou Verdade? *Observador Rural*, 55. Maputo: Observatório do Meio Rural.

Abbas, M., et al. 2021. *Sistemas Alimentares em Moçambique: Rumo a uma Política Alimentar Nacional*. Maputo: Observatório do Meio Rural (OMR), Justiça Ambiental (JA!), La Via Campesina (LVC), Alternactiva – Plataforma para Emancipação Social e União Nacional de Camponeses (UNAC).

Abbas, M. 2022a. A Farming System Approach to Support Food Policies Under Climate Change in Developing Countries: The Case of Mozambique. PhD thesis, Instituto Superior de Agronomia, University of Lisbon.

Abbas, M. 2022b. Caminhos para a Segurança Alimentar em Moçambique: Uma Abordagem de Sistemas de Produção. *Observador Rural*, 120. Maputo: Observatório do Meio Rural.

Abbas, M., and J. Mosca. 2021. *Análise ao Inquérito Agrário Integrado - IAI 2020*. Maputo: Observatório do Meio Rural.

ANAC. 2015. *Plano Estratégico da Administração Nacional das Áreas de Conservação 2015–2024*. Maputo: ANAC.

Borras, S.M., and J.C. Franco. 2018. The Challenge of Locating Land-Based Climate Change Mitigation and Adaptation Politics within a Social Justice Perspective: Towards an Idea of Agrarian Climate Justice. *Third World Quarterly* 39 (7): 1308–1325.

Brito, R., and E. Holman. 2012. *Responding to Climate Change in Mozambique: Theme 6—Agriculture*. Maputo: INGC.

Bruna, N. 2021. Extractivism, Rural Livelihoods and Accumulation in a 'Cimate-Smart' World: The Rise of Green Extractivism. PhD thesis, International Institute of Social Studies, Amsterdam.

Bruna, N., B. Monjane, and E. Samuel. 2021. Para Além do Gás e Carvão: Créditos de Carbono na Corrida Extractivista aos Recursos Naturais em Moçambique. *Destaque Rural*, 135. Maputo: OMR.

Capaina, N. 2019. Titulação e Subaproveitamento da Terra em Moçambique: Algumas Causas e Implicações. *Observador Rural*, 73. Maputo: OMR.

CIAT and The World Bank. 2017. *Climate-Smart Agriculture in Mozambique*. CSA Country Profiles for Africa Series. Washington, DC: International Center for Tropical Agriculture (CIAT)/World Bank.

Clapp, J., P. Newell, and Z. W. Brent. 2018. The Global Political Economy of Climate Change, Agriculture and Food Systems. *Journal of Peasant Studies* 45 (1): 80–88. https://doi.org/10.1080/03066150.2017.1381602.

Corbera, E., K. Brown, and W. Adger. 2007. The Equity and Legitimacy of Markets for Ecosystem Services. *Development and Change* 38 (4): 587–613.

Corbera, E., M. Estrada, and K. Brown. 2010. Reducing Greenhouse Gas Emissions from Deforestation and Forest Degradation in Developing Countries: Revisiting the Assumptions. *Climatic Change* 100 (3/4): 355–388.

Corbera, E. 2012. Problematizing REDD+ as an Experiment in Payments for Ecosystem Services. *Current Opinion in Environmental Sustainability* 4: 612–619.

Corbera, E., C. Hunsberger, and C. Vaddhanaphuti. 2017. Climate Change Policies, Land Grabbing and Conflict: Perspectives from South-East Asia. *Canadian Journal of Development Studies* 38 (3): 297–304.

Dewees, P. 2013. *Bouncing Back: Forests, Trees and Resilient Households*. Washington, DC: PROFOR.

Dixon, J., A. Gulliver, and D. Gibbon. 2001. *Farming Systems and Poverty: Improving Farmers' Livelihoods in a Changing World*. Rome and Washington, DC: FAO/World Bank.

FAO. 2013. *Climate-Smart Agriculture Sourcebook*. Rome: FAO.

FAO, IFAD, UNICEF, WFP and WHO. 2020. *The State of Food Security and Nutrition in the World 2020: Transforming Food Systems for Affordable Healthy Diets*. Rome: FAO.

Fairhead, J., M. Leach, and I. Scoones. 2012. Green Grabbing: A New Appropriation of Nature? *Journal of Peasant Studies* 39 (2): 237–261.

Ford, J., L. Berrang-Ford, A. Bunce, C. McKay, and M. Irwin. 2015. The Status of Climate Change Adaptation in Africa and Asia. *Regional Environmental Change* 15 (5): 801–814.

Foster, J. B., B. Clark, and R. York. 2010. *The Ecological Rift: Capitalism's War on the Earth*. New York: Monthly Review Press.

Giraldo, O.F., and P.M. Rosset. 2017. Agroecology as a Territory in Dispute: Between Institutionality and Social Movements. *Journal of Peasant Studies* 45 (3): 545–564.

Groom, B., and C. Palmer. 2012. REDD+ and Rural Livelihoods. *Biological Conservation* 154: 42–52.

Hausfather, Z. 2018. Explainer: How "Shared Socioeconomic Pathways" Explore Future Climate Change. *Climate Modelling*, April 19. Retrieved from https://www.carbonbrief.org/explainer-how-shared-socioeconomic-pathways-explore-future-climate-change.

Human Rights Watch. 2013. *"What is a House without Food?"* Mozambique's Coal Mining Boom and Resettlements.

Hunsberger, C., E. Corbera, S.M. Borras, J.C. Franco, K. Woods, C. Work, R. de la Rosa, V. Eang, R. Herre, S.S. Kham, C. Park, S. Sokheng, M. Spoor, S. Thein, K.T. Aung, R. Thuon, and C. Vaddhanaphuti. 2017. Climate Change Mitigation, Land Grabbing and Conflict: Towards a Landscape-Based and Collaborative Action Research Agenda. *Canadian Journal of Development Studies* 38 (3): 305–324.

IPCC. 2019. *Research Handbook on Climate Change and Agricultural Law*. Cheltenham: Edward Elgar Publishing.

Irish Aid. 2018. *Mozambique Country Climate Risk Assessment Report*. Maputo.

Jamnadass, R., F. Place, E. Torquebiau, E. Malézieux, M. Iiyama, G.W. Sileshi, K. Kehlenbeck, E. Masters, S. McMullin, J.C. Weber, and I.K. Dawson. 2013. Agroforestry for Food and Nutritional Security. ICRAF Working Paper, 170. Nairobi.

Jindal, R., J.M. Kerr, and S. Carter. 2012. Reducing Poverty Through Carbon Forestry? Impacts of the N'hambita Community Carbon Project in Mozambique. *World Development* 40 (10): 2123–2135.

Knox, J., T. Hess, A. Daccache, and T. Wheeler. 2012. Climate Change Impacts on Crop Productivity in Africa and South Asia. *Environmental Research Letters* 7 (3). https://doi.org/10.1088/1748-9326/7/3/034032.

Lobell, D., M.B. Burke, C. Tebaldi, M.D. Mastrandrea, W.P. Falcon, and R.L. Naylor. 2008. Prioritizing Climate Change Adaptation Needs for Food Security in 2030. *Science* 319: 607–610.

Lötter, D. 2017. Understanding the Vulnerability of Southern Africa to Climate Change: Agriculture and Livestock. In *Climate Risk and Vulnerability: A Handbook for Southern Africa*, ed. C.L. Davis-Reddy and K. Vincent, 54–63, 2nd ed. Pretoria: CSIR.

MADER. 2021. *Inquérito Agrário Integrado 2020*. Maputo.

Mandamule, U. 2016. Tipologia dos Conflitos sobre a Ocupação da Terra em Moçambique. *Observador Rural*, 37. Maputo: OMR.

Mandamule, U., and N. Bruna. 2017. Investimentos, Migrações Forçadas e Conflitos de Terra: Representações Sociais de Populações Reassentadas no Corredor de Nacala. In *Movimentos Migratórios e Relações Rural-urbanas: Estudos de Caso em Moçambique*, ed. J. Feijó, 215–228. Maputo: Alcance Editores.

Manuel, L., and F. César. 2014. Ocupação da Terra por Emprendimentos Económicos no País. *Relatório de Investigação*, 5. Maputo: CTV.

Markwei, C., L. Ndlovu, E. Robinson, and W. Shah. 2011. International Assessment of Agricultural Knowledge, Science and Technology for Development: Summary for Decision Makers of the Sub-Saharan Africa Report. In *Climate Change and Food Systems Resilience in Sub-Saharan Africa*, ed. L.L. Ching, S. Edwards, and N. Scialabba, 59–97. Rome: FAO.

Marques, M., et al. 2009. Assessing Future Impacts of Climate Change on Land Use and Land Cover: Actual and Potential Agricultural Land Use. In *INGC Climate Change Report: Study on the Impact of Climate Change on Disaster Risk in Mozambique*, ed. K. Asante, et al., 89–117. Maputo: INGC.

MITADER. 2016. *Plano de Investimento Florestal (FIP) para Moçambique*. Maputo.

MITADER. 2017. *Environmental and Social Management Framework (ESMF)*. Maputo.

Mohamed-Katerere, J., and M. Smith. 2013. The Role of Ecosystems in Food Security. *Unasylva* 241 (64): 14–22.

Mosca, J., M. Abbas, and N. Bruna. 2016. *Governação 2004 – 2014. Poder, Estado, Economia e Sociedade*. Maputo: Alcance Editores.

ND-GAIN. 2022. *ND-GAIN Country Index: Mozambique*. Retrieved from https://gain.nd.edu/our-work/country-index.

Phelps, J., E.L. Webb, and A. Agrawal. 2010. Does REDD+ Threaten to Recentralize Forest Governance? *Science* 328: 312–313.

Queface, A. 2009. Climate Change Analyses: Climate Change Historical and Baseline Analysis. In *INGC Climate Change Report: Study on the Impact of Climate Change on Disaster Risk in Mozambique*, ed. K. Asante, et al. Maputo: INGC.

Salimo, P. 2017. *Perfil de Terras em Mocambique*. Maputo: ACUT.

Seagle, C. 2012. Inverting the Impacts: Mining, Conservation and Sustainability Claims Near the Rio Tinto/QMM Ilmenite Mine in Southeast Madagascar. *The Journal of Peasant Studies* 39 (2): 447–477.

Tadross, M., et al. 2017. Understanding Climate Risks Over Southern Africa: Projected Climate Change Futures for Southern Africa. In *Climate Risk and Vulnerability: A Handbook for Southern Africa*, ed. C. Davis-Reddy and K. Vincent, 2nd ed., 20–29. Pretoria: CIRS.

Taylor, M. 2018. Climate-Smart Agriculture: What Is It Good For? *Journal of Peasant Studies* 45 (1): 89–107. https://doi.org/10.1080/03066150.2017.1312355.

Thornton, P.K., P.G. Jones, P.J. Ericksen, and A.J. Challinor. 2011. Agriculture and Food Systems in Sub-Saharan Africa in a 4° C+ World. *Philosophical Transactions of the Royal Society a: Mathematical, Physical and Engineering Sciences* 369: 117–136.

Vinceti, B., C. Termote, A. Ickowitz, B. Powell, K. Kehlenbeck, and D. Hunter. 2013. The Contributions of Forest Foods to Sustainable Diets. *Sustainability* 5: 4797–4824.

Warner, K., et al. 2016. *Climate Change Profile: Mozambique*. Wageningen: CDI.

World Bank. 2007. *World Development Report 2008: Agriculture for Development*. Washington, DC.

World Bank. 2010. *Economics of Adaptation to Climate Change: Mozambique*. Washington, DC.

World Bank. 2011. *Climate-Smart Agriculture: A Call to Action*. Washington, DC.

11

Accumulation by Land Rent and Territorial Disputes in a Brazilian Agricultural Frontier

Bernardo Mançano Fernandes, Samuel Frederico, and Lorena Izá Pereira

Introduction

Land grabbing has changed the agrarian question on a world scale, much more intensely from the first decade of the twenty-first century with the financialisation of agriculture, producing the global agrarian question. In Brazil, a new 'administrative region'—MATOPIBA—was created mainly to meet the territorialised interests of financial capital in the production of flexible commodities. Those pressures gave rise to the concept of modern agricultural frontier characterised by capital and technology-intensive agriculture over large areas, aimed at export, where

B. M. Fernandes (✉) · S. Frederico
São Paulo State University (UNESP), São Paulo, Brazil
e-mail: mancano.fernandes@unesp.br

L. I. Pereira
São Paulo State University (UNESP), São Paulo, Brazil

A. A. R. Ioris and B. Mançano Fernandes (eds.), *Agriculture, Environment and Development*, https://doi.org/10.1007/978-3-031-10264-6_11

international trading companies predominate with the purpose of appropriating the rent of land. MATOPIBA—an acronym for the intersection area of the states of Maranhão, Tocantins, Piauí and Bahia (Map 11.1)—comprises of 73,173,485 hectares, including 337 municipalities. The 2010 census registered 590,178 people in this region, with 35% living in the countryside. Established by the Presidential Decree no. 8447 of 2015, the MATOPIBA Agricultural Development Plan is a continuity of agrarian extractivist model (Alonso-Fradejas 2015; McKay 2017), for export as the Japanese-Brazilian Cooperation Programme for the Development of Cerrados (PRODECER), created in the 1970s. MATOPIBA is an example of the land grabbing process in the twenty-first century, reserving an enormous territorial extension to exploit natural resources and produce commodities. PRODECER and MATOPIBA are two examples of the agrarian extractive model at different times.

The agrarian question of the twentieth century consists of a series of conflicts, mostly between indigenous and peasants against landowners, and the participation of international capital was occasional. In the twentieth century, these conflicts between indigenous and peasants are now against new institutions. Such as multinational and national companies that trade various commodities, from land to agricultural production with the use of high technology, with the financial participation of several investment funds, forming a conglomerate complex. There is a fundamental competition over MATOPIBA between (trans)national corporations that promote the territorialisation of land grabbing, and the peasant, indigenous and *quilombola* communities that struggle to defend their territories from the power of agribusiness. The area of the indigenous territories in the region comprises 4157 thousand hectares, the peasant territories comprise 7100 thousand hectares, and the *quilombola* territories totalise 250 thousand hectares. These territories are the result of struggles against the '*grilagem*[1]' representing 16% of the MATOPIBA region. Land concentration is inseparable from income concentration,

[1] A process of fraud of land titles, that include several forms of the falsification of documents, also called 'grilagem industry'.

Map 11.1 Brazil with the MATOPIBA region (*Data source* Pereira and Pauli [2016]; elaboration: Lorena Izá Pereira)

with 6% of the establishments generating 86% of gross income (Miranda 2015), most of it concentrated in the hands of corporations (shown in Table 11.1).

This chapter analyses MATOPIBA from the perspective of the 'agrarian question', that is, highlighting the territorialisation of agribusiness by the territorial income has increased the conflicts and the resistance of the peasant, indigenous and *quilombola* peoples.[2] The so-called agrarian question is a structural problem of the capitalist mode of production in the countryside that aggravates inequality through the concentration of land and wealth. The root of the agrarian question is the capitalist appropriation of land rent from the peasantry that happens when they sell goods in the capitalist market. This appropriation eventually leads to land expropriation. This definition is based on Karl Kautsky's seminal book *The Agrarian Question*, first published in 1899. The long time between the analysis of the agrarian question of late nineteenth-century Germany (Kautsky 1980) and our analysis highlights the relevance of the studies on the rent of land in the Marxist tradition.

The participation of financial capital in the production of commodities has strengthened the process of territorialisation of the agribusiness that relies on the expropriation of the peasantry, indigenous and *quilombolas* (Fernandes 2008). The capitalist appropriation of income produced by the peasantry is a form of expropriation. Another way of expropriation is the enclosure and isolation of communal lands that are surrounded

[2] Three fieldwork campaigns were carried out in the Matopiba region between the years of 2015 and 2017. In the first two, ten days each, technical visits were made to agricultural enterprises and farms, and semi-structured interviews with directors, managers and agronomists of companies, as well as with large capitalist farmers. In total, 13 companies and farms were visited, with 26 interviews. In general, visits and interviews provided primary information on: prospecting, negotiating, leasing and land acquisition; edafoclimatic and pedological conditions necessary for the development of modern agriculture; technical systems used on farms (machinery, cultivars, precision agriculture, etc.); data on productivity, financing, profit and logistics for the outflow of production; methodology (stages) of transforming gross lands into highly productive lands; major technical and edafoclimatic problems for increasing yield; technical and territorial division of the company's labor (number of employees, specialization, hierarchy, location); use of information techniques for business management and for the development of productive activities per se; relationships with financial investors. The third field work was devoted exclusively to visits to traditional communities. Ten communities were visited in the southern states of Piauí and Maranhão, with meetings and semi-structured interviews with peasants and community leaders. From the reports, we obtained a series of data and information on the problems faced by the communities: threats of expropriation of their lands; armed violence (including cases of death of peasants); environmental impacts (contamination with agrochemicals, silting of rivers, deforestation, dissemination of new pests, etc.), impossibility of raising livestock in plains areas [*chapadas*], proliferation of internal conflicts in the community, exploitation of peasants' work by companies. In this way, we were able to counterpose two distinct and conflicting logics of land use and occupation. While the visits to the agricultural companies allowed to verify how the capitalist/financial logic leads to the necessity of constant technical improvement and expansion of the productive area, the visits to the communities allowed to verify how the expansionist logic entails in a series of problems and conflicts for the local population.

by vast capitalist properties. The resistance of the affected communities triggers a territorial dispute against the agribusiness that wants to appropriate the areas to expand the production of commodities. Capitalists, landowners, peasants, indigenous peoples and *quilombola* disputes for these spaces of existence that enable their own ways of life and production. Territorial conflict is also a class struggle (Fernandes et al. 2012). Our departure point is that the creation of MATOPIBA has been associated with the land grabbing that took advantage of several crises, as analysed in Section "Land Grabbing: Land Appropriation and Ensuing Crises". Being territorialised throughout the latifundia and through the expropriation of peasants, indigenous peoples and *quilombola* lands, it has created the, still evolving, modern agricultural frontier. Part of which is already consolidated and part is expanding, as shown in Sections "Agricultural Enterprises and Typology of Financial Capital in the Brazilian Agricultural Frontiers" and "Information Technologies and Ground Rent", where we analyse the process of appropriation of the rent of land by the hegemonic model of development. In the last section, we analyse the conflicts generated by competition over territories and development models.

Land Grabbing: Land Appropriation and Ensuing Crises

At the beginning of the twenty-first century, the increasing appropriation of large areas of land by (trans)national capital was termed 'land grabbing' (Borras Jr. and Franco 2012), 'land rush' (Cotula 2012) or 'land control-grab' (Alonso-Fradejas 2012). Land grabbing is a cyclical process (Edelman and León 2013; Sassen 2013) with each stage marked by new elements. Current land grabbing is a global process, resulting from strategies of capital accumulation based on the convergence of multiple crises: financial, food, energy, environmental and climatic (Borras Jr. et al. 2012).

The recent financial crisis, whose peak occurred in 2007/2008 in the USA and Europe as a result of the overaccumulation crisis begun in the 1970s (Harvey 2003), generated a need for diversification in the financial capital investment portfolio, and land was taken as a profitable alternative, thus introducing a new agent in large-scale land acquisitions—investment funds—and leading to an increased financialisation of agriculture. We are experiencing a period of crisis and fear of a future energy crisis engendered by various factors, particularly global warming due to the use of fossil fuels, to gradual depletion of oil sources and to political instability in producing countries. These factors lead transnational corporations and States to invest in alternative sources of energy production. In this scenario of change in the energy matrix (Fernandes et al. 2010), interest in land grows, behind a conservative discourse of *energy security*. As Sassen (2013) argues, interest in agrofuel is a key element for understanding the current stage of land grabbing. Environmental and climatic crises, associated with the energy scarcity, make use of the discourse of global warming resulting from greenhouse gases emissions to promote the *Green Grabbing*. Through discourses on environmental conservation, carbon credit market, ecotourism, research and conservation practices, they acquire large extensions of land on the grounds that selling nature is a necessary step to preserve it (Fairhead et al. 2012; Corson and Macdonald 2012).

The main factor responsible for the food crisis is the increase in global food prices since 2009. Simultaneously, the food crisis is engendered by neo-Malthusian discourses around increasing world population, and by changes in land use, replacing traditional crops with commodities for export. In 2009, FAO released a document entitled *Global agriculture towards 2050*, projecting the world population to reach 9 billion by 2050. The rationale of more people to be fed contributed to the increase in land acquisition in several countries (Cotula 2012). With the justification of food security, corporations and governments have appropriated large tracts of land for commodity production. It is necessary to remember that the problem of food insecurity does not refer to the amount of food produced, but access to food. In the case of Brazil, corporations, investment funds, sovereign funds and state-owned companies have invested in the appropriation of land for agricultural production with the discourse of food and energy security come from the USA, Japan, United Kingdom and the Middle East. The discourse of resource-rich and resource-poor countries that seek to secure food and energy supply (Wilkinson et al. 2012) shows the geopolitical dimension of land grabbing in Brazil.

According to Borras Jr. et al. (2012) and McKay (2017), what differs from the current land grabbing to that observed in previous periods is the power to control land and its productive resources, that is, to control the territory in its multiple dimensions; the large scale of transactions in terms of land extension and transacted capital and the fact that the current land grabbing is a response to a convergence of multiple crises. In addition to these three characteristics, we argue that the financialisation of agriculture is also a novelty in this process, although the relationship between agriculture and financial capital has a long history (Willians 2014), the intensification of investment funds in both production and speculation is an important element to understand the dynamics of land grabbing. In this way, we understand that land grabbing is an old process, which takes on new characteristics in order to control the territory for the accumulation of capital. Such crises produce concerns among different governmental and multilateral institutions related to food production and environmental problems, and open a door in agriculture for the expansion of financial capital, which—under a crisis of overaccumulation—creates excuses for companies and investment funds to increase capital accumulation, under the guise of environmental conservation and fight against hunger.

Since 2008, the intensification of land grabbing has promoted a global race in scholarly production on the theme (Oya 2013; Sauer and Borras Jr. 2016) and land grabbing became a common headline (Holt-Giménez 2012). According to Borras Jr. and Franco (2010), the term *land grabbing*, understood as the transactions involving large extensions of (trans)national land, was first introduced by activist groups that opposed this process as, for example, Foodfirst Information and Action Network (FIAN). After globally spreading through the media, the theme of land grabbing was appropriated by academia and other agents—companies, investment funds, multilateral institutions and governments—that even proclaim the benefits of this large-scale land transaction.

Multilateral and governmental institutions, nongovernmental organisations, social movements and academia have collected and systematised data from different localities and scales of analysis, therefore steering the debate on land grabbing. The only agreement among researchers is the escalation of land grabbing on a world scale. The own notion of land grabbing is in dispute (Sauer and Borras Jr. 2016), with each researcher defining elements that characterise land grabbing, elaborating different conceptions of a same process through the analysis of various countries, such as India (Levien 2012), Mozambique (Clements 2015), Ethiopia (Moreda 2015), Ukraine (Mamonova 2015), Guatemala (Alonso-Fradejas 2015), among others. Edelman et al. (2013) characterise the earlier assessments of land grabbing as a period of 'making sense', in which most works published between 2009 and 2012 sought to answer broader questions such as: who is involved? Who are the drivers? The focus was then on investment scales, especially regarding the territorial extent of land transactions, which Edelman (2013) later criticised as *fetishisation* of the hectare.

We are currently experiencing the stage called 'deepening our understanding', inaugurated in 2013. In this new stage, researches are more meticulous, prioritising case studies and fieldwork, aiming to understand what happens on a local scale, since the process is not homogeneous. It explores how each community is affected and resists these land transactions; they have a clearer focus, and some methodological and epistemological issues have already been overcome, especially on an international scale. Thus, the year 2013 is considered a milestone because a new phase in land grabbing studies was inaugurated, in which the focus of the research was altered, seeking a better understanding of the process based on its singularities in the local scale, which made possible the analysis from below, that is, land grabbing as a process formed through local, regional and global dynamics. These two periods involve distinct interpretations of land grabbing. Peluso and Lund (2011) use the term 'Land Control', suggesting the existence of new forms of access to land that consist of a recreation of land governance due to great interest from corporate structures acting on a global scale. Borras Jr. and Franco (2010) consider land grabbing as a new cycle of expropriation promoted by capital, in which the need for land governance is increasingly exposed. Land grabbing escalates alongside the convergence of global crises, and results in the expropriation of land by capital, promoting direct and indirect conflict with peasant and indigenous communities.

Philip McMichael (2009, 2013) discusses land grabbing within the context of a shifting food paradigm, that is, subordination to a corporate food system. Edelman and León (2013) and Sassen (2013) emphasise the need to consider the history of the land grabbing process, since it is an old process that presents some new features, such as the recent intervention of financial capital. Borras Jr. et al. (2012) study the Flexible Commodities that offer multiple uses—fibre, food and fuel—and solve a difficult challenge for agriculture: the need for a diversified portfolio of products to avoid the shock of prices. The rise of flexible commodities is related to the financialisation of agriculture, because multiple and flexible uses have the potential to mitigate investment risk, while maximising returns (Borras Jr. et al. 2015).

Land grabbing is a cyclical process (Edelman and León 2013), intensified in the twenty-first century due to the convergence of crises. It reaches all continents (Ploeg et al. 2015) being not exclusive to the global South, where it occurs with greater intensity and speed. It is led by players and drivers who join to get control of the territory, such as (trans)national companies, governments, banks, multilateral institutions and local agents, but the great novelty is the role of financial capital. It happens through land acquisition, lease, concession and partnership contracts, in which the purpose is the control of the territory for accumulating the rent of land. There is a variety of land uses, such as commodity production for export, agro-energy, mining, ecotourism, green and resource grabbing and speculation. Land appropriation takes place in the most profitable areas for capital, either consolidated areas or those in expansion on the agricultural frontier. It creates new territorialities through deterritorialisation processes, promoting conflicts that both directly and indirectly affect the peasant and indigenous communities (Fernandes 2011) that seek to fight land grabbing through resistance from below (Borras Jr. and Franco 2013).

Latin America presents some complex challenges for understanding current land grabbing. The process of land concentration—centuries-old in Latin America—was intensified through land grabbing (Borras Jr. et al. 2012), especially through the production of flexible commodities for export, and the acquisition of lands for territorial control, which produces real estate speculation and increases the rent of the land. This process is not limited to the appropriation of land by foreign capital, since local companies, state and federal governments also participate in it, creating great diversity of ways for multinational capital to access land. Some processes, although recent on the global scale, are older in Brazil. Obviously, there has been an increase in land appropriation in the twenty-first century, but it is not exclusively recent. As an example, we can cite the reports of illegal land grabbing by foreigners in Brazil, which resulted in a parliamentary commission of inquiry which concluded that 28 million hectares, most located in the Amazon, were illegally traded by public agencies (Oliveira 1988). During the process of territorial formation of Brazil, it is possible to identify different phases of land grabbing, each with unique characteristics. At the beginning of the last century, the motivation for land grabbing was the development of infrastructure projects through the granting of land by the Brazilian State to US and European companies (Pereira 2015). Already in the 1960s, the interest was the Brazilian Amazon, with the objective of exploring biodiversity and mineral resources (Garrido Filha 1980), through an agreement between (trans)national companies, local elites and government, that often occurred (Oliveira 1988).

Based on complaints, in 1968 the government of Brazil prepared a commission to verify the allegations of illegal land acquisition in Brazil, where it was found that 20 million hectares of Brazilian territory were under the control of foreign groups (Oliveira 1988; Pereira 2015). In the case of the cerrado (savanna), the expansion of the agricultural frontier began in the late 1960s with the migration of producers from southern Brazil to the North and Central-West regions, especially to the state of Mato Grosso. In 1976, PRODECER began a bilateral cooperation between Brazil and Japan with the objective of promoting the expansion and modernisation of agriculture in the cerrado (Frederico 2010), resulting in 'the miracle of the cerrado'. This expansion of the agricultural frontier with the support of the state attracted landowners, (trans)national enterprises and resulted in the expropriation of the peasantry (Inocêncio and Calaça 2010).

In the study focussed on Latin America, it is still necessary to address the intra-regional character of land grabbing (Borras Jr. et al. 2012), that is, the role played by trans-Latin companies in land grabbing. Brazil is an example, since, while the country is a land grabbing target of (trans)national capital, it also grabs land in other countries of the region, such as Paraguay (Galeano 2012) and Bolivia (MacKey 2011; McKay 2017; Fernandes et al. 2012). Likewise, Argentine companies open to financial capital also play a significant role in promoting land grabbing in other countries, as we show in section three of this chapter. In a context of globally shifting agrarian question, land grabbing is a key element, which redefines social relations in the countryside, under the guise of food production, but actually promoting the consolidation of an agrarian extractivist model. In such model, productive capital is increasingly mixed with speculative capital, through investment funds—what constitutes a new aspect of the current land grabbing.

Agricultural Enterprises and Typology of Financial Capital in the Brazilian Agricultural Frontiers

For Santos and Silveira (2001: 119), the concept of frontier is used in studies of modern agricultural production areas and corresponds to the territories incorporated by modern agriculture since the last quarter of the twentieth century. Modern agricultural frontier is an area of the Brazilian territory that is occupied, since the 1970s, by large scale, capital and technology-intensive monocultures destined to export, which replace original vegetation, peasant cultures and extensive pasture areas. It coincides with the extension of the Cerrado Morphoclimatic Domain[3] that, according to Ab'Sáber (2003: 117), is one of the 'great irregular polygons that form the Brazilian landscape mosaic'. With an area of 2,045,000 km^2, equivalent to ¼ of the Brazilian territorial extension, the cerrado has topographic (plateau, with large flat areas and deep soils) and climatic (regular and adequate rainfall and intense

[3] The Amazon occupation was also done under a modernization discourse, with soy plantation in northen Mato Grosso and Rondônia and cattle breeding in Pará.

luminosity) conditions favourable to mechanisation and large-scale agricultural production. The interest in this region stems mainly from its potential for appreciation of lands, that is, the acquisition of land at relatively lower prices and its possible sale at much higher prices after being transformed into highly productive farms, as discussed in more detail in the next section.

One goal is to find out who are the investors of the agricultural enterprises in the region. During fieldwork, talking with producers, peasants, representatives of civil society and of the state, we observed that enterprises are identified by their market name. Who are the real owners of these companies? What are their intentions? The operational and accumulation strategies? These are issues that worry researchers involved with in-depth analysis of land grabbing in the region. Seeking to identify the financial capital present in the production of grains in agricultural frontier areas, we analysed a group of companies whose extension of controlled land reaches over 2.3 million hectares.

Our selected group included a diverse range of companies with diverse backgrounds, trajectories and forms of connection with financial capital. The most common traits include the control of large tracts of land in areas of modern agricultural frontier expansion, adoption of similar productive models (with intensive use of information technology systems), capitalisation through linkages with financial capital and predominance of strategies of surplus appropriation through the profit from grain yield and, mainly, the capture of the rent of land in a predefined period. Table 11.1 shows the main agricultural grain enterprises operating in the modern agricultural frontier in Brazil, the extension of land they control, the states where the farms are located, the main representatives of the financial capital in their shareholder structure, and the country of origin of the investor.

The main owners/controllers of the companies are highlighted in bold in Table 11.1. While individually the state of Mato Grosso has the highest percentage of land controlled by financial capital, the strong participation of these groups in the MATOPIBA region stands out. Data sum reveals that nearly 60% of the area controlled by the financial capital is located in this region, corresponding to an area of approximately 1.5 million hectares, as shown in Map 11.2.

Table 11.1 Agricultural enterprises and financial capital in the Brazilian agricultural frontier (year 2015)

Financial capital	Country of origin	Agricultural enterprise	Area of grains (1000 ha)	Location of farms (Brazilian state)
Vinci Partners	Brazil	Sollus Capital	30	Maranhão/Piauí/Tocantins/Bahia
Touradji Capital (Hedge)	US			
Altima Partners	England	El Tejar	110	Mato Grosso
The Capital Group	US			
Crestline Investors	US			
Louis Dreyfus Commod	France	Calyx Agro	62	Mato Grosso/Tocantins/Maranhão
AIG Investiments	US			
Brookfield Asset Management Inc	Canada	Brookfield Brasil	243	Maranhão/Minas Gerais/Tocantins/Mato Grosso do Sul/Goiás/Mato Grosso
SLC Participações S.A	Brazil	SLC Agrícola	340	Maranhão/Piauí/Bahia/Mato Grosso do Sul/Mato Grosso
Credit Suisse	Switzerland			
Deutsche Bank	Germany			
Neuberger Berman LLC	US			
Gavea Investimentos	Brazil	Vanguarda Agro S.A	250	Mato Grosso/Piauí/Bahia
Otaviano Pivetta	Brazil			
BTG Pactual	Brazil			
Salo Davi Seibel	Brazil			
Bonsucex Holding	Brazil			
Silvio Tini de Araújo	Brazil			
EWZ investiments LLC (Socopa)	Brazil			

(continued)

Table 11.1 (continued)

Financial capital	Country of origin	Agricultural enterprise	Area of grains (1000 ha)	Location of farms (Brazilian state)
Cresud	Argentina	BrasilAgro	177	Maranhão/Minas Gerais/Tocantins/Mato Grosso do Sul/Piauí/Goiás/Mato Grosso
JP Morgan Whitefrias Inc	US			
Cape Town LLC	US			
CSHG Asset Management S.A	Switzerland			
BRZ Investimentos	Brazil	Agrifirma (Genagro)	94	Bahia
RIT Capital Partners (Lord Rothscild)	England			
Harvard Asset Management (Iochpe family)	Brazil	Insolo Agroindustrial	100	Maranhão/Piauí
Vision Brazil Investments	Brazil	Tiba Agro	348	Bahia/Piauí/Goiás/Mato Grosso
Irmãos Francioni	Brazil			
Grupo Golin	Brazil			
International funds of private equity	US/EU			
SLC Agrícola	Brazil	SLC-MIT	45	Bahia/Maranhão/Mato Grosso
Mitsui	Japan			
Mitsui	Japan	Agricola Xingu	120	Minas Gerais/Bahia
TIAA-CREF	US	Radar	130	Maranhão/Tocantins/Piauí/Goiás/Mato Grosso/Bahia
Cosan	Brazil			
Mitsubishi	Japan	Ceagro Agrícola (Agrex)	94	Maranhão/Tocantins/Piauí/Goiás/Minas Gerais/Bahia
Paulo Fachin	Brazil			
Soros Fund Management LLC	US	Adecoagro	20	Bahia/Tocantins
Qatar Investment Authority	Qatar			

Financial capital	Country of origin	Agricultural enterprise	Area of grains (1000 ha)	Location of farms (Brazilian state)
Stitching Pensionfonds	Netherlands			
Ospraie Management LLC	US			
Management and Directors	US			
Jennison Associates LLC	US			
Brandes Investment Partners	US			
Ridgefield Capital	US	Agrinvest	94	Marnhão/Mato Grosso
Harvest Capital	US	Faz. Dois Rios	7,5	Tocantins
Dutch Businessmen	Netherlands	Kobra Agrícola	15	Bahia
Grupo Iowa	US	Iowa Brasil	10	Bahia
Agribusiness Investments	Brazil	AgriBusiness	8	Mato Grosso
Coteminas/Josué Gomes	Brazil	Cantagalo General Grains	151	Minas Gerais/Piauí/Goiás/Mato Grosso
GFN Agrícola	Brazil			
Agrícola Estreito	Brazil			
Valor Grains LLC	US			
Sojitz Group	Japan			

Source Frederico (2016), based on information available at the websites of the cited companies, at BM&FBovespa website, at the states commercial/trade registers and the Securities and Exchange Commission of Brazil

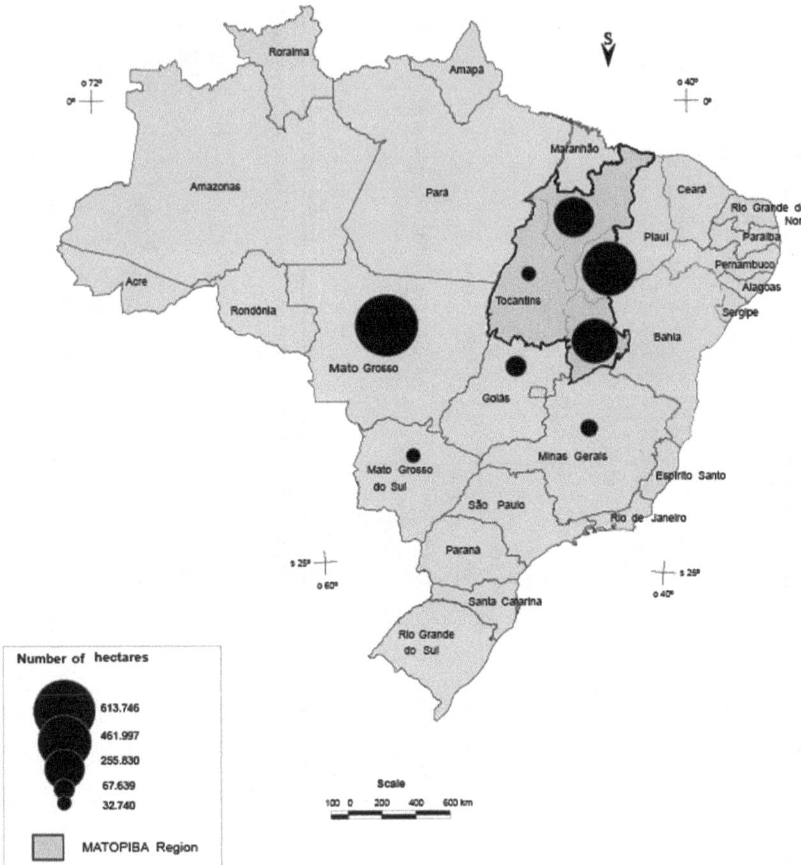

Map 11.2 Land (ha) per state Controlled by Financial Capital at the Agricultural Frontier (2015) (*Data source* Samuel Frederico [2016]; elaboration: Lorena Izá Pereira)

In addition to the identification of the main shareholders, it is important to ascertain the main owners, that is, those who can actually enforce their interests in the organisation and operation of the agricultural enterprises. It included shareholders with over 50% of relative share in the company's bond structure. The relative participation is obtained by the ratio between the square of the shareholding of a particular shareholder and the sum of the square of all shareholdings. This provides a satisfactory measure of the effective capacity of a particular shareholder to interfere in the decisions of the companies, since it strengthens its power in relation to the minority stockholders. This methodology was used by Instituto Mais Democracia, in the project 'Who are the Owners of Brazil?' that aimed to unveil the real controlling shareholders of the 300 largest Brazilian companies.[4] Results show that only the publicly traded companies Adecoagro (New York Stock Exchange) and Vanguarda Agro (BM&F/Bovespa—São Paulo) have more than one controlling shareholder, that is, no single shareholder owns more than 50% controlling interest. All other companies have only one controlling shareholder. For the most part, the controlling shareholders of the above-listed companies comprise investment companies (that manage private equity, mutual funds and pension funds) and other corporations such as the textile company Coteminas, the trading company Louis Dreyfus and the Argentine real estate firm Cresud.

It is often convenient for the control of the company to remain in Brazilian hands, to circumvent restrictions imposed since 2010 by the Presidency of the Republic that approved the Opinion of the General Attorney of the Union, which restricts the purchase of land by foreigners. One illustrative case is that of SLC Agrícola, of which the founding Brazilian family still owns 51% of the shares, even after going public. Other examples are the joint ventures SLC-MIT and Radar. The first constitutes a partnership between SLC Agrícola and the Japanese corporation Mitsui, and, the second, is a partnership between the US pension fund TIAA-CREF and the Brazilian sugar-energy company Cosan. Among the controlling shareholders, there are national and foreign investment companies (Vision Brazil, Brookfield, TIAA-CREF, Altima Partners, Harvard Asset Company, Ridgefield Capital, BRZ Investimentos, Gávea Investimentos, Vinci Partners, EWZ Investimentos, Agribusiness Investments, Harvest Capital, Soros Management Fund, Qatar Sovereign Wealth Fund), individual and family investors (SLC Participações/Logemann Family, Iochpe Family, Hélio Seibel, Otaviano Pivetta, Seibel Family and Sílvio T. de Araújo) and corporations from various economic sectors (Cresud, Coteminas,

[4] http://proprietariosdobrasil.org.br/.

Cosan, Mitsui, Mitsubishi, Louis Dreyfus Commodities, Kobra Group, Bonsucex Holding and Iowa Group). Controlling shareholders listed in Table 11.1 often do not yet represent the ultimate controlling interest, as they are also controlled by other agents such as the company stockholders and the quota holders of the pension funds mentioned. The agrarian question is a set of conflicts, that is, a conflictivity, as stated by Fernandes (2008), which has several components, starting with land rent appropriation and finishing with land expropriation DATALUTA (2017). Therefore, the land rent remains essential to understanding the land expropriation. In the following sessions, we analyse land rent and conflict.

Information Technologies and Ground Rent

Ground rent is one of the most controversial and important categories within the history of political economy. In general, it can be defined as a category referring to the payment by the capitalists to the landlords for the right to use land in the production of grains. In the current land grabbing phenomenon, several authors have adopted the perspective of land as a financial asset, in an attempt to interpret the interest and forms of action of institutional investors in agriculture (Ducastel and Anseeuw 2014; Fairbairn 2013; Gunnoe 2014; Ioris 2016; Li 2014; Ouma 2016). As Visser (2017) observes, the commodification of land is somewhat old, much earlier than the recent financialisation of agriculture. As the last author emphasises, land has been sold, leased, appraised and taxed for many centuries in various parts of the world. Thus, more than a binary transformation of lands of common use into new commodities, the current insertion of financial agents in agriculture represents a new stage of commodification of the land. It is their understanding not only as a commodity but as a financial asset. Thus, in addition to complying with various requirements for comparison and negotiation (standardisation, profit potential, liquidity, legitimacy, specific standards, etc.) (Visser 2017), to be considered an asset, it is not enough to generate positive returns in the future, it is necessary that its expectation of profitability is equal to or higher and the investment risk lower than that of the other financial instruments available in the market (Ouma 2016; Ducastel and Anseeuw 2014).

Although it seems contradictory, due to institutional investors' aversion to capital immobilisation, recent land investments are linked to the 'insatiable' logic of financial accumulation (Chesnais 2005). According to Paulani (2012), it is a rentier capital par excellence, either by the interest earned by the monopoly of money or by income derived from the monopoly of the land. As Harvey argues (2006: 369), for these investors, the land is understood as fictitious capital, that is, it is equivalent to an interest-bearing investment. For the last author, from the perspective of the capitalist, 'what is bought and sold is not the land, but title to the ground-rent yielded by it. The money laid out is equivalent to an interest-bearing investment. The buyer acquires a claim upon anticipated future revenues, a claim upon the future fruits of labour'. The increasing intensification of capital and labour over land assures the rise of future incomes, constituting itself as an attraction for interest-bearing capital.

However, to identify ways of raising future land incomes, it is necessary to analyse how agricultural enterprises operate through the use of technical information systems. Together with the financialisation of the world economy, the possibilities offered by information technologies (Santos 1996) are one of the main distinguishing features of the current land grabbing. Information systems allow the most powerful capital interested in investing in land/agriculture to operate on multiple scales: from the global scale, by facilitating international financial flow—indispensable to acquisitions and mergers, joint ventures, stock purchase, capitalisation of companies and repatriation of profits and dividends—to the detailed mapping of the stock of agricultural land and remote control of production; and to the local scale, by enabling more effective forms of appropriation of land rent, by choosing the best land and introducing informational agricultural technological systems. Thus, after identifying the companies' controlling shareholders, it is important to analyse how information technologies help large-scale agricultural enterprises to capture different forms of income associated, particularly, to land appropriation and modern agricultural production.

Natural Differential Land Rent

The natural rent (Differential Rent 1), according to Marx (1864), derives from the benefits of a monopoly held by a capitalist on a certain part of the terrestrial globe that presents a fertility differential. In this case, the holder of the land with highest natural fertility can get an extra profit derived from the better productivity of the more fertile soil. This generates an extra profit because the market price is determined by the production price of the worst quality land. In explaining the origin of the extra profit, Kautsky (1980: 75) states that 'in agriculture, it is not the production costs necessary for average land but those necessary for the worst terrain that determine the cost price'. In explaining Marx's proposal, Koslov (1981: 228) argues that the price of production in agriculture comes from those established in the worst land because of the scarcity and monopoly of the land, which prevents other capitalists from entering the best land.

This type of rent apparently has no relation to technical availability, since it derives from the control by a certain capitalist of naturally more fertile, and thus more productive and profitable land areas. But the question that arises is: how does a manager of a financial fund, located in the United States or in England, get to find the most fertile lands across the globe? To this end, in addition to establishing relationships with producers, companies and brokers located in the countries and regions targeted by the investments, the use of information technologies, such as satellite images and crosschecking of data on soil quality and historical rainfall records, becomes essential. Today, almost all large Brazilian companies operating in the land market and in modern agricultural production have a technical team focussed on mapping lands that can possibly be incorporated into their investment portfolios. Some of them have even established subsidiary companies aimed at locating, assessing and negotiating lands, such as Land CO, controlled by SLC Agrícola, and the joint venture quite evocatively named, Radar Agrícola. Some companies affirm having a database with georeferenced information on hundreds of thousands of hectares of land scattered throughout Brazil and abroad.[5]

In general, the first stage in assessing a potential purchase or lease of agricultural property involves the use of satellite images and historical climate records of a particular region. This part of the analysis focusses on mapping the property, determining its topography (essential for assessing its potential mechanisation) and the percentage of the land that can be used for agricultural production, with estimates of risks of drought and incidence of pests, as well as soil quality and productivity. Furthermore, a historical survey of satellite images of the property is made, and compared to climate records and soil data, to evaluate performance during periods of cataclysms, including severe weather such as drought or excessive rainfall.

Following the initial evaluation of the images, the second stage involves collecting detailed information on the farmland through visits to the farm. The inspection of the estate generally includes collection of soil samples from its various plots for chemical and physical analysis and assessment of current conditions of crop productivity (in case of an active farm). Data collected are, then, georeferenced and crossed with the general data previously obtained by satellite images, thus allowing to estimate, through computer simulation, the production costs and required capital expenditures throughout the estate. Such expenditures include the necessary initial investment in infrastructure, repairs, drainage and soil correction, with the aim of adapting it to the use of modern production techniques, such as no-tillage and double harvest.

Produced Differential Land Rent

After facilitating selection of the best lands, that is, the appropriation of the natural differential land rent, information technologies are also essential to obtain the produced differential land rent resulting from production (Differential Rent 2). As noted by Paulani (2012), this rent is not derived from the hazards of nature, but, theoretically, from the

[5] Brasilagro (2015), for example, claimed to have carried out, by 2015, the basic survey of about 30 million hectares in Brazil, Colombia and Paraguay.

application of different amounts of capital to plots of land with the same fertility. In lands with greater capital investment, individual price of production falls below the market regulating price, allowing the creation of an additional surplus value. The large agricultural companies—aware that their crops, for being agricultural commodities, could hardly obtain a differentiated pricing, since the price is determined by the international market and is dependent on factors that they cannot control—seek a differential profitability by increasing the efficiency of the production (an increase of yield and reduction of unit costs). In order to demonstrate to shareholders and potential investors that they are increasing their production efficiency, publicly traded agricultural companies emphasise in their reports and public presentations the historical evolution of some indicators showing the rising appropriation of the land rent, such as the amount of hectares per employee (production, processing and administrative), overhead expenses per hectare, use of tractor (cv) per hectare, area of precision agriculture and yield (kg/ha).

The farms with higher density of capital and information technology (use of IT systems), which almost always coincide with those with better natural fertility,[6] can produce an additional surplus value, that is, a produced differential land rent. This latter can be demonstrated, for example, by differences in yield (kg/ha) that result in an individual production price lower than the market regulating price of production. The comparison between indexes of soybean yield for SLC Agrícola (considered one of the most efficient in terms of yield and costs reduction) and the average of the main producing regions and countries provides clear evidence of the yield differentials resulting from investments in agricultural technological systems. The average yield of SLC Agrícola for five harvests (2010/2011 to 2014/2015) was 3006 kg/ha, which was higher than the average of all Brazilian producers (2922), of the United States (2919), Argentina (2673) and of the West of Bahia (2819), the company's main region of operation.

[6] Certainly, capitalists are attracted to making larger investments in agricultural holdings that naturally offer them greater profitability, for example, those with greater water availability, allowing two crops a year without the use of irrigation.

In addition to the production differential, the intensive capital investment in cutting-edge information technologies also promotes higher land appreciation. By artificially raising the quality and yield of land, capital investment also raises its market price. In fact, this is the main strategy used by companies particularly dedicated to the land market. In general, according to information collected from reports of companies such as Adecoagro, BrasilAgro, SLC Agrícola and Tiba Agro, what they call the 'thesis of investments in land development' implies the acquisition, transformation and sale of lands with great potential for appreciation in an average term of five years. To this end, they seek to acquire land at relatively low prices, generally located in areas of expansion of the agricultural frontier, with suitable edaphoclimatic and pedological conditions for dryland farming and, if possible, the development of two crops per agricultural year.

The cycle of land appreciation adopted by the companies has basically five stages: acquisition, licensing, clearing, transformation and sale. The first step is to prospect for areas and negotiate with local brokers and large landowners. We cannot disregard, in many cases, there is *grilagem*. We cannot forget the frontier occupation process runs side by side with the *grilagem* industry, which counts on companies specialised in the land business. It is a process of territorial conflict production. Following this first stage, environmental licensing is requested, indicating the legal reserve areas and authorising deforestation of the remaining areas, and the company's managers prepare, present and obtain the approval for their strategic plan. Then, the area is cleared and the ground levelled, preparing for the fourth stage, consisting in the transformation of the land, through ploughing, soil correction and fertilisation and construction of infrastructure facilities. Finally, it is necessary to find a buyer for the farm, which often maybe not as easy as initially expected due to the high price of farms. According to information provided by SLC Agrícola (2014), after completion of the (five years) cycle, an area of raw cerrado initially acquired at an average price of US$ 2000/ha can be sold at around US$ 6.5 thousand/ha.

Absolute Rent of Land

The strategies of land appropriation and transformation by large agricultural enterprises also illustrate the existence of another type of land rent, which Marx (1864) referred to as absolute land rent. In his own terms, 'the mere legal ownership of land does not create any ground-rent for the owner. But it does, indeed, give him the power to withdraw his land from exploitation until economic conditions permit him to utilise it in such a manner as to yield him a surplus' (p. 225). By acquiring large tracts of land with native vegetation, the conversion of land is carried out gradually, from the areas of greatest natural fertility to those which require greater capital expenditure to adapt it to the requirements of modern agriculture.

The raw land stock, termed land bank by investors, although not providing rent to its owner, as Marx (1864) points out, is used by companies in diverse ways, both in marketing strategies, for demonstrating their growth potential and consequently attracting new investors and for obtaining credit for operational expenditure and investment,[7] without disregarding its function as a reserve and potential future appreciation. All companies surveyed interested in capturing the rent of land through its conversion and trading have their own Land Bank. As part of its business, the continuous transformation of land requires reserve areas for future appreciation as a way of keeping the continuous cycle of capturing the rent of land. In this session, we analysed how land rent is essential to understand the process of land grabbing, being the component that generates the agrarian question and its conflict. The financialisation of agriculture intensified the territorialisation of corporations, increasing impacts and territorial confliction.

[7] Generally, land is pledged as collateral for obtaining loans from financial institutions.

Impacts and Territorial Confliction

Land grabbing changed the agrarian question, keeping and changing the forms of land concentration and ultimate rent appropriation (Pitta and Mendonça 2015). The latifundia, a synonym for unproductiveness, became productive with the purchase or leasing by corporations with either national or foreign controlling capital. They form clusters of agribusiness companies that produce flexible, technology-intensive commodities and simultaneously generate territorial conflicts over the ownership of land, water, labour, and production. Such conflicts constitute the 'conflictivity' (Fernandes 2008) that unveils the class struggle in the disputes over territories and development models. The territorialisation of land grabbing in MATOPIBA threatens peasant, *quilombola* and indigenous communities. In Brazil, technological modernisation of agribusiness, with the use of computerised machines for the production of various commodities, drastically reduced the participation of wage labour and the farmers' contract (Fernandes et al. 2012). The production of agricultural commodities is carried out by large-scale machinery by outsourced companies or by the corporations themselves. These populations have their own development models, practiced in their territories through agroecology and agroforestry. The conflicts are, therefore, territorial disputes and by development model.

Conflictivity (i.e. confliction or antagonism) is an essential concept for understanding territorial disputes over development models and distinct agrarian policies. The overcoming of conflictivity does not happen by consensus, since the models are antagonistic and any possible agreement implies changing both. Land grabbing impacts produce both direct and indirect conflicts, involving multiple dimensions of conflictivity. These new territorialities are the result of the combined actions of latifundia, corporations and state that changed the forms of repression, for example, the gunslingers were replaced by security companies and the *grilagem* became a more intensive process in digital era. This new reality has also widened the scales of peasant resistance, making local communities connected with their and other organisations on national and international scales, making their struggles visible. The direct conflict between these communities and the corporations is an indirect result of public policies, with the State acting as a legitimising agent for the practices of national and transnational corporations and investment funds, through cooperation agreements and robust rural credit policies for agribusiness.

Graph 11.1 shows that the increase in conflicts began in 2003, reaching its peak in 2016. The rise in conflicts is associated with the growing territorialisation of agribusiness, which in turn is associated with the increase in agricultural and livestock credit. In 2003, the agricultural and livestock plan released 27 billion reais and, in 2016, the budget was 187.7 billion reais (MAPA 2015). The growing territorial conflicts are a direct outcome of the impacts of land grabbing territorialisation processes and have shown a growing trend during the last decade. The struggles of peasant and indigenous communities are no longer against farmers who invaded their lands. The impacts of land grabbing are devastating to the territories of peasant, *quilombola* and indigenous communities. Agribusiness territorialisation leaves a trail of deforestation, water pollution thus multiplying the conflicts. Graph 11.1 and Map 11.3 show that, in 1996, 45 conflicts were registered in the municipalities that would come to form the MATOPIBA; in 2016, there were 271 conflicts produced in the territories of expansion of the modern agricultural frontier. In twenty years, the number of land conflicts has increased over five times.

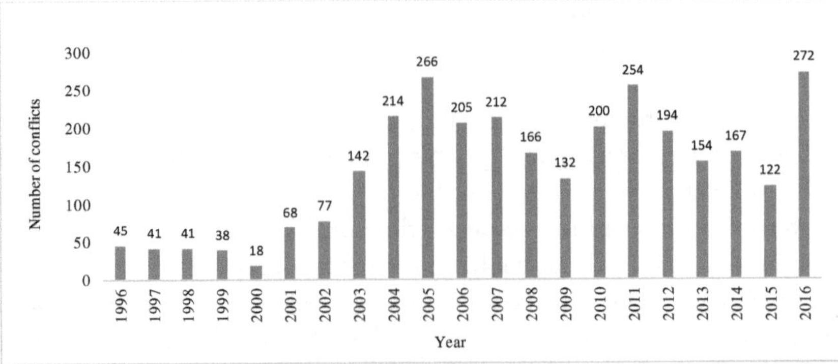

Graph 11.1 MATOPIBA Region—Territorial conflicts (1996–2016) (*Data source* Comissão Pastoral da Terra [2017]; elaboration: Lorena Izá Pereira)

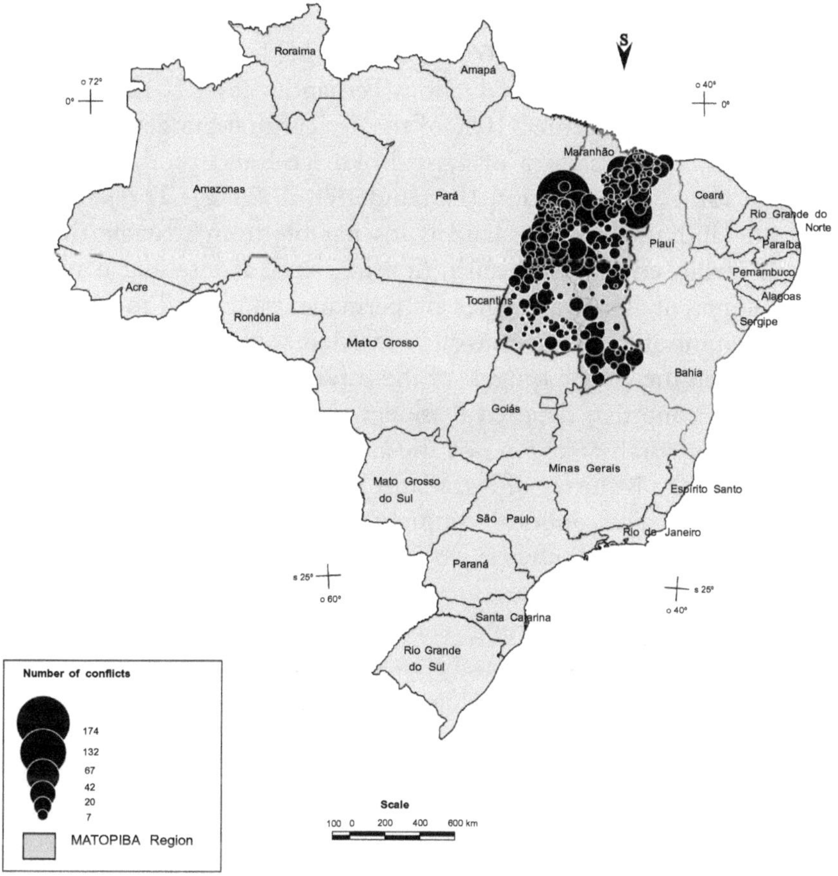

Map 11.3 Territorial conflicts in the MATOPIBA Region (1996–2016) (*Data source* Comissão Pastoral da Terra [2017]; elaboration: Lorena Izá Pereira)

Conclusions

The creation of the MATOPIBA region enabled land grabbing processes to further exploit land and land rent. The territorialisation of agrarian extractivism, intensified by investments from international financial funds, changed the agrarian question, transforming latifundia into productive farmlands, increasing production and engendering territorial conflicts. The State supports this hegemonic model by financing agricultural production, allocating 90% of the agricultural budget to agribusiness, which controls 76% of agricultural land, but produces only 68% of the gross value of production (Fernandes et al. 2012).

This advance represented 10% of the agricultural budget to the peasantry, which controls 24% of agricultural land and produces 32% of the gross value of production (Fernandes et al. 2012). The struggle to resist the land and conquer land is inseparable from struggle the agricultural credit, education, health, housing, etc., as essential conditions for development. These struggles are permanent and tend to grow when the correlation of forces is unfavourable. Maintaining the public policies won has been the recent struggle of the rural movements, especially with the political coup that deposed President Dilma Rousseff in 2016. With the coup, the 'ruralists' composed the allied base of the government that extinguished the Ministry of Agrarian Development and several policies public policies. This political conjuncture contributes to the advancement of agrarian extractivism to the detriment of the development policies of peasant agriculture.

Land grabbing with strong participation of financial capital and agrarian extractivism with use of new technologies are new elements of the agrarian question that develops keeping the same historical process of the inequality and deterritorialisation. This study of MATOPIBA confirms the hegemony trend of the agribusiness development model, which increasingly reveals that manual labour and the farmers' contract are reduced to some crops. The main commodities for export are produced with high mechanisation and on a large scale. But if peasants do not interest corporations as labourers or contracted farmers, these same corporations have a great interest in peasant, indigenous and *quilombola* territories to territorialise the agribusiness. The resistance to maintain the grassroots territories depends, basically, on agroecology and agroforestry. However, the resources received from the State for the development of these crops are insufficient, which reinforces the hegemonic power of agribusiness. So MATOPIBA is an excellent example to better understand the agrarian question of the twenty-first century.

References

Ab'Sáber, A. 2003. *Os Domínios de Natureza no Brasil: Potencialidades Paisagísticas*. São Paulo: Ateliê Editorial.

Alonso-Fradejas, A. 2012. Land Control-Grabbing in Guatemala: The Political Economy of Contemporary Agrarian Change. *Canadian Journal of Development Studies* 33 (4): 509–528.

Alonso-Fradejas, A. 2015. Anything but a Story Foretold: Multiple Politics of Resistance to the Agrarian Extractivist Project in Guatemala. *Journal of Peasant Studies* 42 (3–4): 489–515.

Ball, M. 1976. *Marx's Theory of Rent and the Role of Landed Property*. London: Birkbeck College.

Borras Jr., S., and J. Franco. 2010. La Política del Acaparamiento Mundial de Tierras. Replanteando las Cuestiones de Tierras, Redefiniendo la Resistencia. ICAS Working Paper Series n. 001. Transnational Institute, Land Deal Politics Initiative, Amsterdam.

Borras, S., Jr., and J. Franco. 2012. Global Land Grabbing and Trajectories of Agrarian Change: A Preliminary Analysis. *Journal of Agrarian Change* 12 (1): 34–59.

Borras, S., Jr., and J. Franco. 2013. Global Land Grabbing and Political Reactions 'from Below.' *Third World Quarterly* 34 (9): 1723–1747.

Borras, S., Jr., C. Kay, S. Gómez, and J. Wilkinson. 2012. Land Grabbing and Global Capitalism Accumulation: Key Features in Latin America. *Canadian Journal of Development Studies* 33 (4): 402–416.

Borras, S., Jr., J. Franco, R. Isakson, L. Levidow, and P. Vervest. 2015. The Rise of Flex Crops and Commodities: Implications for Research. *The Journal of Peasant Studies* 43 (1): 93–115.

BrasilAgro (Companhia Brasileira de Propriedades Agrícolas). 2015. *Relatório de Demonstrações Financeiras – Brasilagro*. Retrieved from https://ri.brasilagro.com.

Carcanholo, R.A. 1984. Renda da Terra: Uma Concreção Teórica Necessária. *Revista De Economia Política* 4: 4.

Chesnais, F. 2005. O Capital Portador de Juros: Acumulação, Internacionalização, Efeitos Econômicos e Políticos. In *A Finança Mundializada: Raízes Sociais e Políticas, Configuração e Consequências*, ed. F. Chesnais, 35–67. São Paulo: Boitempo.

Clements, E.A. 2015. Brazilian Polices and Strategies for Rural Territorial Development in Mozambique: South-South Cooperation and the Case of ProSAVANA and PAA. MSc dissertation, Presidente Prudente, Unesp.

Comissão Pastoral da Terra. 2017. *Conflitos no Campo Brasil 2016*. Goiânia: Comissão Pastoral da Terra.

Corson, C., and K.I. MacDonald. 2012. Enclosing the Global Commons: The Convention on Biological Diversity and Green Grabbing. *Journal of Peasant Studies* 39 (2): 263–283.

Cotula, L. 2012. The International Political Economy of the Global Land Rush: A Critical Appraisal of Trends, Scale, Geography and Drivers. *Journal of Peasant Studies* 39 (3–4): 649–680.

DATALUTA. 2017. *Relatório DATALUTA Brasil 2016*. Presidente Prudente: Rede DATALUTA.

Ducastel, A., and W. Anseeuw. 2014. *Agriculture as an Asset Class: Reshaping the South African Farming Sector*. Paper presented at The International Conference "Finance, Food and Farmland." Institute for Social Studies, The Hague.

Edelman, M. 2013. Messy Hectares: Questions about the Epistemology of Land Grabbing Data. *Journal of Peasant Studies* 40 (3): 485–501.

Edelman, M., and A. León. 2013. Cycles of Land Grabbing in Central America: An Argument for History and a Case Study in the Bajo Aguán, Honduras. *Third World Quarterly* 34 (9): 1697–1722.

Edelman, M., C. Oya, and S. Borras Jr. 2013. Global Land Grabs: Historical Processes, Theoretical and Methodological Implications and Current Trajectories. *Third World Quarterly* 34 (9): 1517–1531.

Fairbairn, M. 2013. Just Another Asset Class? Neoliberalism, Finance and the Construction of Farmland Investment. In *The Neoliberal Regime in The Agri-Food Sector: Crisis, Resilience and Restructuring*, ed. S. Wolf and A. Bonanno, 245–261. London: Earthscan.

Fairhead, J., M. Leach, and I. Scoones. 2012. Green Grabbing: A New Appropriation of Nature? *Journal of Peasant Studies* 39 (2): 237–261.

Fernandes, B.M. 2008. Questão Agrária: Conflitualidade e Desenvolvimento Territorial. In *Luta pela Terra, Reforma Agrária e Gestão de Conflitos no Brasil*, ed. A. Buainain, 173–224. Campinas: Unicamp.

Fernandes, B.M. 2011. Estrangeirização de Terras na Nova Conjuntura da Questão Agrária. *Conflitos no Campo Brasil*, 76–83.

Fernandes, B.M., C. Welch, and E.C. Gonçalves. 2010. Agrofuel Policies in Brazil: Paradigmatic and Territorial Disputes. *Journal of Peasant Studies* 37 (4): 793–819.

Fernandes, B.M., C. Welch, and E.C. Gonçalves. 2012. *Land Governance in Brazil: A Geo-Historical Review of Land Governance in Brazil*. Rome: International Land Coalition.

Frederico, S. 2010. *O Novo Tempo do Cerrado: Expansão dos Fronts Agrícolas e Controle do Sistema de Armazenamento de Grãos*. São Paulo: AnnaBlume.

Frederico, S. 2016. *Território, Capital Financeiro e Agricultura: Investimentos Financeiros Estrangeiros no Agronegócio Brasileiro*. Research Report. São Paulo: FAPESP.

Galeano, L. 2012. Paraguay and the Expansion of Brazilian and Argentinian Agribusiness Frontiers. *Canadian Journal of Development Studies* 33 (4): 458–470.

Garrido Filha, I. 1980. *O Projeto Jari e os Capitais Estrangeiros na Amazônia*. Petrópolis: Vozes.

Gunnoe, A. 2014. The Political Economy of Institutional Landownership: Neorentier Society and the Financialization of Land. *Rural Sociology* 79 (4): 478–504.

Harvey, D. 2003. *The New Imperialism*. Oxford: Oxford University Press.

Harvey, D. 2006. *Limits to Capital*. London and New York: Verso.

Holt-Giménez, E. 2012. Land Grabs versus Land Sovereignty. *Food First Backgrounder* 4 (18): 1–3.

Inocêncio, M.E., and M. Calaça. 2010. Estado e Território no Brasil: Reflexões a Partir da Agricultura no Cerrado. *IDeAS* 4 (2): 271–306.

Ioris, A.A.R. 2016. Rent of Agribusiness in the Amazon: A Case of Study from Mato Grosso. *Land Use Policy* 59: 456–466.

Kautsky, K. 1980 [1899]. *A Questão Agrária*, 3rd ed. São Paulo: Proposta Editorial.

Koslov, G. 1981. *Curso de Economia Política*. Lisbon: Editorial Avante.

Levien, M. 2012. The Land Question: Special Economic Zones and the Political Economy of Dispossession in India. *Journal of Peasant Studies* 39 (3–4): 933–969.

Li, T.M. 2014. What Is Land? Assembling a Resource for Global Investment. *Transactions of the Institute of British Geographers* 39 (4): 589–602.

MacKey, L. 2011. Legitimating Foreignization in Bolivia: Brazilian Agriculture and the Relations of Conflict and Consent in Santa Cruz, Bolivia. In *International Conference on Global Land Grabbing*, University of Sussex, Brighton.

Mamonova, N. 2015. Resistance or Adaptation? Ukrainian Peasants' Responses to Large-Scale Land Acquisitions. *Journal of Peasant Studies* 42 (3–4): 607–634.

MAPA. 2015. *Plano Agrícola e Pecuário 2015–2016*. Brasília: Ministério da Agricultura, Pecuária e Abastecimento.

Marx, K. 1986 [1864]. *O Capital, Vol. III*. São Paulo: Abril Cultural.

McKay, B. 2017. Agrarian Extractivism in Bolivia. *World Development* 97: 199–211.

McMichael, P. 2009. A Food Regime Analysis of the 'World Food Crisis.' *Agriculture and Human Values* 26: 281–295.

McMichael, P. 2013. Land Grabbing as Security Mercantilism in International Relations. *Globalizations* 10 (1): 47–64.

Miranda, E.E. 2015. *Caracterização Territorial Estratégica do MATOPIBA*. Campinas: Embrapa.

Moreda, T. 2015. Listening to Their Silence? The Political Reaction of Affected Communities to Large-Scale Land Acquisitions: Insights from Ethiopia. *Journal of Peasant Studies* 42 (3–4): 517–539.

Oliveira, A.U. 1988. *Integrar para não Entregar: Políticas Públicas e Amazônia*. Campinas: Papirus.

Ouma, S. 2016. From Financialization to Operations: Historicizing and Disentangling the Finance-Farmland-Nexus. *Geoforum* 72: 82–93.

Oya, C. 2013. Methodological Reflections on "Land Grab" Databases and the "Land Grab" Literature "Rush." *Journal of Peasant Studies* 40 (3): 503–520.

Paulani, L. 2012. *Acumulação e Rentismo: Resgatando a Teoria da Renda de Marx para Pensar o Capitalismo Contemporâneo*. Rio de Janeiro: ANPEC.

Peluso, N.L., and C. Lund. 2011. New Frontiers of Land Control: Introduction. *Journal of Peasant Studies* 38 (4): 667–681.

Pereira, L.I. 2015. O Processo de Estrangeirização da Terra no Brasil: Estudo de Caso da Empresa Umoe Bioenergy no Município de Sandovalina (SP). BA dissertation. Presidente Prudente, Unesp.

Pereira, L.I., and L. Pauli. 2016. O Processo de Estrangeirização da Terra e Expansão do Agronegócio na Região do MATOPIBA. *Campo – Território* 11 (23): 196–224.

Pitta, F.T., and M.L. Mendonça (eds.). 2015. *A Empresa Radar S/A e a Especulação com Terras no Brasil*. São Paulo: Outras Expressões.

Ploeg, J.D. van der, J. Franco, and S. Borrar Jr. 2015. Land Concentration and Land Grabbing in Europe: A Preliminary Analysis. *Canadian Journal of Development Studies* 36 (2): 147–162.

Santos, M. 1996. *A Natureza do Espaço: Técnica e Tempo, Razão e Emoção*. São Paulo: Hucitec.

Santos, M., and M.L. Silveira. 2001. *O Brasil: Território e Sociedade no Início do Século XXI*. Rio de Janeiro: Record.

Sassen, S. 2013. Lands Grabs Today: Feeding the Disassembling of National Territory. *Globalizations* 10 (1): 25–46.

Sauer, S., and S. Borras Jr. 2016. 'Land Grabbing' e 'Green Grabbing': Uma Leitura da 'Corrida na Produção Acadêmica' sobre a Apropriação Global de Terras. *Campo Território* 11 (23): 6–42.

Visser, O. 2017. Running Out of Farmland? *Agriculture and Human Values* 34: 185–198.

Wilkinson, J., B. Reydon, and A. Di Sabbato. 2012. El Caso del Brasil. In *Dinámicas del Mercado de la Tierra en América Latina y Caribe: Concentración y Extranjerización*, ed. F.S. Baquero and S. Gómez, 105–144. Rome: FAO.

Willians, J.W. 2014. Feeding Finance: A Critical Account of the Shifting Relationships Between Finance, Food and Farming. *Economy and Society* 43 (30): 401–431.

12

Dispossession and Agricultural Commodities: The Case of Oil Palm Farming in the Brazilian Amazon

Adolfo da Costa Oliveira Neto

Dispossession Nowadays: A Brief Introduction

The debate about dispossession is not recent in the academy, but it has increased specially during the last decade. Research done on 29 March 2022 on the Web of Science database has shown that the number of papers published per year regarding dispossession has remarkably increased from 2010 on, as shown in Graph 12.1. From a sample of almost 3200 papers, only after 2013 there were more than 100 yearly papers published about this theme, having the papers by Bouhdiba (1964) been the inspiring one.

However, the current chapter is intended to introduce new approaches on the theme. The dispossession came to be seen as the process through which capitalism expands throughout several world regions, mainly in the southern hemisphere, and features the fight of local individuals

A. da Costa Oliveira Neto (✉)
Federal University of Pará (UFPA), Belém, Brazil
e-mail: adolfo.oliveira.neto@gmail.com

A. A. R. Ioris and B. Mançano Fernandes (eds.), *Agriculture, Environment and Development*, https://doi.org/10.1007/978-3-031-10264-6_12

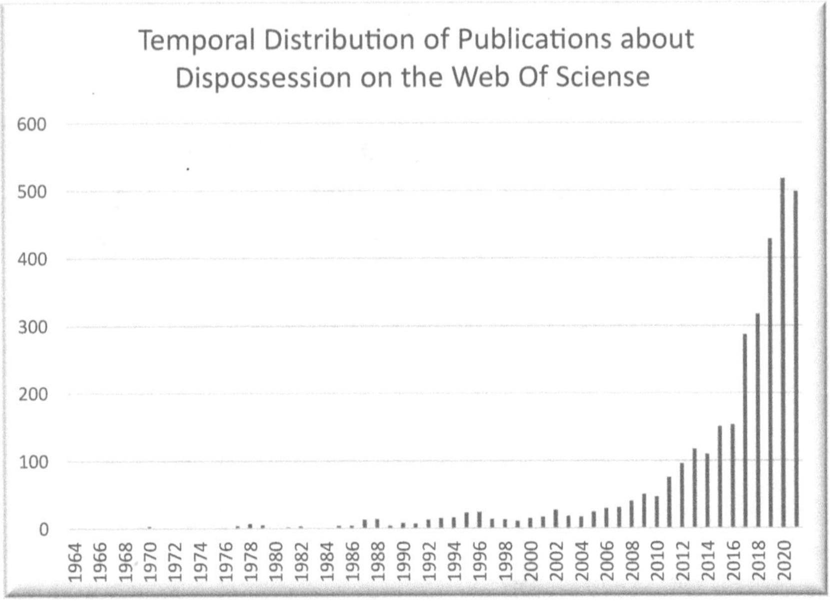

Graph 12.1 Temporal distribution of publications about dispossession on the Web of Science

against the state politics which promote the advance of financed capitalism aiming to take more lands and natural resources by capturing and expelling as a modern approach, both through annexation, subordination or exclusion of local individuals. Issues related to the various kinds of agriculture which opposes to the agrobusiness model, the importance of forests conservation, the climate change, the conflicts, the environmental justice and the ecological politics, the land titling issues, the social movements, the place, the territory and the land rights are some that currently surrounds the debate about dispossession, as shown in Fig. 12.1 (based on the Web of Science search).

The map above has been made from data available on Web Science database, by searching the co-occurrence among the keywords, so that circles are proportional to the frequency of the words and the minimal frequency for a word to be displayed was 10, which reduces the initial sample from 2306 to 169 words. The highlighted terms show that dispossession is a process that occurs in several scales and in different

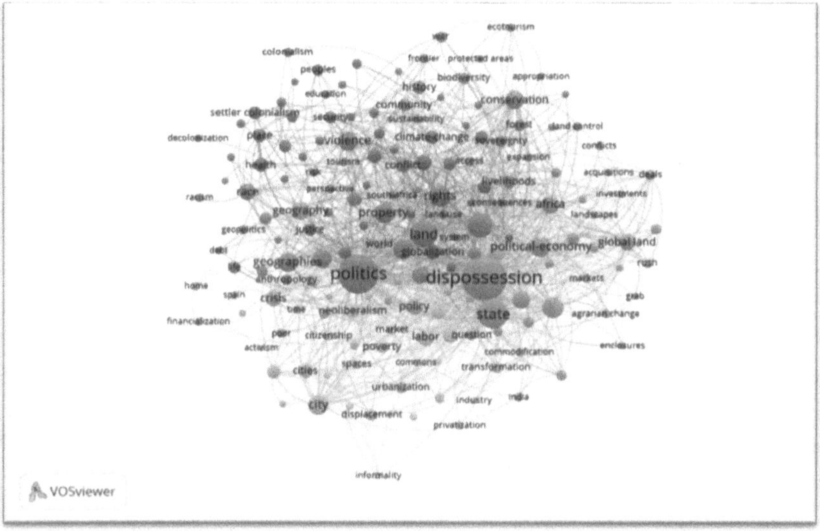

Fig. 12.1 Keywords co-occurrence map

places, comprising processes in rural and urban areas and which are closely related to the structural contradiction among the current forms of capitalist progress and the diversity of resistance.

Agricultural *commodity* and *flex crops*: Oil Palm in the Brazilian Amazon

The expansion of agrobusiness as a system complex and of flexible agricultural commodities are the main basis of the current expansion of oil palm farming in Amazon. On one hand, agribusiness comprises the fusion of agricultural, industrial, commercial, financial and technological systems. On the other hand, the production of commodities and flex crops are the real basis which the financial capital takes over worldwide. The business opportunities for agrofuels have influenced its production as part of the strategy of northern countries to decrease their dependence on fossil fuel. Because of this market, several undeveloped countries make politics for promoting agrofuels based on cereal farming (such as maize and soy), in sugar cane or oily plants (such as palm, sunflower and castor beans). The energy balance of palm is one of the most well-known comparing advantages in relation to other crops which are base for biodiesel. For instance, palm oil has an energy balance 4.6 timer bigger than castor beans.

The Brazilian biodiesel, for instance, which has been initially designed to be based on vegetal oils such as palm, castor beans, sunflower, maize and cotton, ended up being mainly constituted by soy oil and, in a lower quantity, by bovine fat. These two ingredients are more than 90% of the raw material used to make biodiesel in Brazil, according to the Brazilian National Agency of Petroleum (ANP), as shown in Fig. 12.2. There we can see the national profile from October 2008 to December 2014, period which represents the use of Biodiesel 3 (B3) until Biodiesel 7 (B7). The numbers associated with biodiesel indicate the per cent of vegetal or animal oil added on the mineral diesel. Biodiesel 3, for instance, has 3%, Biodiesel 4 has 4% and so on, reaching the mark of 10% in 2021. The main raw material used in biodiesel during that period were soy oil, bovine fat, cotton oil, tallow, reused frying oil, chicken fat, sunflower oil, peanut oil, canola oil and palm oil.

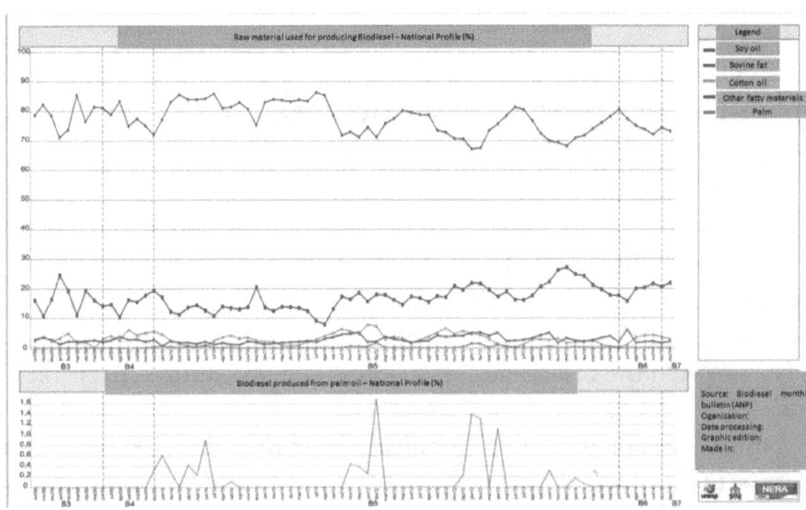

Fig. 12.2 Raw material used in the production of biodiesel–national profile (%)

In the 75 months analysed in Fig. 12.2, in just 18 (24% of the period) there was palm oil in biodiesel and in concentrations rarely above 1% of oils and fatty material used to produce biodiesel. Besides palm oil, castor beans, peanuts and sunflower remained in a low quantity during the period, which means that the initial proposal of biodiesel based on vegetal oil and based on national land development politics was actually replaced by an agrofuel based on agricultural commodities and on sub-products from the traditional rural business from Brazil, specially cattle ranching and soy, based on monoculture and often studied due to their effects on wealth and land concentration, besides the intensive use of agrochemicals.

The increase of biodiesel production in Brazil takes place following the worldwide increase of attempts to replace fossil fuels to conserve the environment, in an attempt to respond to what Houtart (2011) defined as a double crisis, climatic and energetic. According to the author:

> Effectively, on one hand, the double crisis, energetic and climatic, has become an inevitable reality which cannot be ignored anymore, on the other hand, the desire to seek for new sources of energy has turned, face to the price of oil and gas, into a very profitable business to capitalist investors and has an absolutely positive image on the eyes of the public opinion more and more concerned about environmental issues. (Houtart 2011: 13)

On the other hand, the increase in biodiesel production was followed by the consolidation of the productive chain and raw materials which are the basis for the products. The data shows that, nowadays, palm oil is not produced with the aims of serving the agrofuel chain. This is a choice for company which owned industrial plants for producing agri-diesel and other for producing several other products that function as base for food, industry and pharmaceutical areas. This flexibility of palm oil in various areas of industry is one of the main elements to support its valorisation in the commodity market. Flexibility may cause the base industry to quickly change the final product and serve the most convenient market. Houtart highlights the palm as flex crop when he says:

> The use of the acceptance to palm is multiple. At first, it is a basic product for food: margarine, oil, glacé, chocolate, pre-cooked food, animal food, etc., but it is also basic for paints and enamels. Besides, the pharmaceutical industry is also a huge consumer, through the production of a big variety of interesting sub products as, for instance, furfural, antifungal bactericide and natural insecticide, in addition to lignin that comes from wood and is used for plywood. Recently, the acceptance to palm has started turning massively into Agri Diesel, mainly after the increase on the Price of oil. (Houtart 2011: 122)

In the case of Brazil, the addition of soy and bovine fat derivates to agri-diesel caused the biofield politics to strengthen production chains which offered lower costs than palm oil. Analysing this issue Blanca Rublo states:

> This way, the raw material shall convert, since the early 2000's, into an effective shelter for speculative capital, which became a device of United States to strengthen their financial capital and also a hegemonic role in the world market. (…) Thus, a relation among the recurrent capitalist crisis was done, the devaluing of the Dollar, the decrease of financial assets profitability, and their flow used the commodities as a refuge effect. Basic grains thus became a way out of the financial crisis that capital is going through. This situation caused raw materials to be transformed into one of the most financialised products, that is, with the greatest speculative investment in the world. (Vega 2014: 192)

The dispute for grains and oils among food business, agrofuels and pharmaceutical has raised concerns about the food crisis in the world and famine, while it raises the price of commodities under dispute. According to Vega:

> The Price with higher increases in price is those dedicated to agrofuels. As can be seen (...), the price of maize rose 14% between 2006 and 2012, and soybeans 14.1%, while rice and wheat had lower growth. (Vega 2014: 188)

The actions of oil palm farming companies, even when not related to agrofuels production, are typical from global dynamics which aim to boost the capital flow, increase dominance over agricultural lands, to acquire a progressive and more capitalised profit from lands, strengthening the added value and in some cases to earn the price as a monopoly. The criticising elements towards the oil palm agribusiness dynamics worldwide are present in important papers such as: Houtart (2011), Fabrini (2010), Oliveira (2011), Bernardes (2011), Bernardes and Aracri (2011), Puttilli and Tecco (2012), Brandão (2010), Oliveira Neto and Fernandes (2017), Oliveira Neto (2021) and Laschefski and Zhouri (2010), among others.

Dispossession and Conflicts in the Amazon Related to Oil Palm Farming

Dispossession in added value the Amazon, when related to the production of agricultural commodities, especially palm oil, is directly related to another process: the land grabbing, which may or not be associated to the subordination of farmers' lands, and as a critical element for the comprehension of the oil palm dynamics. These processes have been responsible for countless conflicts since its implementation in Amazon in the 1940s, until today. Figure 12.3 shows some recent news about conflicts involving oil palm farming in the Amazon. The excerpts show recent articles which demystify the production and palm oil as a green commodity able to bring income associated to local development, with respect to local stakeholders, their territories and environment. Actually, the oil palm has revealed major environmental contradictions related to monocultures and also social ones, regarding the way the primitive and permanents capitalism thrives in Amazon taking lands and workforce.

These conflicts are taking place in at least six fronts presented here in a summarised and random way. The first one is the front for land control. Data from Nahum et al. (2020), presented in Table 12.1, shows how oil palm farming has expanded during the first two decades of twenty-first century, mainly in Pará, turning from 38.9 thousand hectares of farming area in 2001–100.8 thousand hectares in 2018, which indicates how the land disputes for expanding farming areas has taken place in the region, mainly in areas which have better edaphoclimatic conditions for farming.

Fig. 12.3 News about conflicts related to oil palm farming

This expansion happened through a broad process of land acquisition by acquiring small, medium and large farms to compose large farms in the surroundings of industries, causing a sort of allotment as shown in Fig. 12.4 by Nahum and Santos (2015). The land conflicts caused, on

Table 12.1 Harvested area of oil palm in Para and Bahia (in thousand hectares), 2001–2018

Ano	Para	Bahia	Brasil
2001	38.9	45.7	84.6
2002	36.6	41.7	78.4
2003	44.5	41.4	85.9
2004	46.0	41.6	87.5
2005	46.7	41.2	87.9
2006	51.7	44.8	96.5
2007	49.1	52.9	102.0
2008	49.5	53.6	103.2
2009	50.3	53.5	103.9
2010	52.2	53.7	106.4
2011	54.0	54.7	109.1
2012	58.8	53.9	113.1
2013	54.5	53.8	108.6
2014	72.4	54.0	126.6
2015	85.9	54.0	140.1
2016	99.4	42.3	142.2
2017	101.8	8.7	111.2
2018	100.8	8.2	109.0

Source Nahum et al. (2020)

one hand, a huge speculation and an increase of land prices, making many farmers sell their lands in order to buy new properties somewhere else or to try other sorts of business, creating a new front of expansion for oil palm farming, including the reorganisation of the regional productive base. This search for lands caused the monoculture of palm to turn into the prevailing landscape, both through large crops with adult palm trees, which reminds earlier capitalist expansions in this region, and through planting and replanting, as in Fig. 12.5, and show how much expanding and renewing power the monoculture of palm still has in the region.

However, this dispute for lands took place under a very fragile legal basis. Despite of the state attempts to identify the ownership of lands and to regularise them, the capturing of lands for palm farming thrived, often making use of dubious illegitimate titling documents, a process known in Amazon as "land fraud", in which one attempts to make a legit title for lands from a false one, since the land ownership chain in Brazil and, particularly in the Amazon is pointed out by Oliveira (2005).

Fig. 12.4 Replanting of palm tree in the municipality of Tailândia, Pará

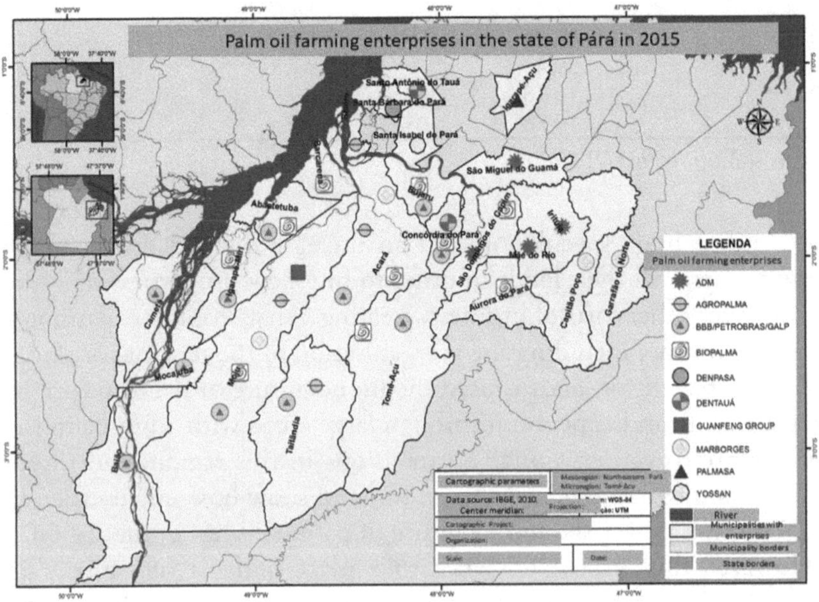

Fig. 12.5 Oil palm farming companies in the municipalities of Pará in 2015 (Adapted from Nahum and Santos 2015)

The second front of conflict typical from the dispossession is for the territory. This dispute is not constrained to the ownership of land, but also comprises the comprehension of what a territory is, understood from the typology of territories (Fernandes 2009), it is where the most profound social conflicts take place, and as such conflicts:

> Capitalism stablishes itself from the consolidation of the capitalist territory. To state that capitalist social relations generate non capitalist social relations is also state that capitalist territories produce non-capitalist territories. This production occurs in an inequal and conflicting way generating permanent land conflicts. Territorial disputes are not limited to the economic dimension. Because the territory comprises a multidimensional totality, the territorial disputes occur in all dimensions; thus, also occurring on the political, theorical and ideological levels, which allows understanding the material and immaterial territories. The politics of dominance and resistance make use of the concepts of territory to delimit both the disputed geographical spaces, and the pleaded ones. (Fernandes 2009)

That is the aspect which causes the original people to retake their lands. People who had their ancestors expelled from their lands for the establishment of large farms in a process of expulsion. The dispute over territories comprises indigenous, riverside and quilombo people, as well as small farmers who currently demand their collective land rights based on the ancestry, memory, work and on farming methods. Palm also disputes territory with other branches of agrobusiness, especially medium and large farmers who work with cattle ranching. However, differently from these ones, the aforementioned original people resist for understanding that their lives and projects are neither monitorable nor reduced to compensations, raising the authenticity of the territory as a common (Dardot and Laval 2017), built from territorial learning (Oliveira Neto et al. 2021).

The third dimension is the control over the agrarian system (Oliveira Neto 2017), which manages not only to make its logical seem as the truth, but also the conversion of several crops and structures to oil palm, both through the rise of new businesses, the decrease of other farming areas, the work monopolisation, the rise of indirect activities related to palm and the rising of structures controlled by big enterprises. It is in this scope that the municipalities impacted by oil palm farming nowadays have their agrarian systems directly related to this business, and it is what defines most of the work dynamics, the resources available and the circulation of capital.

The fourth dimension is the very contradictions regarding work and the many forms of appropriation of it, both salaried, cooperative, hired or in partnership. The change in the work relations in oil palm farming and the changes in the capitalist and non-capitalist relations of production aim to increase the added value, while aims to soften social tensions with local individuals by re-editing the myth of progress, trying to use salary and income as social mobilisers to serve the interest of enterprises. The fifth dimension is the environmental and the proportional impact of palm as a large project in the Amazon. The landscape homogenisation caused by the removal of biodiversity and the impacts on fauna and flora, the contamination of streams, rivers and the soil and the use of agrochemicals are one of the most reported conflicts by local individuals, impacting not just the habitants, but also all the collectivity, causing entire communities to leave their territories and activities.

The last dimension of the dispossession is the territorial subordination, expressed through: (a) the strengthening of the North–South dominance relation which occurs through the imposition to southern countries (via market) of a range of farming products necessary to sustain the development model of northern countries; (b) the way this production must be conducted regarding agrobusiness; in this business, the merging of farming, industrial, financial, commercial, technological and political systems build a territorial system which allows a verticalised production and increase of profit; (c) the valorisation of farming business which follows this logic. Such crops comprise, preferably, grains or vegetal oils with high flexibility for usage, and which can be dealt in the market by several branches, thus causing disputes for who shall afford the best prices; and (d) the fact that this agricultural model survives from big exchanges of public money, both from financing devices, interest rates and tax devices.

Due to such strategies, they give up autonomy over their territories, which then get controlled under the models and relation of production, based on the companies' interests. Nevertheless, lacking any sort of device which grants protection to farmers against the constant fluctuation of market prices. In this relation, the peasantry gives up autonomy for an income promised by the enterprise and this one becomes a productive medium between the financial capital and the territory, which actually becomes subordinate not to the enterprise, but to the speculative capital.

Lessons Learned

The debate about the large-scale farming of oil palm in Amazon is not recent. It started in the 1940s and had ups and downs. However, the moment in which oil palm farming thrived faster in Amazon was between 2004 and 2013. In that period there happened an alliance among companies and the state in an attempt to re-edit the narrative of oil palm as a synonym of development, disregarding local stakeholders and territorial dynamics. Having public capital and private decisioning, the oil palms has expanded throughout the region, monopolising lands through expanding large pieces of land, assimilating small farmers or territorialising itself in peasant territories, both from dwelling processes and land reforms.

If the enterprise's estimates are accurate, the process observed from 2004 to 2013 tends to intensify, since in that period a set of infrastructures have been built which allowed oil palm farming companies to remarkably increase their action power. Estimates show that the farming areas might reach 316,000 hectares by 2019, what surpasses by far the 54,475 hectares identified in 2013 on IBGE database. It is important to highlight that the impact of oil palm farming on the regional productive structure differed among municipalities and has changed along the time, comprising a diverse set of actions and different levels of intervention power. There are cases in which oil palm farming ended in some municipalities. In other ones, it has lost space to other branches of agrobusiness, such as cattle ranching, or it has lost space to the peasantry in cases where their lands were set to land reforms. In other municipalities, it managed to progress but neither establishing itself as a hegemonic business nor monopolising the territory. However, in most of the cases from the last decade, the oil palm farming expanded and reconfigured territories, relations and productive structures.

The diversity of processes regarding these dynamics indicates that its critical aspect is not the oil palm itself, but the force with which agrobusiness imposes itself on the territories and having a critical support of the state. The expansion of oil palm farming and the cases of success are directly related to the amount of capital invested in this process. In places where oil palm agrobusiness imposed itself more fiercely, mobilising more capital and monopolising lands and workforce, oil palm prevailed over other cultures. However, in planes where there was less capital invested, it also progressed less. Agrobusiness based on a globalised production of flex crop commodities is expanding, making use of the narrative of combat to climate and energy crisis, both on world scale, and which could be mitigated by the agrofuels. The production model is the agrobusiness and the trade is centred in the stock market, which indicates that grains and oily plants are assimilated by the financial capitalism, which thrives faster in the short and long term.

The verticalised production process of agrobusiness, the subordinated assimilation process of the peasantry to the palm oil industries and the level of speculation under which those cultures are reinforce the thesis that the expansion of agricultural commodities in southern countries is marked by subordination of the territory to the financial market and by the increase of land grabbing that takes over southern lands to reinforce the development model of the northern countries.

References

Bernardes, J.A. 2011. Políticas Públicas, Selo Combustível Social e Território Usado. In *Novas Fronteiras do Biodiesel na Amazônia: Limites e Desafios da Incorporação da Pequena Produção Agrícola*, ed. J.A. Bernardes and L.A. Aracri, 65–76. Rio de Janeiro: Arquimedes.

Bernardes, J.A., and L.A. Aracri (eds.). 2011. *Novas Fronteiras do Biodiesel na Amazônia: Limites e Desafios da Incorporação da Pequena Produção Agrícola*. Rio de Janeiro: Arquimedes.

Bouhdiba, A. 1964. Review of Jacques Berque. Dépossession du Monde. *Cahiers Internationaux De Sociologie* 37: 178–179.

Brandão, C. 2010. Acumulação Primitiva Permamnente e Reiteração do Capitalismo Dependente no Brasil. In *Capitalismo Globalizado e Recursos Territoriais*, 9–38. Rio de Janeiro: Lamparina.

Dardot, P., and C. Laval. 2017. *Comum: Ensaio sobre a Revolução no Século XXI*. São Paulo: Boitempo.

Fabrini, J.E. 2010. O Campesinato Frente à Expansão do Agronegócio e do Agrocombustível. In *Geografia Agrária, Território e Desenvolvimento*, ed. M.A. Saquet and R.A. Santos, 55–88. São Paulo: Expressão Popular.

Fernandes, B.M. 2009. Sobre a Tipologia de Territórios. In *Territórios e Territorialidades: Teorias, Processos e Conflitos*, ed. M.A. Saquet and E.S. Sposito, 197–215. São Paulo: Expressão Popular.

Houtart, F. 2011. *El Escándalo de los Agrocombustibles para el Sur*. Quito: Ediciones la Tierra.

Laschefski, K., and A. Zhouri. 2010. Conflitoa Ambientais Norte-Sul: Agrocombustíveis para Quem? In *Capitalismo Globalizado e Recursos Territoriais*, 257–310. Rio de Janeiro: Lamparina.

Nahum, J.S., and C.B. Santos. 2015. O Boom do Dendê na Microrregião de Tomé-Açu, na Amazônia Paraense. *Confins - Revista Franco-brasileira de Geografia* 25: 1.

Nahum, J.S., L.S. Santos, and C.B. Santos. 2020. Formação da Dendeicultura na Amazônia Paraense, 19, 1–14.

Oliveira Neto, A., and B.M. Fernandes. 2017. A Produção de Óleo de Palma a Partir da Subordinação de Territórios Camponeses em Assentamentos de Reforma Agrária na Amazônia. In *A Geografia do Pará em Múltiplas Perspectivas: Políticas Públicas, Gestão e Desenvolvimento Territorial*, ed. A. C. Leal, C. A. Bordalo, and J.O. Nunes, 99–124. Tupã: ANAP.

Oliveira Neto, A. 2017. Territórios Subordinados: Análise da Política de Desenvolvimento Territorial a Partir da Produção de Óleo de Palma pela Agropalma em Assentamentos de Reforma Agrária no Pará. PhD thesis, Presidente Prudente, Unesp.

Oliveira Neto, A. 2021. Questão Agrária e Conflitualidade em Territórios Subordinados: O que Revelam os Usos e Abusos do Território? In *Ofensivas e Resistências: Ação Coletiva, Movimentos Socioespaciais e Movimentos Socioterritoriais*, ed. J. Sobreiro Filho, A. Oliveira Neto, S.H. Cruz, and C.Q. Paula, 41–52. Belém: UFPA.

Oliveira Neto, A., J. Sobreiro Filho, C.Q. Paula, and S.H. Cruz. 2021. *Aprendizagem Territorial, Educação e Resistências no Campo e na Cidade*. Belém: UFPA.

Oliveira, A.U. 2005. Cuiabá-Santarém: Geopolítica, Grilagem, Violência e Mundialização. In *Amazônia Revelada: Os Descaminhos ao Longo da BR-163*, 67–183. Brasília: CNPq.

Oliveira, A.U. 2011. Os Agrocombustíveis e os Dilemas da Agricultura Camponesa e Familiar. In *Novas Fronteiras do Biodiesel na Amazônia: Limites e Desafios da Incorporação da Pequena Produção Agrícola*, ed. J.A. Bernardes and L.A. Aracri, 47–64. Rio de Janeiro: Arquimedes.

Puttilli, M., and N. Tecco. 2012. Biocarburanti e Territorio: Il Biodiesel tra Industria e Sviluppo Rurale: Case Empirici e Modelli Interpretativi. In *Geografia da e para a Cooperação ao Desenvolvimento Territorial: Experiências Brasileiras e Italianas*, ed. M.A. Saquet, E. Dansero, and L.Z. Candiotto, 251–266. São Paulo: Outras Expressões.

Vega, B.R. 2014. *El Dominio del Hambre: Crisis de Hegemonia y Alimentos*. México: Universidad Autónoma Chapingo/Universidad Autónoma de Zacatecas/Juan Pablos Editor.

13

Three Pillars of the Global Governance of Coffee Production

Luiza Dulci

Introduction

Coffee. One of the most consumed beverages in the world, it has become part of the food culture of East and West, North and South. Currently, coffee is sold worldwide in the form of green beans, ground and roasted, instant coffee and in capsules, as well as various caffeinated, ready-to-drink beverages. Its production network comprehends a big diversity of actors, from small farmers to major exporters, retailers and financial agents, firms, research institutes, social organisations, and other public and private institutions. Although spatially dispersed, all those actors are functionally integrated into the network, which is based on a division of competences between the regions that cultivate it and those that transform and commercialise it. To better understand the main features and dynamics of what is happening in the coffee market today, this study

L. Dulci (✉)
Perseu Abramo Foundation, São Paulo, Brazil
e-mail: luiza.dulci@gmail.com

A. A. R. Ioris and B. Mançano Fernandes (eds.), *Agriculture, Environment and Development*, https://doi.org/10.1007/978-3-031-10264-6_13

identified three key pillars of its global network governance: (i) private regulation of the coffee market; (ii) consolidation-internationalisation of firms; and (iii) financialisation. The analysis of each pillar and their interrelations aims to answer the following question: considering that Brazil is the main producer and the second largest coffee consumer, why has it been unable to capture a bigger portion of the rent generated throughout the network?

Our investigation departs from the main coffee producer region in Brazil: the southern section of the Minas Gerais state, which responds for about 30% of Brazilian coffee and 10% of the world's coffee. The region concentrates small, medium, and large coffee growers, the operation of warehouses, processing plants, and roasting industries, in addition to brokers and exporters, both national and transnational. Because of that, it is a privileged territory for observing dynamics, limits, and why not possibilities of reversing existing inequalities along the network. Inspired by the approaches of Knowles (2014) and Maluf (2021), the study presented in this chapter takes coffee as the central element of relationships and, consequently, analysis. Coffee is thus the central character of this story and its key investigative tool. By following its journey from farm to cup, it is possible to observe the coexistence of a process of valorisation and 'revival' of coffee in the countries of the North while there is an increasing devaluation of coffee growers in the South. That leads us to the so-called *Coffee Paradox* (Daviron and Ponte 2005). Despite being responsible for more than 70% of the world's coffee production, smallholders retain only 5–10% of the financial profit (FJP 2018). The share retained by producing countries, which usually already includes initial processes of separation and transformation of grains, is still below 15%. All the rest is captured by traders, roasters, and retailers. Therefore, strategies to expand the commercialisation of green coffee through export do not reach—and even less control—crucial issues and arenas where most of the rent is being created and captured (Fitter and Kaplinsky 2001).

A fruitful approach to frame and explain the features and dynamics of today's extended network of actors and landscapes, as well as the functioning of the Coffee Paradox, is the Global Production Network (Gereffi et al. 2005; Henderson et al. 2002). It is known that many theoretical perspectives and concepts have attempted to explain the complex set of interconnections related to development studies. They go back to the world-systems theory or to dependency approaches and include framings such as global value chains; *filière*; global commodity chains; and actor-network theory. Benefiting from all those conceptions, there is the Global Production Network (GPN). According to this methodology, the network's characteristics and governance are collectively produced by its participants, nevertheless one must note power and scale inequalities between them. The GPN comprehends as "an effort of grasping the myriad complexities of economic globalisation [...] capable of delivering a better analytic purchase on the changing international distribution of production and consumption - and the viability of different development strategies to which they relate" (Henderson et al. 2002: 5). Situated in the field studies of Sociology and Political Economy, the present piece will analyse the GPN of coffee to understand the relationships between the actors of the network on the production and distribution of coffee. Hence, we pay special attention to governance structures that enable certain actors to capture wealth and rents at the expense of others. According to Gereffi et al. (1994: 97) governance structures are defined as "authority and power relationships that determine how financial, material, and human resources are allocated and flow within a chain".

The examination of the ongoing processes around coffee market utilised a varied group of oral and textual sources, along with quantitative and qualitative methods, including fieldwork conducted in the region of South of Minas Gerais and 44 interviews with farmers, representatives of public and private institutions, traders, and coffee shop owners. These primary sources allowed a deep dialogue with the specialised bibliography on the subject and the possibility to raise questions and innovative perspectives on Brazil's condition in the global network; the relational approach between commodity and specialty coffees; and the construction of productive and financial arrangements between national and private capital. Therefore, the study contributes to deepening and refining the understanding of the main features and dynamics of the coffee market and contemporary global capitalism.

The chapter is organised into three sections, besides this Introduction and Final Remarks. Each one of them addresses one specific pillar of the governance structure. The second section analyses change among the regulation realm, especially regarding the transition between public and private regulation in the coffee network. Then we move on to the analysis of the consolidation strategy, controlled by big coffee roasters and retailers, identified as the main leaders of the network's dynamics since the 1990s. The fourth section looks at what has been called financialisation of agriculture and aims to understand how finance shapes agriculture and coffee's activities in the present time and contributes to the growing distance and inequalities between actors and territories.

Transitioning from Public to Private Regulation

Regulation is a key feature of political economy analysis, for it considers the production and distribution of goods, power, and rents between and within nations and classes (Lipietz 1984). It encompasses the set of rules, procedures, and instances that define patterns of reproduction and distribution of the product associated with coffee activity in the world. The coffee regulation system dates to the beginning of the twentieth century and until the 1980s was strongly influenced by producing countries—with a leading role of Brazil. The very first attempt to regulate the coffee market started at a sub-national level in Brazil, comprising São Paulo, Rio de Janeiro, and Minas Gerais states in 1906, known as *Convênio de Taubaté* (the Taubaté Agreement). Indeed, as the country produced a high proportion of worldwide's coffee,[1] it could influence its overall price. "By maintaining international prices single-handedly, Brazil was in fact providing all coffee growers in the world with protection" (Daviron and Ponte 2005: 84). But this policy had dramatic consequences, worsened by the 1929 crisis and its oversupply result (Furtado 1969).

[1] Around 1830s Brazil produced about 25% of the world's coffee; three decades later, it produced half of the world's total, and by the end of the nineteenth century, it was responsible for about 75% of coffee production worldwide (Daviron and Ponte 2005).

The 1930s inaugurated a new phase, which lasted until 1962, distinguished by fragmentation in the world coffee market. Anglophone and Francophone African countries were stimulated to grow coffee, augmenting even more supply worldwide. To establish common rules and goals, since the 1940s a few international regulating experiences took place in the American continent, such as the Inter-American Coffee Agreement signed in 1940, and the Latin American Agreement in 1958, following the Mexican Agreement signed in the previous year (Daviron and Ponte 2005). Finally, in 1962, it was signed the first International Coffee Agreement (ICA), a commitment between producing and consuming countries, which agreed to establish a new regulatory policy based on price band and export quotas of coffee beans (Lucier 1988; Talbot 2004). Since 1967, seven ICA have been signed, the last one still in force. Despite that, 1989 is often referred to as the ending year of the ICA, for since then all economic rules related to quotas and prices have been excluded from the Agreements. From then on, the international coffee market has been regulated by so-called 'market forces'.

The ICA has also encouraged many coffee producing countries to build their own national regulatory systems. Working together, these national and international sets of rules and processes have resulted in a more stable coffee market. "On average, from the end of the 1950s to the end of the 1980s, the international price was more or less double what it would have been on the free market" (Daviron and Ponte 2005: 120). Despite the success in maintaining prices and revenues for producing countries, the ICA may not have been able to address a critical issue of the GPN of coffee: its structural oversupply (Talbot 2004). Other problems had to do with the many ways producing and consuming countries were found to circumvent the rules of the Agreements. Along with the general liberalisation trend of the 1980s economy (Chang 2002; Shafaeddin 1998), these factors were decisive for the end of the political arrangement guiding the coffee market. Thereafter, the power imbalance between producing and consuming countries became unprotected and a much more unequal rent distribution is now in place within the network. Lower coffee prices were accompanied by a higher level of price volatility, financialisation increased, as well as market consolidation. Regarding market prices, the average of the International Coffee Organisation indicators for the period immediately following the Agreement's extinction (1990–1994) is US$ 0.77 per pound of coffee, whereas it had been US$ 1.34 for the five years prior (1984–1988).

Between 1990 and 2003, the average price of a pound of coffee was only US$ 0.62—even considering the price spikes observed in 1994/95 due to frosts and droughts in Brazil and 1997, when a speculative wave caused a rise in prices (Daviron and Ponte 2005: 88). Considering market instability, in the last eight years of operation of the Agreement, the nominal monthly variation in prices was 14.8%. Between 1990 and 1997, the variation had increased to 37%, reaching 43% between 1998 and 2000. Contrast is also significant considering price series performance during the AIC period. On average, from the late 1950s to the late 1980s, coffee's international price was twice the market price (Daviron and Ponte 2005). Income's share captured by farmers reduced at about one third, from an estimated 20% in the period 1971–1988 to 13% in the years 1989–1995 (Talbot 1997 *apud* Grabs and Ponte 2019).

Across the world, privatisation of social protection was extended to public policies dedicated to agriculture. In the 1990s, while opening its economy to the global market, Brazil extinguished its main public agency dedicated to supporting and regulating its coffee market and production, the *Instituto Brasileiro do Café* (IBC). At the same time, it also extinguished its national extension services agency (*Empresa Brasileira de Assistência Técnica e Extensão Rural*—Embrater), leaving farmers with the single option to buy those services from private sources. The process of dismantling public policies and weakening public regulation on national and international spheres did not lead to a blank space sort of regulation. On the contrary, what has been happening over the last four decades may be better analysed as a continuous transference of state functions and powers to private actors. From then on, firms and civil society organisations have assumed a much more active role in establishing and monitoring agricultural norms, procedures, and rationales (Breger-Bush 2012; Clapp 2017; Grabs and Ponte 2019; Raynolds 2014). An important feature of this new dynamic is the advent of third-party certifications, which involve voluntary agreements, code of conducts, and social and environmental certifications, among others. Those "non-state market-driven" novelties "dictate business standards above legal requirements, establish mechanisms to foster compliance, and often award labels to differentiate and promote participating enterprises and products" (Raynolds 2014: 500).

In the coffee sector, the main third-party certificated in place are Organics, Fair Trade, Rainforest Alliance, Utz, and IBD, all of them present in Brazil. Some firms have their own code of conduct, such as Starbucks (Coffee and Farmer Equity—C.A.F.E) and Nestlé (Nespresso AAA)—those last two programmes function as points systems, ranking producers according to their alignment to the company's requirements. These strategies are embedded and support what the literature calls 'the quality turn', which involves moving the focus from price (Coutinho et al. 2002; Saes and Farina 1999) to quality (Murdoch et al. 2000; Niederle 2013; Wilkinson 2016). They are usually identified with notions of embeddedness, trust, and place, and may express tensions and conflicts both at local and global levels (Goodman 2003). In 2016, worldwide coffee production aligned with sustainability standards was around three million tons, amounting US$ 7.2 billion. Approximately 70% of that comes from Latin America, especially Brazil, Colombia, and Peru; which are followed by Vietnam and Indonesia, in Asia; and Ethiopia, Tanzania, and Uganda in Africa (Voora et al. 2019).

Despite practical and ideological differences between certifiers, there is a growing convergence among them. This approximation has led to an increasing standardisation of processes and products (Daviron and Vagneron 2011). As it allows substitution possibilities between suppliers, it also re-edits the principle that gave rise to commodities at the end of the nineteenth century. In that sense, it contrasts with the original purposes that led to the formation of organic and fair-trade movements in the 1960s, which sought to highlight particular processes and producers, in the opposite direction of the anonymity resulting from commoditisation. Given this scenario, Controlled Denomination of Origin and Geographical Indications are alternatives that preserve distinctions of place and, to some extent, local producers, for they link a particular product and/or production process to a specific location. Overall, third-party certifications and geographical indications are part of a set of contradictory dynamics in food markets, as they move between the so-called 'commodity markets' and 'unique markets'. Thus, instead of opposing global and sectoral dynamics, those mechanisms of differentiation operate within a business-to-business (B2B) configuration, while, at the same time, may open possibilities for a change or, at least, for mitigating present inequalities between actors and territories.

The Consolidation Strategy

The coffee industry, especially roasters, traders, and other firms that process coffee beans, are the main beneficiaries of the Coffee Paradox. Roasters in particular act as leading firms in the global coffee production network, as they dictate its dynamics and control the distribution of rents generated by it. A key aspect of the Second Food Regime (Friedmann and McMichael 1989) is the intense process of a firm's consolidation, supported by mergers and acquisitions of small and medium-sized traders, roasters, and retailers by large, generally multinational, companies. Such a strategy is not unique to the agri-food sector, nor is it a historical novelty. Concentration is part of a general movement of the capitalist economy, already noted and analysed in the literature since the end of the nineteenth century (Luxemburg 1985; Marx 2011 [1867]). It is interesting to note that oligopolistic markets allow the creation of specific rents (from oligopolies). Thus, one of the main consequences of the consolidation strategy is inequalities increase among and between network participants (Davis et al. 2018). The overall international division of labour in the GPN of coffee may be better visualised through the following data (Table 13.1).

World coffee production in the 2017/18 harvest totalised around 158 million tons. Regarding domestic consumption, the Philippines, Venezuela, and Thailand also stand out, all three with coffee consumption exceeding 1.3 million tons. In terms of exports, other relevant countries, (more than one million bags), are Guatemala, Nicaragua, Papua New Guinea, and Costa Rica. It can be seen, therefore, that coffee is mostly produced in countries of the South. Among them, Brazil stands out as the largest producer, exporter, and consumer; followed by Vietnam and Colombia in terms of volume of production (Table 13.2).

Table 13.1 Coffee production, domestic consumption, and exports of the 10 biggest producers on the harvest of 2017/18

	Production		Domestic consumption		Exports	
Country	mil ton	%	mil ton	%	mil ton	%
Brazil	51.000	32.16	21.997	43.62	33.467	27.98
Vietnam	29.500	18.60	2.500	4.96	23.540	19.68
Colombia	14.000	8.83	1.800	3.57	13.488	11.28
Indonesia	10.902	6.88	4.700	9.32	6.891	5.76
Honduras	8.349	4.82	375	0.74	7.290	6.09
Ethiopia	7.650	4.82	3.750	7.44	3.497	2.92
India	5.840	3.68	2.350	4.66	6.371	5.33
Uganda	5.100	3.22	245	0.49	4.605	3.85
Peru	4.280	2.70	250	0.50	3.973	3.32
Mexico	4.000	2.52	2.400	4.76	2.853	2.39
Other countries	17.939	11.77	10.064	19.94	13.647	11.40
Total	158.560	100	50.431	100	119.622	100

Source FAO-Stat

Table 13.2 The 10 biggest importers and re-exporters of coffee in 2013

Importers	mil ton	%	Re-exporters	mil ton	%
USA	27.016	23.30	Germany	12.020	33.13
Germany	21.174	18.26	Belgium	4.257	11.73
Italy	8.823	7.61	USA	3.248	8.95
Japan	8.381	7.23	Italy	3.183	8.77
France	6.173	5.32	Netherlands	1.781	4.91
Belgium	5.502	4.75	Spain	1.636	4.51
Spain	5.137	4.43	Poland	1.615	4.45
Russia	4.410	3.80	Switzerland	1.547	4.26
United Kingdom	4.206	3.63	United Kingdom	1.380	3.80
Netherlands	3.407	2.94	France	1.014	2.79
Other countries	21.702	18.73	Other countries	7.784	12.70
Total	115.931	100	Total	36.282	100

Source FAO-Stat

Re-exporting countries are those that import coffee beans and export them after processing and transformation. The most recent data from FAO, from 2013, show that 36,282 thousand tons of coffee were re-exported, of which 30,605 thousand (84.3%) left the borders of European Union countries—with Switzerland, it amounts 32,152 thousand tons—which is the equivalent to 88.6% of the total coffees' re-exports in the world. The last two tables show that despite controlling more than 30% of global coffee production and being the second largest consumer market, Brazil struggles to capture larger portions of the rents generated in the network. This fact shows that participating in international trade is not a sufficient condition for the development of producing countries—their companies and farmers. Thus, "the key issue is not that these producing countries are not trading, but rather that they are not gaining much from trade" (Daviron and Ponte 2005: xviii). In fact, coffee producing countries have exported more over the years, as shown in Graph 13.1.

In the period considered (1990/91–2018/19), Brazilian exports more than doubled, from 17,863 thousand to 37,614 thousand bags, while world exports increased from 74,862 thousand to 129,028 thousand bags.

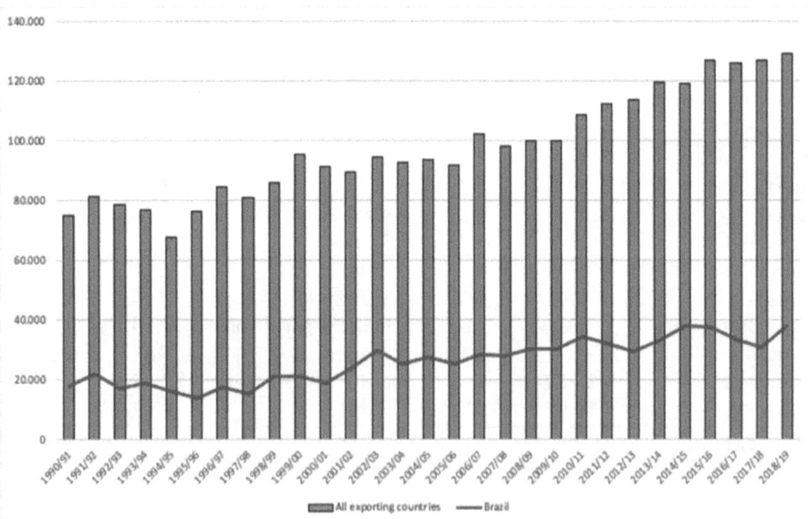

Graph 13.1 Export coffee from Brazil and total exports between the 1990/91 and 2018/19 harvests (thousands of 60 kg bags) (*Source* Data from ICO)

Graph 13.2 shows the price fall in the aftermath of the ICA economic clauses extinction in 1989, as well as the spikes associated with regulatory changes in financial markets in the 2000s and 2010s. Between 2002 and 2005 the average price went from US$ 62.75 to $ 111.52; and between 2010 and 2011, from US$ 166.50 to US$ 284.72. Since then, there has been a price fall. That raises a crucial question about how to solve or soften the inequality issue in the network? Considering the mechanism that governs the Coffee Paradox, the key "is not to make consumers pay, but how to transfer a greater proportion of the final price to producers" (Daviron and Ponte 2005: 263). When we look at Brazil with all those specific and locational advantages (Wilson 2013) and installed capacity in terms of firms, human resources, and knowledge, the struggle has to do with the adding value process value, especially in the stages of transformation and commercialisation. Industry denationalisation and the increasing control of commercialisation by foreign roasters and retailers are a central aspect of interest.

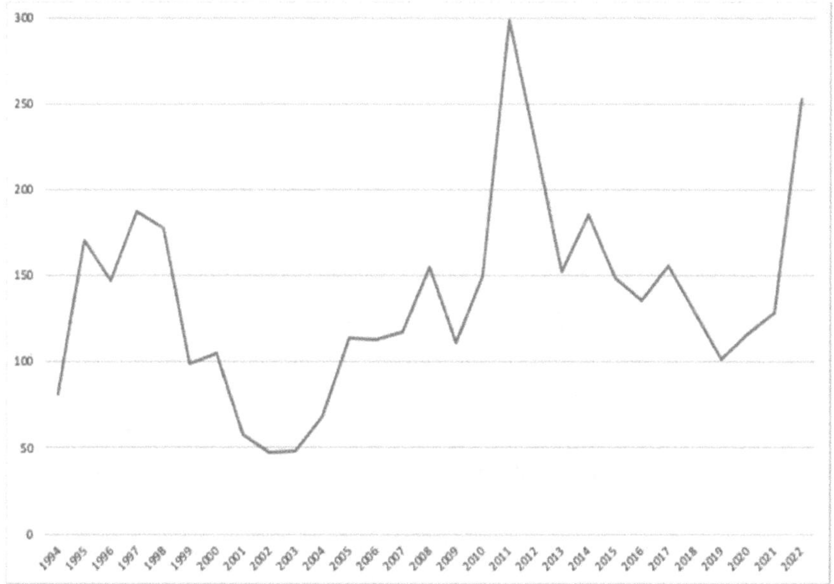

Graph 13.2 Average price of coffee bags sold by Cooxupé in the months of March between 1994 and 2022 (values in US$) (*Source* Data from Cooxupé)

It should be noted, however, that the presence of a national industry does not resolve inequalities between classes—industrialists and farmers—or intraclasses—family farming and agribusiness, for example. Therefore, changes are needed in the governance structure that allows a better rent distribution along the network. In this sense, we envision two possible paths. The first concerns vertical integration, with the increasing control of the transformation and commercialisation stages by small farmers. The second relates to mechanisms of income transfer from the final price back to farmers: "it is difficult for producers of tropical products to capture value from them, unless consumption is brought closer to them (as in agro-turism or ecotourism) unless the sites of service provision are more directly controlled by producer organizations" (Daviron and Ponte 2005: 46). As an example of this second strategy, there are the Juan Valdez coffee shops, present in the USA, controlled by the Colombian FedeCafé (Lucier 1988; Molleda and Roberts 2008). They are a success story, involving public and private sectors and prioritising the organisation of coffee growers. It is one of the possible alternatives for reversing the distances between growth and commercialisation landscapes and, therefore, to provide farmers and other local actors with greater control over material and immaterial assets demanded by consumers.

In recent decades, concentration among traders has largely stemmed from roasters' demands and dynamics. Until then, traders used to operate mostly within national markets. Apart from two of the big agri-food companies (Germany's Volkart Bros., and Switzerland's Nestlé), most traders had family backgrounds and were very much connected to their operating territories. Things began to change with the growth of roasters and the surge in demand for greater green beans volumes and the requirement that traders themselves store them. As a result, only large traders, with bigger logistical capacity could survive. For all these reasons and the end of the ICA, many traders went bankrupt, both small and large ones. In the early 1990s, only five of them (Neumann, Volcafé, ED&F, Econ, and Goldman Sachs) controlled 40% of the world coffee market (Talbot 2004); see Fig. 13.1. Data from 1998 indicated that the two main ones, Neumann and Volcafé, both German, controlled 29% and the six main ones, 50% of the world market. In the early 2000s, with the mergers between Volcafé and ED&F; and Esteve with Cargill, the latter giving rise to Ecom; the three main groups controlled 45% of the global coffee market (Daviron and Ponte 2005). Currently the top five traders are Neumann Kaffee Gruppe; ED&F Man Volcafe; Louis Dreyfuss; ECOM (which operates in Brazil through EISA Empresa Interagrícola S.A.) and Sucafina. Neumann alone holds around 10% of the market. In 2017, it sold 15 million bags of coffee, equivalent to the total coffee production in Colombia (Panhuysen and Pierrot 2018).

Consolidation between roasters happens at even more intense levels. Considering the US market, Talbot (2004) points out that the market share of the four main roasting companies grew from 46 to 69% between 1958 and 1978, while the number of independent roasters dropped from 261 in 1963 to 162 in 1972. Globally, in 1998, the top two groups (Nestlé and Philip Morris) controlled 49%, and the top five controlled 69% of the world coffee market. As it is more capital-intensive, the instant market is even more susceptible to concentration. In the same year, 1998, Nestlé alone controlled 56% of the world market for instant coffee (Daviron and Ponte 2005). In 2016, the market share of two

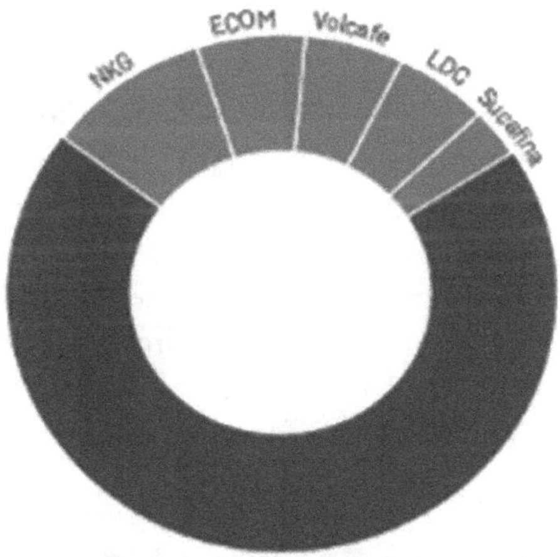

Fig. 13.1 Five main world coffee traders (2018) (*Source* Panhuysen and Pierrot [2018])

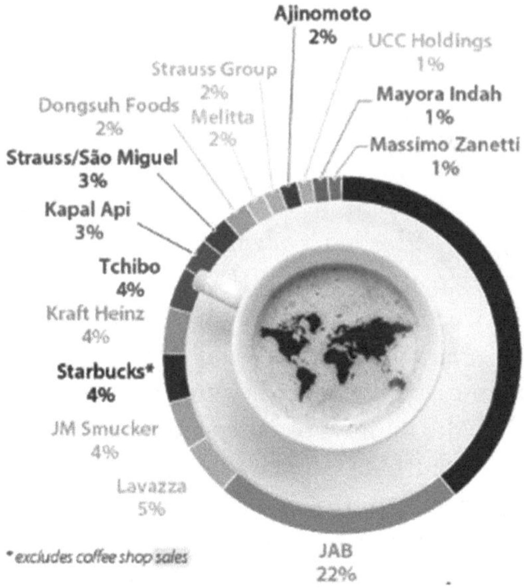

Fig. 13.2 Coffee roasters' market share (2016) (*Source* Euromonitor [2016], *apud* Rabobank [2017])

largest companies (Nestlé and JAB) corresponded to 61% of the market, as shown in Fig. 13.2.

The level of market concentration continues to rise and currently the ten largest roasters account for 35% of the world market. Two of them occupy the isolated leadership, Nestlé (Switzerland) and JAB Coffee (USA), followed by Lavazza (Italy), which occupies a slightly more distant third position. The other big seven are Smuckers (USA); Starbucks (USA); Três Corações/Strauss Group (Brazil, more recently with Israeli capital) (in Fig. 13.2, as Strauss/São Miguel); UCC (Japan); Tchibo (Germany); and Massimo Zanetti (Italy).

Consolidation in the Brazilian Market

The contemporary portrait of the Brazilian coffee industry is one of a highly concentrated market. In 2018, Abic (Brazilian Coffee Industry Association) recorded in its database the existence of 1393 industries in the sector. Of these, 405 were members and accounted for 73.5% of domestic consumption of roasted and ground coffee. In the same year, the production of the 10 largest companies totalled 78.03% of national production (ABIC 2019). Back in 2002, the 10 largest roasters were responsible for 41.69% of production (Silva et al. 2017) (Table 13.3).

Things have changed since then. Sara Lee, national market leader in Brazil in 2004, transformed into JDE and currently belongs to JAB. Cia Cacique was also acquired by JAB in 2017. Santa Clara, Mitsui, and Café 3Corações, which occupied the 2nd, 6th, and 8th positions in 2004, respectively, currently belong to Grupo 3Corações. Another recent change concerns the acquisition of the Café Pacaembu brand by Italian Massimo Zanetti in 2019. It should be noted that the presence of multinationals among the five largest companies present in Brazil went from two (Sara Lee and Melitta) to four (Grupo 3Corações; JDE; Melitta and Mitsui) between 2004 and 2018 (in 2020, Grupo 3Corações acquired Mitsui's coffee operations in Brazil). Table 13.4 registers data regarding market shares of the five main Brazilian companies in terms of volume and values transacted.

Table 13.3 Ten largest member companies of the Brazilian Coffee Industry Association (2004 and 2018)

2004	2018
Sara Lee Cafés do Brasil Ltda	Grupo Três Corações
Santa Clara Ind. e Com. de Alimentos Ltda	Jacobs Douwe Egberts Br Com. De Cafés Ltda
Mellita do Brasil Ind. e Com. Ltda	Inds. Aliments. Marata Ltda
Café Damasco Ltda	Melitta Do Brasil Ind. E Com. Ltda
Cia. Cacique de Café Solúvel	Mitsui Alimentos Ltda
Mitsui Alimentos Ltda	São Braz S/A Ind. E Com. De Alimentos S.A
Café Bom Dia Ltda	Coop. Regional de Cafeicultores em Guaxupe Ltda.—Cooxupe
Café Três Corações S/A	Café Bom Dia Ltda
São Braz s/a Ind. e Com. de Alimentos S.A	Café Pacaembu Ltda
Moka Trading Company Ltda	Café Rancheiro Agro Indl. Ltda

Source Abic (2018)

Table 13.4 Five main roasters in Brazil—market share (2019)

Companies	Volume (%)	Companies	Values (%)
Grupo 3Corações	29	Grupo 3Corações	23
JDE	26	JDE	19
Marata	11	Nestlé	15
Melitta	7	Marata	7
Mitsui	3	Melitta	6
Outras	24	Outras	30

Source Data from Rabobank (2020)

Grupo 3Corações is the national leader in terms of volume traded and value transacted, followed by JDE (currently JAB Holding). Nestlé appears as the third company in terms of values but does not appear in the list of the five largest in relation to the volume of sales. A possible explanation has to do with the fact that it sells instant coffee and capsules, while it is not present in the roasted and ground coffee market. Regarding internationalisation, the study by Silva et al. (2017) identified 30 mergers and acquisitions operations between the 1990s and March 2016. Of these, 17 involved foreign corporations, as we may see in Table 13.5.

Overall, to the average observer, the coffee market can seem highly diverse, with thousands of options from independent coffee shops or chains like Starbucks and similar ones. However, these many options occult the underlying intense oligopoly process going on over the last decades. As we have seen, the agri-food sector is thus highly populated by intermediary agents, positioned between producers and consumers—very active promoters of distancing in agricultural landscapes and the weakening of rural belonging. For Legun and Bell (2016: 105) those are conducers, who "might broadly be considered middle industries (…) and participate in the food economy through *conduction*: transporting, processing, warehousing, advertising, retailing, cooking, serving, and more". They are agents who play a bridging role and fill structural holes between producers and consumers in agri-food networks, and, although maintaining a persona of invisibility, are very much present in the making of food, both materially and ideologically, as well as capture bigger shares of the rent generated across the food network.

Table 13.5 List of mergers and acquisitions in the Brazilian coffee sector (1990 to March 2016)

Buyer	Buyer's origin	Firm	Year
Mitsui & Co. Ltd.	Japan	Mitsui-Yoshioka	1990
Grupo Branco Peres	Brazil	Café do Centro	1994
Café Damasco	Brazil	Café Negresco	1996
Sara Lee	USA	Café do Ponto	1998
Sara Lee	USA	Café Seleto	1998
Café Damasco	Brazil	Café Pacheco	1999
Sara Lee	USA	Ativos de café da Cia. União	2000
Nhá Benta Alimentos	Brazil	Café Tiradentes	2000
Strauss	Israel	Café Três Corações	2001
Segafredo Zaneti	Italy	Café Nacional	2001
Minas Export	Brazil	Café Moka	2003
Café Damasco	Brazil	Café América	2003
Café Damasco	Brazil	Café Palheta	2004
Santa Clara	Brazil	Café Pimpinella	2004
Strauss	Israel	Santa Clara (50%—joint venture)	2006
Café Toko	Brazil	Café Minas Rio	2006
Melitta	Germany	Café Bom Jesus	2007
Lavazza	Italy	Café Grão Nobre	2008
Sara Lee	USA	Café Moka	2008
Strauss/Santa Clara	Israel/Brazil	Café Letícia	2009
Lavazza	Italy	Café Terra Brasil	2009
Sara Lee	USA	Minas Export	2009
Sara Lee	USA	Café Damasco	2010
Grupo 3 Corações	Israel/Brazil	Café Fino Grão	2011
Foods Alimentos	Brazil	Marca Café Seleto	2012
Grupo 3 Corações	Israel/Brazil	Café Itamaraty	2014
Brasil Espresso	Brazil	Café Astro	2014
Café Três Marias	Brazil	Café Solúvel Brasília	2015
Grupo 3 Corações	Israel/Brazil	Marcas de café da Cia. Iguaçu	2016
Brasil Espresso	Brazil	Madame D'Orvilliers	2016
Jacobs Douwe Egberts	Netherlands	Foods Alimentos	2016
AC Café	Brazil	Café do Centro	s.d

Source Silva et al. (2017)

The growth of the specialty market brought some novelties into this dynamic. It did not interrupt the consolidation movement, but strengthened new marketing routes, especially the so-called direct trade of small quantities of coffee—microlots. The sale of microlots has been carried out by both large and small traders, as well as by coffee shops that negotiate directly with producers. As we have seen, by following connections from a small farmer we may reach varied territories and landscapes, some of which are very distant from those commonly associated with coffee beans. An increasingly important arena is one of the financial markets, where fundamental decisions are made, both regarding coffee and agriculture in general. It is also a key promoter of the growing distance between farmers and consumers, as it is analysed in the next section.

Finance and the Expansion of the Agriculture's Frontier

Although the relationship between finance and agriculture is not recent, there are historical novelties to be analysed, especially regarding the mutual support of the three pillars of governance since the end of the economic clauses of the International Coffee Agreement in 1989, it was only from the 1970s onwards that financial dynamics began to overlap certain agricultural organisational and productive processes (Breger-Bush 2012; Balestro and Lourenço 2014; Clapp 2015; Isakson 2013). Since then, weakening regulation and financialisation have been both promoting and driving themselves together. "In 1980, the amount of coffee traded in the futures market was only four times the coffee traded in the physical market. By the early 1990s the ratio had risen to eleven times" (Dijk et al. 1998 *apud* Daviron and Ponte 2005: 90).

One of the most widespread definitions of financialisation in the literature points to it as the "financialization means the increasing role of financial motives, financial markets, financial actors and financial institutions in the operation of the domestic and international economies" (Epstein 2005: 3). His diagnosis therefore involves the identification of an essential distinction between the historical processes that linked agriculture and industry to finance (Hilferding 1985). Financialisation begins when the financial sector elevates its self-financing to its priorities—especially from the FIRE triad: finance, insurance, and real estate (Mazzucato 2021). That is, "speculators may do no harm as bubbles on a steady stream of enterprise. But the position is serious when enterprise becomes the bubble on a whirlpool of speculation" (Keynes 1985: 116).

More recently, Talbot (2004) observes in coffee what Gras and Hernandez (2016) and others point to agricultural markets in general: the entry of new agents. Until the 1980s, futures contracts were mainly used by importers and roasters, mainly for the purpose of securitisation and protection against price instability. While in the mid-1970s almost all coffee traded globally was sold at a fixed price, two decades later about 90% of coffee sales were based on futures market prices. Finance is present in the coffee business in the south of Minas Gerais and Brazil in general in different ways. Public and private credit is one possible entry point. Another way is through transactions linked to the stock exchange. Each of these possibilities gives materiality to what Knowles (2014) calls the 'hyper locality of globalisation', that is, the composition of small 'hyperlocal' sections, distinct but connected and strategically aligned with the varied globalisation's practices, processes, and project (Williams et al. 2013).

Data from the National Ministry of Agriculture shows that credit destined for coffee in Brazil comes from both public and private sources. In 2016, funding for the coffee sector amounted to R$ 4.736 billion, distributed among 72,634 contracts. Minas Gerais was responsible for 67.69% of the resources (R$ 3.206 billion) and 48,822 contracts. Of the total credit for coffee, 19.05% was financed by the Medium Size Rural Producers National Support Programme (Pronamp); 15.58% by the Family Farming Strengthening Programme (Pronaf); 11.29% by the Coffee Economy Defense Fund (Funcafé); while 54.08% had no public institutional identified link (Mapa 2017).

Funcafé's resources are managed by public and private financial institutions. According to Funcafé's 2014 Activity Report, the amount of R$ 3.8 billion invested that year was passed on to creditors through 26 financial agents, 19 of which were banks, and seven were credit unions, which served 11,662 beneficiaries. The five main agents were Banco do Brasil (19.5%)—which historically manages most of the funds—, Banco Cooperativo do Brasil S.A. (Bancoob) (9.1%), Itaú Unibanco S.A. (7.6%), Banco Santander Brasil S.A. (6.5%), and Banco Fibra S.A. (5.5%). Also noteworthy is the 4.1% managed by Banco RaboBank S.A., a global financial institution headquartered in the Netherlands, very much present in the coffee business in the South of Minas Gerais region.[2]

A hyperlocal analysis of coffee finance in the South of Minas Gerais points to the existence of different forms of penetration and operation of financial logic in the region. There are, for example, cases such as Ipanema Coffees company, headquartered in the municipality of Alfenas, which grows and exports specialty coffee, trades most of its production on the New York Stock Exchange, and has foreign and national shareholders, including banks and investment funds from Brazil, Japan, and Germany. Likewise, there are cases such as the Montesanto Tavares Group, which in 2018 signed a US$ 326 million partnership agreement with the Public Sector Pension Investment Board of Canada in order to raise funds to invest in its coffee plantations situated in the states of Minas Gerais and Bahia. Finally, financial logic is also getting closer to small and medium size farmers through its relations with coffee cooperatives. The case of Cooxupé, the world's largest coffee cooperative, with more than 15,000 members and 2500 employees, is emblematic. Cooxupé intermediates the relationship between farmers and banks, as well as it sells inputs and machinery through in the forms of credit and barter. By doing so, it operates as a kind of hub of the New York Stock Exchange among farmers in the south of Minas, as it translates financial language for its cooperative members, shares market information, and assists in carrying out futures market operations.

Finally, financialisation is also present in the land market, as it is well analysed by land grabbing literature (Edelman 2013; Sauer and Borras Jr. 2016; Wilkinson et al. 2012). Detachment between commodity and land prices in Brazil is one evidence of how land has become a financial asset (Fairbairn 2014), capable of generating income from speculation. Since 2011, agricultural commodity prices are falling. However, land

[2] In 2016, Rabobank had US$ 160 million invested in Brazilian coffee, distributed among around 150 customers (Rabobank 2016).

prices did not stop growing. Flexor and Leite (2017) show that between 2010 and 2015 the average price of a Brazilian hectare jumped from R$ 4756.00 to R$ 10,083.00, an increase of 112%. Data from the municipality of Patrocínio, the largest coffee grower territory in Minas Gerais state, shows that a similar decoupling process is underway. Between 2008 and 2016, coffee prices increased from R$ 251.22 to R$ 482.24 (a bag of 60 kg) (it went up 92%) In turn, land prices variation between 2008 and 2017 was 183.7%. For comparison purposes, inflation in the period, measured by the Brazilian Broad National Consumer Price Index (IPCA), increased by 71% between 2008 and 2016.

Conclusions

This study analysed the global coffee production network from the perspectives of Political Economy and Economic Sociology, supported by primary and secondary data, interviews, and fieldwork, as well as contributions from specialised literature on the subject from various disciplines of the social sciences. The research basically aimed to answer the question of 'Why has Brazil and the region of the South of Minas been unable to capture a bigger portion of the rent generated in the global coffee production?' In dialogue with the Coffee Paradox thesis, we demonstrated how capital consolidation and internationalisation are essential aspects of the governance model that favours actors in the North. That is, how the globalising impulse, driven by large corporations of the agri-food system, promotes consolidation in the market, weakens public regulation, and intensifies financialisation—the three governance pillars of the global production network of coffee. Regarding regulation, our aim was to show how different regulation models are associated and give support to different patterns of accumulation. Changes observed at the turn of the present century highlight the prevalence of a private logic, combined with the strengthening of third-party certificates, to the detriment of the previous arrangement, which almost exclusively centred on States' power and action.

It then moved on to understand the consolidation logic, that encompasses the agri-food system, although with a focus on the coffee market. We have seen how roasters, followed by traders and retailers, predominantly based in the North, manage to guide rent and power distribution and influence the organisation of all other actors in the network. In the South, particularly in Brazil, these dynamic produces denationalisation of local and national firms and increases even more inequalities between actors and landscapes. The third and final pillar of governance is financialisation. As we have seen, it affects many different actors and relationships of the South of Minas Gerais and connects them to the global network, promoting both distance and approximations. Through the hyperlocal approach we could identify distinct patterns of financial penetration: Ipanema Coffees, Grupo Montesanto Tavares, and Cooxupé. Although there are marked differences between them, it is noted that the prices of coffee bags derived from the New York Stock Exchange prevail as a hegemonic reference in the region. Certainly, the examination of these configurations and the meanings attributed to what has been called financialisation deserve a more detailed analysis and remain as an agenda for future work.

A valuable approach to devise and promote alternatives to these unequal patterns is a systemic perspective, which, if properly applied, may broaden the understanding of problems with the potential to enhance the position of certain neglected actors. In the context of food systems, that perspective leads us to *food politics*, instead of being limited to *food policies* (Maluf 2021). Therefore, it is necessary for a new governance arrangement committed to deconcentrating of rents and power, embodied with a different scale composition; focused on conducers and intermediaries' reduction; decreasing information and power asymmetries; and committed with increasing farmers control over transformation, commercialisation, and marketing strategies and power.

References

Abic (Associação Brasileira da Indústria do Café). 2018. *Produção por Porte de Empresas Associadas*. Retrieved from https://www.abic.com.br/estatisticas/indicadores-da-industria/indicadores-da-industria-de-cafe-2018-2.

ABIC. (2019, February 4). Associação Brasileira da Indústria de Café. Consumo de café no Brasil aumenta 4,8% e chega a 21 milhões de sacas. Available at http://consorciopesquisacafe.com.br/arquivos/consorcio/consumo/Press_release_consumo_final_vs_04_02_19.pdf. Acessed 10 September 2022.

Balestro, M.V., and L.C. Lourenço. 2014. Notas para uma Análise da Financeirização do Agronegócio: Além da Volatilidade do Preço das Commodities. In *O Mundo Rural no Brasil do Século 21: A Formação de um Novo Padrão Agrário e Agrícola*, ed. A.M. Buainain, E. Alves, and Z. Navarro, 241–266. Brasília: Embrapa.

Breger-Bush, S. 2012. *Derivatives and Development: A Political Economy of Global Finance, Farming, and Poverty*. New York: Palgrave Macmillan.

Chang, H.-J. 2002. *Kicking Away the Ladder: Development Strategy in Historical Perspective*. London: Anthem Press.

Clapp, J. 2015. Distant Agricultural Landscapes. *Sustainability Science* 10: 305–316.

Clapp, J. 2017. The Trade-ification of the Food Sustainability Agenda. *The Journal of Peasant Studies* 44 (2): 335–353.

Coutinho, L., et al. 2002. Cadeia do Café. In *Estudo da Competitividade de Cadeias Integradas no Brasil: Impactos das Zonas de Livre Comércio*. Campinas: NEIT-IE-UNICAMP.

Daviron, B., and S. Ponte. 2005. *The Coffee Paradox*. London and New York: Zed Books.

Daviron, B., and I. Vagneron. 2011. From Commoditization to De-Commoditization... and Back Again: Discussing the Role of Sustainability Standards for Agricultural Products. *Development Policy Review* 29 (1): 91–113.

Davis, D., R. Kaplinsky, and M. Morris. 2018. Rents, Power and Governance in Global Value Chains. *Journal of World-Systems Research* 24 (1): 43–71.

Edelman, M. 2013. Messy Hectares: Questions About the Epistemology of Land Grabbing Data. *Journal of Peasant Studies* 40 (3): 485–501.

Epstein, G., ed. 2005. *Financialization and the World Economy*. Cheltenham: Edward Elgar.

Fairbairn, M. 2014. 'Like Gold with Yield': Evolving Intersections Between Farmland and Finance. *The Journal of Peasant Studies* 41 (5): 777–795.
Fitter, R., and R. Kaplinsky. 2001. Who Gains from Product Rents as Coffee Becomes More Differentiated? A Value Chain Analysis. *IDS Bulletin* 32 (3).
FJP (Fundação João Pinheiro). 2018. *A Produção de Café em Minas Gerais: Desafios para a Industrialização*. Belo Horizonte: Fundação João Pinheiro.
Flexor, G., and S. Leite. 2017. Land Market and Land Grabbing in Brazil During the Commodity Boom of the 2000s. *Contexto Internacional* 39 (2): 393–420.
Friedmann, H., and P. McMichael. 1989. Agriculture and the State System: The Rise and Decline of National Agricultures, 1870 to the Present. *Sociologia Ruralis* 29: 93–117.
Furtado, C. 1969 [1959]. *Formação Econômica do Brasil*. São Paulo: Editora Nacional.
Gereffi, G., J. Humphrey, and T. Sturgeon. 2005. The Governance of Global Value Chains. *Review of International Political Economy* 12 (1).
Gereffi, G., M. Korzeniewicz, and R. Korzeniewicz. 1994. Introduction: Global Commodity Chains. In *Commodity Chains and Global Capitalism*, ed. G. Gereffi and M. Korzeniewicz, 1–14. Westport: Praeger Publishers.
Goodman, D. 2003. The Quality 'Turn' and Alternative Food Practices: Reflections and Agenda. *Journal of Rural Studies* 19 (1): 1–7.
Grabs, J., and S. Ponte. 2019. The Evolution of Power in the Global Coffee Value Chain and Production Network. *Journal of Economic Geography* 19: 803–828.
Gras, C., and V. Hernandez. 2016. *Radiografía del Nuevo Campo Argentino: Del Terrateniente al Empresario Transnacional*. Buenos Aires: Siglo Veintiuno.
Henderson, J., P. Dicken, M. Hess, N. Coe, and H. Yeung. 2002. Global Production Networks and the Analysis of Economic Development. *Review of International Political Economy* 9 (3): 436–464.
Hilferding, R. 1985 [1910]. *O Capital Financeiro*. São Paulo: Nova Cultural.
Isakson, R. 2013. Financialization and the Transformation of Agro-Food Supply Chains: A Political Economy. In *Food Sovereignty: A Critical Dialogue*. International Conference, 14–15 September.
Keynes, J.M. 1985 [1936]. *A Teoria Geral do Emprego, do Juro e da Moeda*. São Paulo: Nova Cultural.
Knowles, C. 2014. *Flip-Flop: A Journey Through Globalisation's Backroads*. London: Pluto Press.
Legun, K., and M. Bell. 2016. The Second Middle: Conducers in the Agrofood Economy. *Journal of Rural Studies* 48: 104–114.

Lipietz, A. 1984. *La Mondialization de la Crise Genérale du Fordisme: 1967–1984*. Paris: CEPREMAP.

Lucier, R. 1988. *The International Political Economy of Coffee: From Juan Valdez to Yank's Diner*. New York: Praeger Publishers.

Luxemburg, R. 1985 [1913]. *Acumulação do Capital*. Rio de Janeiro: Nova Cultural.

Maluf, R. 2021. Decentralized Food Systems and Eating in Localities: A Multi-Scale Approach. *Revista de Economia e Sociologia Rural* 59 (4).

Mapa (Ministério da Agricultura, Pecuária e Abastecimento). 2017. *Sumário Executivo: Café*. Out/2017. Retrieved from http://www.sapc.embrapa.br/arquivos/consorcio/informe_estatistico/SumarioExecutivoCafe__Outubro2017.pdf.

Marx, K. 2011 [1867]. *O Capital*. São Paulo: Boitempo.

Mazzucato, M. 2021. *Mission Economy: A Moonshot Guide to Changing Capitalism*. London: Penguin.

Molleda, J.C., and M. Roberts. 2008. The Value of 'Authenticity' in Global Strategic Communication: The New Juan Valdez Campaign. *International Journal of Strategic Communication* 2 (3): 154–174.

Murdoch, J., T. Marsden, and J. Banks. 2000. Quality, Nature, and Embeddedness: Some Theoretical Considerations in the Context of the Food Sector. *Economic Geography* 76 (2).

Niederle, P. 2013. Indicações Geográficas e Processos de Qualificação nos Mercados Alimentares. In *Indicações Geográficas: Qualidade e Origem nos Mercados Alimentares*, ed. P. Nierdele. Porto Alegre: UFRGS.

Panhuysen, S., and J. Pierrot. 2018. *Coffee Barometer*. Retrieved from https://america-latina.hivos.org/assets/2018/07/Coffee-Barometer-2018.pdf.

Rabobank. (2016, November). *Banking for coffee*. Available at https://research.rabobank.com/far/en/sectors/beverages/Infographic-Banking-for-Coffee.html. Acessed 10 September 2022.

Rabobank. (2017). *Banking for coffee*. Available at https://static1.squarespace.com/static/584f6bbef5e23149e5522201/t/5d7f82abd90bed78ce407cb8/1568637622720/Coffee+infographic+2017+-+Rabobank.pdf. Acessed 10 September 2022.

Rabobank. (2020, January). *The Brazilian coffee market*. RaboResearch.

Raynolds, L. 2014. Fair Trade, Certification, and Labor: Global and Local Tensions in Improving Conditions for Agricultural Workers. *Agriculture and Human Values* 31 (3): 499–511.

Saes, M.S., and E. Farina. 1999. *O Agribusiness do Café no Brasil*. São Paulo: Editora Milkbizz.

Sauer, S., and S. Borras, Jr. 2016. Land Grabbing e Green Grabbing: Uma Leitura da Corrida na Produção Acadêmica. *Campo – Território* 11 (23 July).

Shafaeddin, M. 1998. How Did Developed Countries Industrialize? The History of Trade and Industrial Policy: The Cases of Great Britain and the USA. UNCTAD Discussion Paper, 139.

Silva, E., A. Azevedo, and L.G. Júnior. 2017. Transformações Recentes na Indústria Brasileira de Café. *Gestão e Desenvolvimento Em Revista* 1 (1): 35–47.

Talbot, J. 2004. *Grounds for Agreement: The Political Economy of the Coffee Commodity Chain.* Lanham: Rowman & Littlefield Publishers.

Voora, V., S. Bermudez, and C. Larrea. 2019. Global Market Report: Coffee. *IISD: Sustainable Commodity Marketplace Series.*

Williams, S., H. Bradley, R. Devadason, and M. Erickson. 2013. *Globalization and Work*. Cambridge: Polity Press.

Wilkinson, J. 2016. Os Mercados não Vêm mais do 'Mercado'. In *Construção de Mercados e Agricultura Familiar*, ed. F. Marques, M.A. Conterato, and S. Schneider. Porto Alegre: UFRGS.

Wilkinson, J., B. Reydon, and A. Di Sabbato. 2012. Concentration and Foreign Ownership of Land in Brazil in the Context of Global Land Grabbing. *Canadian Journal of Development Studies* 33 (4): 417–438.

Wilson, J. 2013. *Governing Global Production: Resource Networks in the Asia-Pacific Steel Industry*. Cham: Palgrave Macmillan.

14

(De)institutionalising Agroecology: A Historical-Relational-Interactive Perspective on the Evolution of Brazil's Agri-Environmental State

Dana James, Antonio Ademir Cazella, Evan Bowness, Natal João Magnanti, and Hannah Wittman

Introduction

Scientists, policymakers, and farmers increasingly agree on the need to transform the industrial food system using an agroecological approach (FAO 2018; HLPE 2019; IPES-Food 2016; Nyéléni International Forum for Agroecology 2015; Rosset and Martínez-Torres 2012; Val et al. 2019). While described as both a science rooted in ecological approaches to agriculture and a set of practices for sustainably managing agroecosystems, agroecology is also a social movement due to its linkages with the transnational food sovereignty movement (Wezel et al. 2009), which advocates for food systems that benefit people and

D. James (✉) · H. Wittman
Centre for Sustainable Food Systems, University of British Columbia, Vancouver, BC, Canada
e-mail: dana.james@ubc.ca

A. A. Cazella
Federal University of Santa Catarina (Ufsc), Florianópolis, Brazil

A. A. R. Ioris and B. Mançano Fernandes (eds.), *Agriculture, Environment and Development*, https://doi.org/10.1007/978-3-031-10264-6_14

nature rather than corporations (Desmarais 2007; Nyéléni International Forum for Agroecology 2015; Wittman 2011). A key debate in the agroecology and food sovereignty literature relates to the role of the state in promoting—or institutionalising—agroecology. On one hand, state support is seen as necessary for enforcing rules and minimising negative agri-environmental externalities, incentivising agroecological production, and supporting farmers during the transition. On the other, social movements and scholars articulate concerns over the possible co-optation of transformative approaches to agroecology by political and economic elites, as well as the risks associated with losing movement power (demobilisation) once policy gains are made (Desmarais et al. 2017; Giraldo and McCune 2019; Mestmacher and Braun 2021; Sauer and Mészáros 2017).

This tension is evident in Brazil, where decades of social movement and civil society advocacy led to the institutionalisation of agroecology-related programming and policy, particularly under the 2003–2010 administration of Luiz Inácio Lula da Silva ("Lula"). These perceived successes have led to Brazil being viewed as a "pioneer" or reference point for agri-environmental policy globally (Sabourin et al. 2020a, b). However, subsequent government administrations have begun to systematically dismantle this institutional infrastructure. This chapter asks: How did the institutionalisation of agroecology develop in Brazil and what have been the effects? Which historical processes, paradigms, and key actors have contributed to the institutionalisation and de-institutionalisation of agroecology in the Brazilian context?

E. Bowness
Food and Agriculture Institute, University of the Fraser Valley, Abbotsford, BC, Canada

N. J. Magnanti
Centro Vianei de Educação Popular, Lages, Brazil

Conceptual Framework

To answer these questions, we draw upon Schiavoni's Historical-Relational-Interactive (HRI) framework (Schiavoni 2017) and the literature on the environmental state (Duit 2016; Mol and Buttel 2002). An HRI framework can be used to highlight the *historical* processes which shape and condition current efforts for advancing agroecology; the ways in which these efforts unfold as an ongoing struggle *in relation* to competing agricultural "development" discourses and paradigms; and the ways in which *interactions* between state and non-state actors influence key agroecology outcomes. We combine the HRI perspective with insights from the literature on the environmental state, defined by Duit et al. (2016: 5–6) as "a state that possesses a significant set of institutions and practices dedicated to the management of the environment and societal-environmental interactions". Conceptually, the status of a real-world environmental state is assessed *in relation* to the ideal-type "ecological state", which systematically prioritises environmental sustainability above the state's economic considerations, and to an ideal-type "neoliberal state", which prioritises economic concerns over ecological sustainability (Duit 2016; Eckersley 2004; Meadowcroft 2005; Mol and Buttel 2002).

We build upon the concept of the environmental state by explicitly incorporating agricultural development concerns, herein referred to as the "agri-environmental state". We characterise the agri-environmental state as "a state that possesses a significant set of institutions and practices" to support environmentally and socially sustainable agriculture (based on Duit et al. 2016: 5). While the environmental state literature alludes to the threats posed by agriculture to the environment, explicitly integrating agriculture into the environmental state concept highlights the fundamental role that states play in mediating the relationship between the environment, agriculture, and rural well-being through land governance, development, environmental, and social welfare policy. Duit (2016) further notes that there are four faces, or functions, of the environmental state—Administration, Regulation, Redistribution, and Knowledge Production—which we adapt to the agri-environmental state concept and define as follows:

1. Administration/organisation: The creation of agri-environmental public administration entities.
2. Regulation: The development of agri-environmental regulations, institutions, and policies.
3. Redistribution: The ability to redistribute agri-environmental resources and risks/benefits.
4. Knowledge generation: The production and dissemination of agri-environmental knowledge and norms.

Having cross-nationally compared 28 countries according to these four dimensions, Duit finds that countries range across a spectrum from "weak" to "established" environmental states (Duit 2016); see Fig. 14.1.

Both the concepts of agroecology and neoliberalism are important in shaping the agri-environmental state, and our typology places an ideal-type agribusiness state (linked to neoliberalism) and an ideal-type agroecological state (advocated by agrarian movements) as opposing paradigms that shape the faces of real-world agri-environmental states (Fig. 14.2).

Following economic imperatives, an agribusiness state prioritises economic growth and profit through deregulation and privatisation, enabling the concentration of land, wealth, and power in its pursuit. Following the demands of social movements, the vision for the agroecological state prioritises socioecological rights and responsibilities, with the purpose of scaling out food systems that work for people and planet. Therefore, an ideal-type agroecological state has institutions that

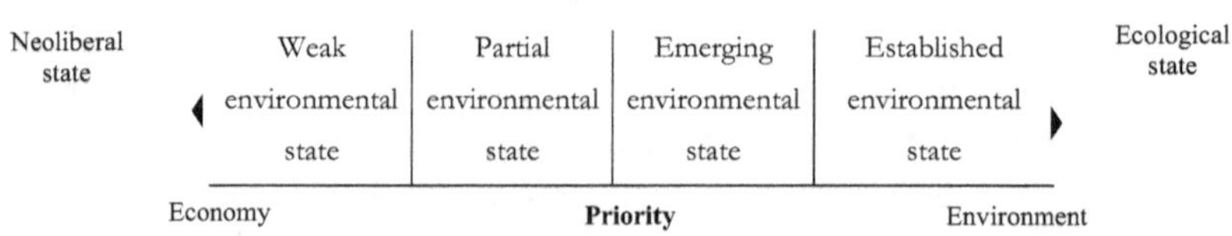

Fig. 14.1 Economy-ecology typology. A typology for understanding the relationship between economy and environment in the environmental state literature. An ideal-type "neoliberal state" would always prioritise economy over environment, and an ideal-type "ecological state" would always prioritise environment over economy. Typology based on Duit (2016)

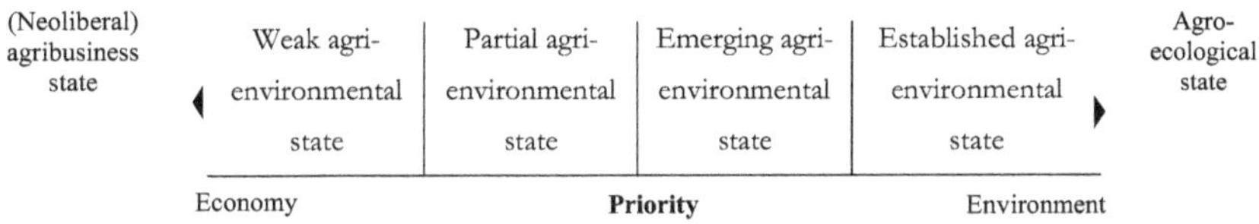

Fig. 14.2 Agribusiness-agroecology typology. A typology for understanding how the economy and environment interact in the environmental state when agriculture (food, fibre, and fuel for people) is taken into account. An ideal-type "agribusiness state" would always prioritise agribusiness interests over agroecology, and an ideal-type "agroecological state" would always prioritise agroecology over agribusiness interests

guarantee the right to food and a healthy environment; meet social movement demands for peoples' participation in decision-making; redistribute resources for the development of sustainable rather than industrial food systems; and foster knowledge exchange across ways of knowing. The spectrum of governance from an ideal-type "agribusiness state" to an ideal-type "agroecological state" informed the creation of our criteria for assessing the four faces of Brazil's real-world agri-environmental state(s) (Table 14.1). By incorporating an HRI perspective with this framework, we assess how historical processes, competing agricultural development paradigms, and state–society interactions have shaped Brazil's agri-environmental state, and the degree to which agroecology has been institutionalised over time.

In what follows, we describe the political-economic factors that influenced the Brazilian state's development of agroecology-oriented entities, policies, redistributive efforts, and knowledge. We do so through a review of the academic literature by agri-food policy scholars; key laws and policy materials (e.g. Law N° 10.831—*A Lei da Agricultura Orgânica*; Decree N° 7.794—*Política Nacional de Agroecologia e Produção Orgânica*, Law N° 11.947—*Programa Nacional de Alimentação Escolar*, etc.); and relevant governmental, civil society, and social movement websites (e.g. *Ministério do Desenvolvimento Social* (dissolved), *Brasil Agroecológico, Articulação Nacional de Agroecologia, Associação Brasileira de Agroecologia, Movimento dos Trabalhadores Rurais Sem Terra*, etc.). We contextualise

Table 14.1 A comparison of the ideal-type "Agribusiness State" and ideal-type "Agroecological State" according to the four faces of the agri-environmental state

Four faces of the agri-environmental state	Characteristics of the agribusiness state	Characteristics of the agroecological state
1. Administration	The minimisation of public administration entities with respect to implementing, monitoring, and enforcing environmental and social welfare policies	The creation and organisation of public administration entities that implement, monitor, and enforce environmental and social welfare policies
2. Regulation	Deregulation and dismantling of laws and policies that protect human rights and the environment in the name of efficiency; creation of a policy environment favourable to investment; externalisation of negative costs	The participatory creation of agri-environmental regulations, institutions, and policies that uphold ecological integrity and the right to food; true-cost accounting
3. Redistribution	Laissez-faire approach to resource distribution, fuelling concentration of resources, land, and power; unequal distribution of agri-environmental harm	Redistribution of resources, land, and power to growers and eaters by eliminating unsustainable subsidies and incentives; equal distribution of agri-environmental harm
4. Knowledge generation	Top-down knowledge transfer of agri-environmental knowledge in line with the modernisation paradigm	The co-creation and dissemination of agri-environmental knowledge with civil society

this review with insights from fieldwork carried out from 2018 to 2019, which included participant observation at agroecology policy events and meetings as well as interviews with 51 agroecological actors, including political advocates, farmers, and NGO leaders. Interviews were coded in NVivo to identify political-economic factors that enable(d) and hinder(ed) the institutionalisation of agroecology.

With these data, we trace the evolution of the Brazilian agri-environmental state across various periods—from landed captaincies during colonisation and imperial empire; to agricultural modernisation during the dictatorship and post-dictatorship period of neoliberalism; to the development of agroecology policy in the Workers' Party (PT) era; to the current regime of agroecological policy rollback. Tracing this terrain helps us to understand how agri-environmental institutions have evolved dynamically in relationship to Brazilian social movements' particular visions for a highly participatory, decentralised, and democratic agroecological state, raising questions for how the agroecology movement might create new relationships and institutions to counter the increasingly unfavourable political situation in Brazil.

An Historical-Relational-Interactive Perspective on the Agri-Environmental State in Brazil

Coffee with Milk: Colonização, Coronelismo, and Clientelismo (1500–1960)

Dating back to the Portuguese invasion of Brazil in 1500, historical and ongoing processes of colonial capitalism led to the consolidation of land, resources, and power in the hands of a select few (Fernandes 2013; Robles 2018). In the early 1500s, 12 men chosen by the Portuguese monarchy were "gifted" with 14 territories[1] known as "captaincies" (Carvalho 2015; Johnson 1987; Wolford 2010a). The "captains" or recipients of these hereditary grants (which would pass down to their sons) were then responsible for dividing up these tracts of land for private interests (Tarlau 2014), forming the basis for Brazil's land inequality. Land became heavily concentrated in the hands of rural elites, known as

[1] Two of the 12 captains were each awarded two captaincies, leading to this discrepancy between the number of captains (12) and captaincies (14) (Johnson 1987).

ruralistas, for "one overriding purpose: the settlement and development of 'vacant' territory" (Johnson 1972: 211, emphasis added)—a common colonial narrative used to justify the dispossession and eradication of Indigenous Peoples, who were subjected to disease, evangelism, enslavement (for example, on rubber baron landholdings in the Amazon), and genocide. Throughout the next three centuries, millions of African people were violently brought to Brazil via the trans-Atlantic slave trade to become enslaved labourers, fuelling what Johnson calls "*latifundia* capitalism" (Johnson 1972: 214).

Following Brazilian independence in 1822, a limited representative system was put in place by the Brazilian empire via the constitution of 1824 (Tranjan 2016). As Tarlau notes, during this time "much of [the previously distributed] private land was supposed to be 'returned' to the Brazilian state…However, through a process of land grabbing and the production of false land deeds—known as *grilagem*—rural elites ensured that they could maintain ownership over their lands" (2014: 4–5; see also Paulino 2014). The empire's abolishment of slavery following the leadership of Afro-Brazilian/quilombo communities such as Palmares (Bowen 2021; Fernandes 2013; O'Dwyer 1995; Reis and dos Santos Gomes 1996) in 1888 led to growing resentment harboured by rural elites, many of whom were involved with the export-oriented production of sugar, coffee, and cotton. This resentment culminated in a *ruralista*-backed coup in 1889 that officially ended the empire period and resulted in the formation of the Brazilian republic.

Throughout the late 1800s and early 1900s, political-economic power was mainly concentrated among the agrarian elites of Minas Gerais (MG) and Sao Paulo (SP). MG was the most populous state at the time, with the economy heavily based on dairy cattle; SP was the second-most populous and most prosperous state, with an economy heavily based on coffee plantations. Together, agrarian elites from MG and SP formed a strong alliance to maintain and solidify the status quo, which allowed them to greatly benefit from the export of dairy products and coffee. The absolute political and economic control exercised by this alliance during this period, which fuelled the expansion of export-oriented agriculture at the expense of workers and natural ecosystems, has led to it being dubbed the 'coffee with milk' period (Meade 2010; Tarlau 2014;

Tranjan 2016). With an increased lack of opportunity in the countryside for peasants and workers as a direct consequence of land and policy capture by the agrarian elite, the state promoted a strategy of industrialisation and urbanisation. By 1960, the majority of the southeast's population—57%—was already urban (Tranjan 2016). The continuous flow of people from rural to urban areas put ever more pressure on the need to standardise agricultural production to improve labour efficiency and facilitate mechanisation.

The intersecting forces of colonialism, *latifundia* capitalism, and industrialisation during this period contributed to patterns of inequality that continue to shape the present-day agri-environmental governance regime in Brazil, as many people found themselves pushed off the land while relatively few elites benefitted from land consolidation and concentration. This period set a deeply entrenched foundation for an agribusiness-oriented, weak agri-environmental state: The expansion of agriculture on supposedly "empty" land served as a tool of dispossession, and the sector rapidly became characterised by concentrated land and wealth ownership, labour and environmental exploitation, and commodity export for the financial gain of landed elites. Public institutions were in a fledgling state, so there was little-to-no regulation, oversight, or enforcement by public entities, accompanied by a high degree of corruption due to the collision of landed elites and public officials, who maintained the political and economic exclusion of non-elite workers (for example, through processes like *grilagem*) (Table 14.2).

"Anos de Chumbo": The Takeover of the Military Dictatorship (1961–1985)

The emergence of João Goulart—a member of the Brazilian Labour Party (PTB)[2] who intended to take on a number of social reforms, including land reform—as president in 1961 signalled some changes that were favourable to the (settler) rural poor. Goulart passed a Statute on Rural Labour in 1963, which granted rural people the right to organise

---

[2] Note that the current PTB has no relationship to the PTB referenced here, which collapsed during the 1964 military coup.

Table 14.2 An assessment of early Brazil's agri-environmental state according to the four faces of the agri-environmental state

Four faces of the agri-environmental state	Characteristics of the evolution of Brazil's agri-environmental state (1500–1961)
1. Administration	Public administration entities were in a nascent stage, and public officials were prone to corruption
2. Regulation	Little-to-no environmental and social welfare policies. Costs of growth are virtually entirely externalised to exploited and/or enslaved people and native ecosystems
3. Redistribution	Land inequality is extreme, with the entire country initially "owned" by only 12 men. Land redistribution and democratising efforts were blocked by economic elites. Resource colonialism is pervasive as commodities are produced for export and peasants are pushed off the land. Racialised and poor workers bear the agri-environmental harms
4. Knowledge generation	State-promoted industrialisation contributed to a rural exodus, beginning to erode land-based knowledge

in unions as part of a confederation, giving birth to the National Confederation of Agricultural Workers (CONTAG) later that same year (Tarlau 2015). Not long after, Goulart was ousted in a US-backed military coup in 1964, marking the start of a repressive and violent military dictatorship that would last 21 years (Puzone and Miguel 2019; Tarlau 2014; Tranjan 2016).

It was during the military dictatorship, at a time where "the state had an almost exclusive role in rural development", that the US-funded Green Revolution (GR) took off globally (Schneider et al. 2010: 225). In Brazil, state-sponsored research and extension services assumed a strong position in supporting agricultural modernisation. For example, the National Rural Credit System provided farmers with a financing scheme for input investment (Garrett and Rausch 2016; OECD-FAO

2015), and in 1972 the state-owned Brazilian Agricultural Research Corporation (Embrapa) was created to further advance and institutionalise the Green Revolution approach (Sencébé et al. 2020). The research of Embrapa in particular was instrumental in opening up frontier areas that were previously thought to be uncultivable, most notably the vast wooded savanna biome known as the *Cerrado*, where Embrapa's top-down plant breeding efforts combined with GR technologies (principally synthetic fertilisers and limes and promotion of mechanisation) have largely transformed the region to be characterised by large-scale, industrial, export-oriented commodity production of soy, maize, and sugarcane monocultures and extensive cattle grazing (Brannstrom et al. 2008; Carvalho et al. 2009; Jepson et al. 2010; Lathuillière et al. 2014).

The dictatorship's pro-modernisation position led to heavily subsidised rural credit tied to the use of GR technologies, such as synthetic pesticides and fertilisers, mechanised infrastructure, and high-yielding varieties of cereal crops (Caporal and Costabeber 2004; da Costa et al. 2017; Franco Lerrer and Servolo de Medeiros 2014; Petersen et al. 2013). Because of their poorer access to resources, technical assistance, and markets—and under the threat of violence—many small-scale farmers continued to be pressured off their lands (Martine 1991). These now-landless farmers migrated to urban areas at a rate that far outpaced job creation, leading to the development of *favelas*. Others invaded Indigenous reserves; for example, scholars have described the military dictatorship's complicity when more than 8000 farming families occupied Indigenous reserves in Brazil's south[3] (Branford and Rocha 2002; Tarlau 2014). Rather than dealing with the politics of unequal land distribution in Brazil's south and elsewhere, the military regime invested in infrastructure and created the National Institute for Colonisation and Agrarian Reform (INCRA) in 1970 to incentivise farmers to settle the "agricultural frontier" along the Amazon border (Sencébé et al. 2020). Scores of farmers then headed north only to be (violently) forced off of these lands by large-scale cattle ranchers (Hochstetler and Keck

[3] Many of these farmers and others would go on to organise occupations of large and unused landed estates, eventually forming the Landless Workers' Movement (the MST).

2007) and, more recently, soy plantation owners (le Polain de Waroux et al. 2017). Tensions between agricultural expansion and environmental protection along the Amazon frontier highlight how throughout most of settler Brazilian history, there has been "direct conflict between the aims of the Ministries of the Environment and Agriculture: with the former seeking to conserve natural resources and the environment, and the latter being responsible for destroying forests and other resources to encourage production" (Schneider et al. 2010: 226).

Yet globally, the environmental and human rights movements were taking off, and Brazil's ecological and sociocultural diversity drew international attention. Brazil's National Foundation for Indigenous Peoples (FUNAI) was founded in 1967 following the publication of the Figueiredo report, which extensively documented and publicised the abuses perpetrated by the Brazilian state against Indigenous Peoples (de la Peña 2005; de Sant'Anna et al. 2018). In addition, scholar-activists with connections to rural leaders in Brazil (such as rubber-tapper organiser and environmentalist Chico Mendes) advocated in the halls of development institutions in the Global North (like the World Bank) for environmental protection and for the rights of traditional and Indigenous Peoples to maintain their way of life (Keck 1995). While still under the military regime, the appointment of the slightly more moderate Ernesto Geisel to the presidency in 1974 created a political opening for social movements to begin building alliances around on-the-ground land conflicts, including in the agricultural frontier region.[4] New satellite imagery depicting Amazonian deforestation, increasing talk of climate change, and Chico Mendes's assassination in 1988 came together to make Brazil "the emblematic case for discussion of third world environmental problems" (Keck 1995: 417). Under global pressure, the government created the *Nossa Natureza* (Our Nature) programme to disincentivise ranching by increasing surveillance of forest burning and removing prior economic incentives to colonise the Amazon frontier (such as subsidised credit), yet monitoring and enforcement remained weak and the "withdrawal of fiscal incentives was no match for the

[4] For example, see Keck (1995) for an overview of the alliance between rubber tappers, CONTAG, and the fledgling Workers' Party (PT).

growing profitability in the 1990s of the ranching sector" (Hochstetler and Keck 2007: 149).

Thus, interactions between local and global actors in combination with changing political dynamics shifted the terrain just enough such that grassroots organising around human rights and the environment was able to spring up, even under military rule. Towards the end of the military regime, mounting public pressure led to government entities taking a more active position with respect to regulation and monitoring (for instance, through the *Nossa Natureza* programme), but there was no real effort to challenge the rural elite or to redistribute the wealth and resources they had accrued; in fact, in 1985, the largest land-holders—representing under 1% of establishments—still owned almost 44% of Brazil's agricultural land, while 50% of establishments owned less than 3% of agricultural land (Robles 2018). Knowledge transfer was based on the GR paradigm and was highly technocratic and top-down. The Brazilian agri-environmental state during this period could be classified as weak (Table 14.3).

From Re-democratisation to the Rise of a Popular Government (1985–2002)

Following the 21-year-long military dictatorship, Brazil's seventh and newest Constitution was created in 1988. At the time, there was "a deep-rooted mistrust of the formal political institutions [including political parties] that had been permeated and corrupted...Direct citizen involvement in political processes was considered the only truly democratic option" (Tranjan 2016: 68). Scholars note that during this process of democratisation, "the progressive regaining of public freedoms, the resumption of the organisational processes of popular movements, and the intensification of the debate on alternatives for the democratic development of society" led to a political climate that was "exceptionally favorable" for advancing alternative agriculture (Petersen et al. 2013: 106). An NGO leader involved in agroecology policy development in Brazil echoed this in an interview:

Table 14.3 An assessment of Brazil's agri-environmental state during the *Anos de Chumbo* according to the four faces of the agri-environmental state

Four faces of the agri-environmental state	Characteristics of Brazil's agri-environmental state: *Anos de Chumbo* (1961–1985)
1. Administration	Public administration bodies related to agriculture, the environment, and human rights (e.g., FUNAI) were emerging, but in the context of heavy state repression
2. Regulation	A pro-modernisation and pro-agribusiness position led to heavily subsidised rural credit in support of GR technologies. Environmental and human rights activism, legislation, and policy were mostly repressed until the end of the regime, and programmes like *Nossa Natureza* were limited due to the enormous profitability of agribusiness
3. Redistribution	Efforts to improve land access did not address the root causes of land inequality via redistribution and instead incentivised the dispossessed to colonise agricultural frontiers
4. Knowledge generation	State-sponsored research and extension services, such as Embrapa, assumed a strong position in supporting agricultural modernisation and top-down knowledge transfer

A very important moment to open the space for the agroecological perspective is the 1988 Constitution. It is important to understand that the entire process of re-democratisation of society and the state has an arc, coinciding with the process of democratic reopening after the military dictatorship…The emergence of agroecology during the dictatorship happened exactly in that period when agricultural modernisation was implemented, the Green Revolution was implemented in Latin America and worldwide. The Green Revolution process would not have been implemented in the way it was if it had not coincided with the authoritarian process. So, the Green Revolution, the big agribusiness as we call

> it today, is the result of a very authoritarian process that imposes technical and economic logics on the rural world. I would say that we have to demarcate the end of the dictatorship and the 1988 Constitution as the moment that created new conditions for agroecology as an idea.

It was against this backdrop—key to interpreting the institutionalisation of agroecological policy—that movements like the Landless Rural Workers Movement (*Movimento dos Trabalhadores Rurais Sem Terra*, MST) and the Movement of People Affected by Dams (*Movimento dos Atingidos por Barragens*, MAB) were born, in 1984 and 1991,[5] respectively. Worth noting is that the most pressing initial concern for groups like the MST was access to land; the question of *how* land should be managed emerged among the MST a decade later, as it consolidated an overall vision for agriculture and rural life that was compatible with its aim to challenge neoliberalism's effects in the countryside (Wittman 2010; Wolford 2010b). Nonetheless, ongoing pressure from movements united around these common issues (most notably, land reform and the right to land), along with the state's desire to quell social unrest in the countryside, contributed to the Brazilian government creating policies to address some of the needs of small-scale and family farmers. One such example is the National Programme for the Strengthening of Family Farming (PRONAF), a rural credit policy created in 1995 under the Cardoso administration to help family farmers access resources and markets through low-interest, state-sponsored loans (Grisa and Schneider 2015; Sencébé et al. 2020). Yet only a small fraction of family farmers—typically those who were already in a position to access loans for capital investment—were able to access PRONAF, whose ultimate purpose was aligned with the state's modernisation goals of economically and technologically advancing the family farming sector.

Even though the family farming sector was beginning to receive political attention, funding for agribusiness-oriented public entities, research institutes, policies, and programmes greatly outpaced funding for family farming-oriented equivalents. This was partly due to the importance of the industrial agriculture sector for economic growth, and partly due

[5] The MAB has been active since the 1970s but was institutionalised in 1991.

to the continued blurred lines between the agrarian elite and government: In 1995, *ruralistas* (also known as the Rural Bloc) still occupied 23% of seats in the Chamber of Deputies (Brazil's lower Congressional house), and their opposition to land redistribution resulted in virtually no changes to land inequality (Robles 2018). In response to this stagnation, the MST dramatically escalated land occupations, and a turning point occurred in 1996 when Brazil's military police infamously murdered 19 MST members who had occupied a landed estate at Eldorado do Carajás (Fernandes 2016; Robles 2018). Growing international attention and scrutiny as a result of the violence led to an increase in agrarian reform settlements and the 1999 creation of a public administration entity—The Ministry of Agrarian Development (MDA)—to support family farming and land redistribution (Fernandes 2016; Sauer and Mészáros 2017). In contrast, the Ministry of Agriculture, Livestock, and Supply (MAPA) has existed (under various names) since 1860 to support export-oriented agriculture and agribusiness (Cazella et al. 2016; Sencébé et al. 2020). These contradictions highlight what Sencébé et al. (2020) call a "parallel path" for agricultural development, where fledgling entities (such as MDA) and policies to support family farming (such as PRONAF) existed alongside agribusiness-oriented growth; therefore, this period of re-democratisation corresponds to a partial agri-environmental state, where a dual strategy was institutionalised for the development of Brazilian agriculture (Cazella et al. 2016; Sencébé et al. 2020); see Table 14.4.

The "Golden Decade": Toward an Established Agri-Environmental State? (2002–2010)

The growing political capital of social movements following re-democratisation contributed to, and benefited from, a new institutional opening in 2002 with the election of Luiz Inácio Lula da Silva, a former factory worker, trade union organiser, and co-founder of the Workers' Party (PT), which had "close historical ties and joint struggles [with] agrarian movements for the democratization of Brazil in the 1980s" (Sauer and Mészáros 2017: 408). This relationship between the PT

Table 14.4 An assessment of Brazil's agri-environmental state during the re-democratisation period according to the four faces of the agri-environmental state

Four faces of the agri-environmental state	Characteristics of Brazil's agri-environmental state: Re-democratisation (1985–2002)
1. Administration	New public administration bodies focused on family farming, such as MDA, were created, but the power of long-standing agribusiness-oriented bodies like MAPA grew as Brazil's economic development strategy depended on agricultural commodity exports
2. Regulation	Family-farming-oriented programmes like PRONAF were developed, but they favoured farmers in more industrialised agricultural contexts and often supported activities for agricultural modernisation
3. Redistribution	Social movement activity and international scrutiny pushed the government to increase land reform settlements. Civil society actors pushed for participatory mechanisms to redistribute power in policymaking following the dictatorship and had some success, but public offices were still highly influenced by *ruralistas*
4. Knowledge generation	The Brazilian state institutionalised separate government bodies for family farming and agribusiness, bifurcating its approach to agricultural development, research, extension, and resourcing

and agrarian movements informed social inclusion policies advanced under the Lula administration and increased avenues for civil society participation in policymaking. For example, in 2003, pre-existing FSN and poverty-reduction initiatives were consolidated under the overarching *Fome Zero* (Zero Hunger) policy. Led by MDA and the newly created Ministry of Social Development and the Fight against Hunger (MDS), *Fome Zero* brought together policies and programmes to address social welfare through income generation, improved food access and public health, and a strengthened family farming sector (Chappell 2018; Chmielewska and Souza 2011; Wittman and Blesh 2015). That same year, the Lula administration also reinstated[6] the National Council on Food and Nutritional Security (CONSEA) to provide an avenue for civil society to play a role in policymaking related to FSN. Comprised of one-third government officials and two-thirds civil society actors, CONSEA served as a direct advisory body to the presidency on FSN (Chmielewska and Souza 2011; Sonnino et al. 2014). In essence, CONSEA acts as what Fox (1993, quoted in Schiavoni 2017) calls an "institutional access route", linking sympathetic state officials with proactive societal actors, thereby opening up space for new political possibilities that advance representation, participation, and joint policy articulation for food security, nutrition, sustainable agriculture, and rural development more broadly.

Civil society's influence and involvement in CONSEA was central to the development of a suite of food security, agroecology, and sustainable rural development initiatives, like the Food Acquisition Programme (PAA), the National School Feeding Programme[7] (PNAE), and the National Policy for Agroecology and Organic Production (PNAPO). The PAA (created in 2003) was a structured demand, or mediated market, intervention that targeted both urban consumers and family farmers: The government committed to purchasing from family farmers as a social inclusion strategy to combat poverty, and then directed the purchased food to food insecure people and to institutions like hospitals and food banks. Another structured demand programme was established in 2009 (Law 11.947), when the government revamped the PNAE and committed to sourcing 30% of food for school cafeterias from family farmers enrolled in the programme, thus linking these producers to a stable market (Veras Soares et al. 2013). Notably, through the use of these formal structured demand programmes, the Brazilian government began

[6] CONSEA was originally founded in April 1993 (Franco administration) through Decree 807; however, it was cut in 1995 (Cardoso administration) and only re-created in 2003 with the Lula administration (see Grisa and Schneider 2015).

[7] PNAE has been around since the 1950s, but was modified in 2009 by Law 11.947 to include specific support to local family farmers.

to incentivise a transition to agroecology by providing price premiums as well as access to extension and other services to farmers who were certified organic or agroecological. For PAA the contract ceiling ranged between $1500 and 2500 *reais* more for certified products than for non-certified products, and certified organic/agroecological PNAE producers received contract priority over non-certified farmers (Guerra et al. 2017; Wittman and Blesh 2015). The aforementioned National Policy for Agroecology and Organic Production (PNAPO) came into effect in 2012 after a policymaking process that involved not only government officials but also over 150 non-governmental organisations, networks, and social movements (McKay and Nehring 2014), further expanding the government's commitment to support joint policymaking and transitions to organic and agroecological farming on family farms.

As a result of the interplay between civil society and key state actors, the decentralised governance structure of new public entities like CONSEA represented an important innovation allowing for civic participation in formal policymaking spaces at multiple (national, state, and municipal) levels, a feature that stems directly from Brazil's complicated history with democratic participation and representation (or lack thereof). Signalling its commitment to also decentralise regulation, participatory processes for organic/agroecological certification became officially recognised by the state in Lula's first term. In the words of an organic policy consultant interviewed,

> The country was in a very democratic process – all this happened under the [first term of the] PT government. There were spaces of democracy in which family farmers were greatly strengthened and non-governmental organisations that represented the interests of family farmers were also strengthened, and it was in this moment that participatory certification arrived as a response and as a testament to the strength of the family farmers, because it was a system of certification created by them.

At first glance, Brazil's agri-environmental state during this period might be classified as established. However, gains from the institutionalisation of agroecology may be more discursive than empirical. For example, citing 2010 data from FNDE, McKay and Nehring state that in 2011, family farmers should have been provided with R$ 930 million through the administration of PNAE based on the fact that family farmers are, by law, supposed to receive at least 30% of the total PNAE budget, which was R$ 3.1 billion in 2011 (McKay and Nehring 2014). However, in the case of PNAE, few states and municipalities actually allocated 30% of the PNAE budget to family farmers, and the law has remained decidedly unenforced. People participating in the agroecology movement (in both civil society and government spaces) point to a lack of enforcement and accountability protocols, a lack of associated funding to ensure legal compliance, and a lack of accessible data about programme expenditures as ongoing hurdles for agroecology policy (Petersen et al. 2013).

With respect to PAA, less than 3% of Brazil's family farmers participated in PAA, and uptake has not been equal across regions; while almost half of organic farms in Brazil are in the northeast, the majority of products sourced through PAA that are certified as agroecological or organic are produced in southern Brazil, indicating "a severe regional bias in PAA organic procurement" (McKay and Nehring 2014: 21). This is tied to long-standing patterns of racial inequality and the racialised distribution of resources in Brazil, where the majority of the Afro-Brazilian population lives in the northeast and the majority of the white population lives in the South and Southeast (Lovell 1993). Farmers interested in participating in PAA have expressed concerns about the lack of information shared about the opportunity to enrol (Guerra 2016; Nehring et al. 2017), and about "institutional complexity, patronage relations, and a failed distribution infrastructure, as well as the inability of the small-scale farming sector to meet food safety regulations and compete with larger scale producers" (Wittman 2015: 177; see also Araújo 2010). In addition, although certified organic foods would be able to receive a price premium through PAA, "in 2012 (the year in which the PAA had its best performance), [only] around 2% of the total volume of purchases" was certified (Schmitt et al. 2017: 60).

In discussing the distribution of state resources, Nowak (2019) reports how public finances received by small-scale and family farming still pale in comparison to funding received by Brazil's agribusiness sector: "In 2003, when Lula came to power, support for agribusiness was five times higher than that for family agriculture. By 2015, one year before [Lula's successor] Rousseff left office, it rose to an amount that was six times higher". In addition, land reform efforts dwindled dramatically towards the end of Lula's term and during Rousseff's term (Fernandes 2019). While agrarian reform peaked under the Lula administration in 2005 with 117,484 settled families, by the time Lula left office in 2010 only 15,536 families had been settled that year (DATALUTA 2017). By the time Rousseff was ousted, the number of settled families had fallen to 5490 (DATALUTA 2017). In addition to this decreasing number of land redistribution efforts, the quality of redistributive reforms also declined. As Giraldo and McCune note, "despite distributing more than 51.2 million hectares to 721,442 families, the Workers' Party (PT) governments of Brazil mostly allocated public and marginal lands, attempting the least possible impact on landlords" (Giraldo and McCune 2019: 10; see also Sauer and Mészáros 2017). For instance, most families were settled in agricultural frontier areas in the Amazon, where the government could buy land cheaply and provide land titles without having to confront the land-owning class or expropriate land—a kind of spatial fix that resulted in an overall failure to diminish structural inequality (Sauer and Mészáros 2017).

In sum, while representing important advances towards an established agri-environmental state, the Golden Decade era could be characterised as partial-to-emerging (Table 14.5). While government-sponsored policies and programmes that aimed to support family farmers and facilitate the transition towards agroecology, such as public procurement, were important innovations, they were also significantly underfunded in comparison to the agribusiness sector. The Lula administration never broke rank with agribusiness in Brazil, and investments in and exports from industrial agriculture largely fuelled Brazil's economic growth under his watch (Sauer and Mészáros 2017). In fact, throughout Lula's term (2003–2010), agricultural exports on average accounted for almost 40% of Brazil's total exports and the agribusiness sector accounted for, on average, almost a quarter of Brazil's GDP (Robles 2018). The administration's contradictory strategy opened Brazil to a commodity boom while simultaneously weakening state power to regulate, control, and enforce agri-environmental rules in the face of growing transnational agribusiness power. The commodity boom in turn "led to a boom in both demand for land and land prices" such that "the state's financial margin of action for land reform was substantially diminished, since market price was the basis upon which the compensation value of expropriated land was determined" (Sauer and Mészáros 2017: 402). The increase in the value of rural land as a result of the commodity boom has negatively affected possibilities for structural land reform. The Lula administration's inability to reconcile the contradictions between the agribusiness and agroecology paradigms and the competing interests of their proponents (*ruralistas* on the one hand and social movements and civil society on the other) ultimately led to a legitimacy crisis for the PT, the consequences of which are described in the following section.

The Agri-Environmental State "Under Pressure": Underestimating Contradictions

The described political-economic contradictions boiled over during the presidency of Lula's successor, Dilma Rousseff (also PT) from 2011 to 2016. In struggling with economic downturn and growing social discontent (including very public and widespread strikes), the party again allied itself with industrial, financial, and agrarian elites as well as "centre-right parties at all levels of government in order to guarantee its electoral viability…leaving the status-quo intact. It did not challenge the ideological dominance of neoliberalism to any significant extent" (Braga and Purdy 2019: 205–206), ultimately eroding the support of their original voter base (Carter 2011; Grisa and Schneider 2015; Saad-Filho 2015).

Political and economic turmoil and media-fuelled public outrage[8] over Operation Car Wash (*Lavo Jato*), the largest corruption scandal in Latin

[8] As multiple sources note, the "corporatized" media particularly singled out the PT for blame in Operation Car Wash, despite virtually every political party and numerous prominent politicians being implicated (Saad-Filho 2015; Braga and Purdy 2019).

Table 14.5 An assessment of Brazil's agri-environmental state during the Golden Decade period according to the four faces of the agri-environmental state

Four faces of the agri-environmental state	Characteristics of Brazil's agri-environmental state: The "Golden Decade" (2002–2010)
1. Administration	Important bodies allowing for civil society participation in FSN, such as CONSEA, were reinstated, democratising policymaking. Ministries like MDS and MDA supported family farming and sustainable agriculture programming but were drastically underfunded compared to MAPA
2. Regulation	FSN policies were consolidated under *Fome Zero* and there was the expansion of innovative public procurement programmes with mechanisms to support transitions to agroecological/organic management, such as PAA and PNAE, but effects were limited. At the same time, the Lula administration heavily relied on agribusiness growth as part of its neo-developmentalist strategy, dramatically subsidising/investing in this sector
3. Redistribution	Land redistribution efforts peaked and then steeply declined, and never challenged the structural inequality embedded in the Brazilian countryside; the land reform strategy involved settlement in agricultural frontiers rather than expropriating land from the landlord class
4. Knowledge generation	The democratisation of policymaking spaces allowed for inclusion of more diverse forms of knowledge, including the knowledge and experience of social movement actors and civil society

American history,[9] bolstered the political factions that lost the tight 2014 election—including the *ruralistas*, who held over 30% of seats in the lower house of Congress during Rousseff's term (Robles 2018)—to build a movement pursuing Rousseff's impeachment (Braga and Purdy 2019; Saad-Filho 2015; Sauer et al. 2019). President Rousseff was impeached in 2016 under the pretences of inappropriate budgetary conduct and replaced by Michel Temer of the opposition party. To avoid corruption charges himself and to "repay" the Rural Bloc (which was instrumental in orchestrating Rousseff's impeachment), Temer advanced a deeply neoliberal political platform that heavily favoured agribusiness (Leite et al. 2019; Sauer et al. 2019), aiming at the "wholesale privatization of Brazil's remaining state-owned enterprises and public resources" (Pahnke 2017). This marked the beginning of a radical decline of what were tenuous agroecology policy gains to begin with.[10] For example, in 2016 the Temer administration extinguished MDA,[11] which housed programmes that encouraged the promotion of agricultural development in line with the principles of agroecology and social participation, like the Technical Assistance and Rural Extension programme (ATER).

Another example relates to *grilagem*, or the theft of public land through the falsification of private property deeds (Paulino 2014; Schmink et al. 2017), a practice which dates back to colonial era. It has been particularly problematic in the Amazon frontier region, where extensive cattle and soy production, logging, and land speculation are the leading drivers of deforestation, but where various past government policies have also incentivised colonisation by small-scale farmers. During the Lula administration, the federal government led an inquiry into this process, and proposed a new law (Law 11.952) to, in theory, regularise the land claims of small-scale farmers who had settled in the area. But a number of scholars have criticised the law for loopholes that

[9] Operation Car Wash was a corruption scandal that involved construction executives defrauding and laundering money through the state-owned oil company Petrobras, and bribing public officials and Petrobras executives (Andrade 2019; Saad-Filho 2015).

[10] For a comprehensive review of the policy rollback during this time, see Leite et al. (2018) and Sabourin et al. (2020a, b).

[11] MDA was extinguished and its programmes were ultimately downgraded via a transfer to a Secretariat, the Special Secretariat for Family Agriculture and Agrarian Development (Sabourin et al. 2020a, b).

also legalised—and even favoured—the claims of large-scale landowners at the forefront of land speculation and *grilagem*, essentially "granting amnesty" to the *ruralistas* (Caetano 2019; May et al. 2016; Paulino 2014). As Paulino (2014: 136) describes,

> Since 2009, with the enactment of the Law 11.952, [67 million hectares] of inappropriately occupied public land have been recognized as private property… Peasants occupied only around an eighth of this area – some 8.3 million hectares. Their gain with 'regularization' allowed legislators and the state to spin a discourse of having advanced agrarian reform goals by legalising the land claims of these smallholders, while transferring the vast majority of the land involved – the remaining 58.7 million hectares – to large land grabbers.

The passing of Law 11.952 under Lula set the precedent for the Temer administration to propose 2016's Provisional Measure 759 (MP 759/16), which (again under the guise of regularising land claims of family farmers who migrated to the Amazon Frontier in the 1970s–1980s) further legalised the privatisation and financialisation of land, extended the time period within which land grabbers could apply for title, increased the size of properties that were eligible from 1500 to 2500 hectares, and in essence accepted deforestation as legal proof of occupation, and thus ownership (1° Defensoria Pública Especializada em Atendimento Fundiário do Estado do Amazonas et al. 2017; Caetano 2019; Leite et al. 2019). Despite strong pushback, MP 759 was amended and passed into Law 13,465/2017. As a result of policies (and political lock-ins) like these, land grabbing and concentration in Brazil has continued to increase, with recent figures for Brazil's Gini coefficient for land inequality reported as 0.86 and trending upwards (DATALUTA 2017; GRAIN 2014), further fuelling a vicious cycle of deforestation and inequality.

Political and economic turmoil following the parliamentary coup and Operation Car Wash set the stage for the authoritarian Bolsonaro administration to come into power in 2019. This "anti-institutional" and "anti-state" administration has sowed confusion about public data as part of its overall dismantling strategy (Sabourin et al. 2020a, b) and has continued to systematically roll back agroecological policy in favour of agribusiness as it pushes "an ultra-neoliberal economic agenda, which includes the scrapping, dismantling and mischaracterization of the state apparatus" (Sauer et al. 2020: 286; see also Fernandes et al. 2020). The first of Bolsonaro's attacks on the state apparatus occurred on the first day of his presidency, when the administration introduced Provisional Measure (MP) 870 and eliminated the National Council on Food and Nutritional Security (CONSEA), the primary mechanism for civil society participation in food-related policymaking. Immediately after MP 870 was introduced, CONSEA representatives and civil society networks began actively denouncing and protesting the move through petitions and rallies, working to get it reinstated. Through this political struggle, CONSEA became not just an "object" of social movements' demands for increased participation in FSN policymaking, but also a "field of action" or site of struggle in and of itself (Silva and Schmitt 2012: 23).

On 2 January 2019—the day after the provisional measure was introduced—the national councillors who represent CONSEA on behalf of civil society wrote a public letter declaring:

> The institutionalisation of the participation of representatives of different sectors of civil society in a direct advisory body of the Presidency of the Republic, such as CONSEA, has been an important instrument for listening to civil society *in order to improve public policies and strengthen the Brazilian State*...It is necessary to strengthen and consolidate the National Council of Food and Nutritional Security (CONSEA) *as a democratic space of the Brazilian State – and not only as a strict government forum* – in order to give voice to representative social organisations so that public policies can respond to the problems of the most vulnerable sectors of Brazilian society. (Councillors of CONSEA 2019, emphasis added)

Here, agroecology civil society actors discursively and strategically highlight the difference between the *government* and the *state*. This framing makes it clear that agroecology advocates in Brazil view civil society participation as a crucial part of "the state" and its authority, and that the function of the state should be to institutionalise policy that protects citizens' rights to food, a healthy environment, and dignified work. In the words of another agroecology policy leader in Brazil,

> The example is very clear: Why did they extinguish CONSEA? CONSEA was exactly the space where civil society and government representatives participated to monitor policies and propose new measures. This means ruling out the possibility of social participation. It is a blow to citizenship, to democracy.

This vision for a deeply democratised state apparatus is rooted in an understanding of the need to protect against the highly variable changes that can take place across government administrations, as seen through the evolution of Brazil's agri-environmental state. Historically in Brazil, the institutionalisation of agroecology policy and programming has advanced in direct relationship to processes of democratisation *outside* of the state (through social movement mobilisations for redistributing land and power), *within* the state (through mechanisms for civil society participation in policymaking, such as CONSEA), and as a result of *interactions and dialogue between state and non-state actors* (for instance, through the relationship of agrarian reform movements to the PT). Historical patterns, competing paradigms, and state–society interactions have contributed to a particularly Brazilian vision for an agroecological state that is deeply tied to democratisation processes and direct civil society participation in politics and policymaking (Schmitt et al. 2017). As stated by an NGO leader and policy advocate,

> Our last [national] agroecology meeting [in 2018] had the motto "agroecology and democracy: uniting the countryside and the city." Because so many people understand that agroecology is an approach to democratise society and the state; the democratisation of the state and society is a necessity for us in agroecology.

> We talk about radicalising democracy, and democratising agri-food systems…to institutionalise agroecology, then, it's necessary to break this vision that was built with modernisation and which was a top-down process, very structuralist, very vertical, where the role of farmers and their organisations are passive recipients of policies. We don't understand

> it that way, we understand that not only must local organisations design policies, they must also execute and monitor policies, together with the state…Public policy is not exclusive to the government, public policy is a co-production between the government with civil society…The government has no way, it has no strength alone to overcome the power of corporations, especially in the world of neoliberal globalisation…So, one way to face this is exactly to create [state-society] partnerships and create territorial mechanisms for regulating agri-food systems.

The agroecology movement's emphasis on democratisation and decentralisation of the state apparatus has emerged directly out of long-standing and legitimate concerns over how centralised state power has led to historical and ongoing inequities, repression, clientelism, and corruption. As noted in an interview by an NGO coordinator and agroecology organiser,

> Here in Brazil, if we're not careful, we can lose, right? It's not that all is conquered and consolidated. Nothing is consolidated. At least not in Brazil. And we see this setback that we are experiencing worldwide [the turn toward authoritarian neoliberalism], we see we can lose a lot.

Yet despite these major, looming challenges, this participant also went on to articulate a kind of radical optimism of the intellect *and* of the will (Gramsci 1973; Panitch 2017; Wright 2010), rooted in an understanding of the politics of social struggle and social change:

> But if we see what we've advanced in the last 30 years – we've advanced a lot. I don't think we achieved little, considering the financial conditions and opportunities we had…we were always few. There was always everyone against us. Even so, we managed to materialise some things…This is a hell of a legacy, right. I would say that it is no small thing.

Conclusions

How and to what degree states successfully institutionalise agroecology remains in need of deeper theoretical and empirical engagement. However, there are limitations to state-centric analyses, particularly amid the emergence of authoritarian state regimes and in settler-colonial contexts, where jurisdiction and the legitimacy of the state and any state policy are unsettled and highly contested. Similarly, creating economy-ecology dichotomies can obscure the fact that any economy is always embedded in an ecology, and runs the risk of subsuming heterogeneous economic modes into a homogeneous, globalised, and neoliberalised market. Nonetheless, linking an HRI approach with the concept of the agri-environmental state can be useful to understand dynamic processes of agroecology (de)institutionalisation, with the understanding that progress is structured by the historical context in which competing agricultural development paradigms dialectically emerged (neoliberal/agribusiness and agroecological), and how state and non-state actors interact and shape (and are shaped by) their relationship to one another and these paradigms (Schiavoni 2017).

Certainly, iterations of Brazil's agri-environmental state have fallen short of prioritising agroecology over agribusiness. While re-democratisation and the rise of the PT provided institutional openings for agroecology, policy gains were tenuous even during the PT era due to alliances with agrarian elites, leading to a kind of "institutional fragility" (Sabourin et al. 2020a, b). Yet despite the Lula regime's ongoing ties to the neoliberal economic order, state–civil society relations qualitatively changed during the Lula administration, where progressive actors who were once side-lined by state processes found new opportunities to articulate their demands (Grisa and Schneider 2015). Thus, the political moment that arose with the election of Lula, while not representing a large political-economic transformation, did open up space for social movement and civil society actors to co-construct public policies that institutionalised a (fragile) agroecological approach.

Particularly instrumental to constructing and institutionalising agroecological approaches have been democratisation processes *outside of* (through movements to democratise land and the means of production) and *within* (through struggles to democratise access to public resources, knowledge, and policymaking) the state. In the current political moment—where basic democratic principles are again under attack and human rights and agri-environmental policy are being rolled back and hollowed out—it remains to be seen how social movement and civil society actors will adapt their advocacy efforts. As Schiavoni (2017: 28) states, "nothing is predetermined" in food sovereignty and agroecology construction, and the policy landscape in Brazil has already experienced some remarkable shifts—"no small thing", indeed.

References

1° Defensoria Pública Especializada em Atendimento Fundiário do Estado do Amazonas et al. 2017. *CARTA AO BRASIL. Medida Provisória N° 759/2016: A Desconstrução da Regularização Fundiária No Brasil*. Retrieved from https://docs.google.com/document/d/1dPHipNy_aJxGkjwk2PiRGXu_8mklCO5BGTw9J1W8Les/edit.

Andrade, D. 2019. Populism from Above and Below: The Path to Regression in Brazil. *The Journal of Peasant Studies* 47 (7): 1470–1496.

Araújo, S. 2010. The Promise and Challenges of Food Sovereignty Policies in Latin America. *Yale Human Rights and Development Journal* 13 (2): 493–506.

Bowen, M.L. 2021. *For Land and Liberty: Black Struggles in Rural Brazil*. Cambridge: Cambridge University Press.

Braga, R., and S. Purdy. 2019. A Precarious Hegemony: Neo-Liberalism, Social Struggles, and the End of Lulismo in Brazil. *Globalizations* 16 (2): 201–215.

Branford, S., and J. Rocha. 2002. *Cutting the Wire: The Story of the Landless Movement in Brazil*. London: Latin America Bureau.

Brannstrom, C., W. Jepson, A.M. Filippi, D. Redo, Z. Xu, and S. Ganesh. 2008. Land Change in the Brazilian Savanna (Cerrado), 1986–2002: Comparative Analysis and Implications for Land-Use Policy. *Land Use Policy* 25 (4): 579–595.

Caetano, C.A.P. 2019. *Regularização Fundiária Rural Na Amazônia Legal: Análise de Uma Política de Contrarreforma Agrária*. Goiânia: Universidade Federal de Goiás.

Caporal, F.R., and J.A. Costabeber. 2004. *Agroecologia e Extensão Rural: Contribuições Para a Promoção Do Desenvolvimento Rural Sustentável*. Brasília: MDA/SAF/DATER/IICA.

Carter, M. 2011. The Landless Rural Workers Movement and Democracy in Brazil. *Latin American Research Review* 45: 186–217.

Carvalho, B. 2015. The Modern Roots of Feudal Empires: The Donatary Captaincies and the Legacies of the Portuguese Empire in Brazil. In *Legacies of Empire: Imperial Roots of the Contemporary Global Order*, ed. S. Halperin and R. Palan, 128–148. Cambridge: Cambridge University Press.

Carvalho, F.M.V., P. de Marco, and L.G. Ferreira. 2009. The Cerrado Into-Pieces: Habitat Fragmentation as a Function of Landscape Use in the Savannas of Central Brazil. *Biological Conservation* 142 (7): 1392–1403.

Cazella, A.A., A.J. Capellesso, M. Medeiros, A. Tecchio, Y. Sencébé, and F.L. Búrigo. 2016. Políticas Públicas de Desenvolvimento Rural No Brasil: O Dilema Entre Inclusão Produtiva e Assistência Social. *Política e Sociedade* 15: 49–79.

Chappell, M.J. 2018. *Beginning to End Hunger: Food and the Environment in Belo Horizonte, Brazil and Beyond*. Oakland: University of California Press.

Chmielewska, D., and D. Souza. 2011. *The Food Security Policy Context in Brazil*. Brasília: International Policy Centre for Inclusive Growth.

Councillors of CONSEA. 2019. *Carta Sociedade Civil de CONSEA*. Brasília.

da Costa, M.B.B., M. Souza, V. Müller, Jr., J.J. Comin, and P.E. Lovato. 2017. Agroecology Development in Brazil Between 1970 and 2015. *Agroecology and Sustainable Food Systems* 41 (3–4): 276–295.

DATALUTA. 2017. *Land Struggle Database Report: Brazil 2016*. São Paulo.

de la Peña, G. 2005. Social and Cultural Policies Toward Indigenous Peoples: Perspectives from Latin America. *Annual Review of Anthropology* 34: 717–739.

de Sant'Anna, A.L.O., A.C. Castro, and A.M. Jacó-Vilela. 2018. Military Dictatorship and Disciplinary Practices in the Control of Indigenous People: Psychosocial Perspectives on the Figueiredo Report. *Psicologia e Sociedade* 30: 1–10.

Desmarais, A.A. 2007. *La Vía Campesina: Globalization and the Power of Peasants*. Halifax: Fernwood.

Desmarais, A.A., P. Claeys, and A. Trauger (eds.). 2017. *Public Policies for Food Sovereignty: Social Movements and the State*. New York: Routledge.

Duit, A. 2016. The Four Faces of the Environmental State: Environmental Governance Regimes in 28 Countries. *Environmental Politics* 25 (1): 69–91.

Duit, A., P.H. Feindt, and J. Meadowcroft. 2016. Greening Leviathan: The Rise of the Environmental State? *Environmental Politics* 25 (1): 1–23.

Eckersley, R. 2004. *The Green State: Rethinking Democracy and Sovereignty*. Cambridge, MA: MIT Press.

FAO. 2018. *Scaling Up Agroecology Initiative: Transforming Food and Agricultural Systems in Support of the SDGs*. Rome: FAO.

Fernandes, B.M. 2013. Re-Peasantization, Resistance and Subordination: The Struggle for Land and Agrarian Reform in Brazil. *Agrarian South: Journal of Political Economy* 2 (3): 269–289.

Fernandes, B.M. 2016. Development Models for the Brazilian Countryside: Paradigmatic and Territorial Disputes. *Latin American Perspectives* 43 (2): 48–59.

Fernandes, B.M. 2019. Land Grabbing for Agro-Extractivism in the Second Neoliberal Phase in Brazil. *Revista Nera* 22 (50): 208–238.

Fernandes, B.M., J. Cleps, Jr., J. Sobreiro Filho, A.Z. Leite, and R.B. Sodré. 2020. *A Questão Agrária no Primeiro Ano do Governo Bolsonaro*. Presidente Prudente: Unesp.

Fox, J. 1993. *The Politics of Food in Mexico: State Power and Social Mobilization*. Ithaca, NY: Cornell University Press.

Franco Lerrer, D., and L. Servolo de Medeiros. 2014. Food Sovereignty and Struggle for Land: The Experience of the MST in Brazil. In *Alternative Agrifood Movements: Patterns of Convergence and Divergence*, ed. D.H. Constance, M.-C. Renard, and M.G. Rivera-Ferre, 111–135. Bingley: Emerald Group Publishing Limited.

Garrett, R.D., and L.L. Rausch. 2016. Green for Gold: Social and Ecological Tradeoffs Influencing the Sustainability of the Brazilian Soy Industry. *The Journal of Peasant Studies* 43 (2): 461–493.

Giraldo, O.F., and N. McCune. 2019. Can the State Take Agroecology to Scale? Public Policy Experiences in Agroecological Territorialization from Latin America. *Agroecology and Sustainable Food Systems* 43 (7–8): 785–809.

GRAIN. 2014. Hungry for Land. Retrieved from https://www.grain.org/article/entries/4929-hungry-for-land-small-farmers-feed-the-world-with-less-than-a-quarter-of-all-farmland.

Gramsci, A. 1973. *Letters from Prison*. New York: Harper & Row.

Grisa, C., and S. Schneider (eds.). 2015. *Políticas Públicas de Desenvolvimento Rural No Brasil*. Porto Alegre: Universidade Federal do Rio Grande do Sul.

Guerra, J. 2016. *Pathways to Agroecology: Mediated Markets and Credit Access in Santa Catarina, Brazil*. Vancouver: University of British Columbia.

Guerra, J., J. Blesh, A.L. Schmitt Filho, and H. Wittman. 2017. Pathways to Agroecological Management through Mediated Markets in Santa Catarina, Brazil. *Elementa* 5 (67).

HLPE. 2019. *Agroecological and Other Innovative Approaches for Sustainable Agriculture and Food Systems That Enhance Food Security and Nutrition.* Rome.

Hochstetler, K., and M.E. Keck. 2007. *Greening Brazil: Environmental Activism in State and Society*. Durham, NC: Duke University Press.

IPES-Food. 2016. *From Uniformity to Diversity: A Paradigm Shift from Industrial Agriculture to Diversified Agroecological Systems*. IPES-Food.

Jepson, W., C. Brannstrom, and A. Filippi. 2010. Access Regimes and Regional Land Change in the Brazilian Cerrado, 1972–2002. *Annals of the Association of American Geographers* 100 (1): 87–111.

Johnson, H.B. 1972. The Donatary Captaincy in Perspective: Portuguese Backgrounds to the Settlement of Brazil. *The Hispanic American Historical Review* 52 (2): 203–214.

Johnson, H.B. 1987. Portuguese Settlement, 1500–1580. In *Colonial Brazil*, ed. L. Bethell, 1–39. Cambridge: Cambridge University Press.

Keck, M.E. 1995. Social Equity and Environmental Politics in Brazil: Lessons from the Rubber Tappers of Acre. *Comparative Politics* 27 (4): 409–424.

Lathuillière, M.J., M.S. Johnson, G.L. Galford, and E.G. Couto. 2014. Environmental Footprints Show China and Europe's Evolving Resource Appropriation for Soybean Production in Mato Grosso, Brazil. *Environmental Research Letters* 9 (7): 074001.

le Polain de Waroux, Y., R.D. Garrett, J. Graesser, C. Nolte, C. White, and E.F. Lambin. 2017. The Restructuring of South American Soy and Beef Production and Trade Under Changing Environmental Regulations. *World Development* 121: 188–202.

Leite, A.Z., L.F.P. Castro, and S. Sauer. 2018. A Questão Agrária No Momento Político Brasileiro: Liberalização e Mercantilização Da Terra No Estado Mínimo de Temer. *OKARA: Geografia Em Debate* 12 (2): 247.

Leite, A.Z., N.L.G. Tubino, and S. Sauer. 2019. Políticas Públicas Para Terra e Território: Um Olhar Prospectivo Sobre Os Próximos Quatro Anos No Campo Brasileiro. In *Brasil Incertezas E Submissão?*, 171–189. São Paulo: Fundação Perseu Abramo.

Lovell, P.A. 1993. The Geography of Economic Development and Racial Discrimination in Brazil. *Development and Change* 24:83–101.

Martine, G. 1991. A Trajetória Da Modernização Agrícola: A Quem Beneficia? *Lua Nova: Revista de Cultura e Política* 23: 7–37.

May, P.H., M.F. Gebara, L.M. Barcellos, M.B. Rizek, and B. Millikan. 2016. *The Context of REDD+ in Brazil: Drivers, Actors and Institutions*, 3rd ed., 160. Bogor, Indonesia: CIFOR.

McKay, B., and R. Nehring. 2014. *Sustainable Agriculture: An Assessment of Brazil's Family Farm Programmes in Scaling Up Agroecological Food Production.* Brasília.

Meade, T.A., 2010. *A brief history of Brazil*. Infobase Publishing.

Meadowcroft, J. 2005. From Welfare State to Ecostate. In *The State and the Global Ecological Crisis*, ed. J. Barry and R. Eckersley, 3–23. Cambridge, MA: MIT Press.

Mestmacher, J., and A. Braun. 2021. Women, Agroecology and the State: New Perspectives on Scaling-Up Agroecology Based on a Field Research in Chile. *Agroecology and Sustainable Food Systems* 45 (7): 981–1006.

Mol, A.P.J., and F.H. Buttel. 2002. The Environmental State Under Pressure: An Introduction. In *The Environmental State Under Pressure*, ed. A.P.J. Mol and F.H. Buttel, 1–11. Elsevier.

Nehring, R., A. Miranda, and A. Howe. 2017. Making the Case for Institutional Demand: Supporting Smallholders Through Procurement and Food Assistance Programmes. *Global Food Security* 12: 96–102.

Nowak, J. 2019. Bolsonaro: The First 100 Days. *The Bullet*: 1–6.

Nyéléni International Forum for Agroecology. 2015. Declaration of the International Forum for Agroecology, Nyéléni, Mali: 27 February 2015. *Development* 58 (2–3): 163–168.

O'Dwyer, E.C. (ed.). 1995. *Terra de Quilombos*. Rio de Janeiro: Associação Brasileira de Antropologia.

OECD-FAO. 2015. *Brazilian Agriculture: Prospects and Challenges.*

Pahnke, A. 2017. The Brazilian Crisis: Corruption, Neoliberalism, and the Primary Sector. *Monthly Review* 68 (9): 43–54.

Panitch, L. 2017. On Revolutionary Optimism of the Intellect. *Socialist Register* 53: 356–563.

Paulino, E.T. 2014. The Agricultural, Environmental and Socio-Political Repercussions of Brazil's Land Governance System. *Land Use Policy* 36: 134–144.

Petersen, P., E.M. Mussoi, and F. Soglio. 2013. Institutionalization of the Agroecological Approach in Brazil: Advances and Challenges. *Agroecology and Sustainable Food Systems* 37 (1): 103–114.

Puzone, V., and L.F. Miguel (eds.). 2019. *The Brazilian Left in the 21st Century: Conflict and Conciliation in Peripheral Capitalism.* Cham, Switzerland: Palgrave Macmillan.

Reis, J.J., and F.S. Gomes (eds.). 1996. *Liberdade Por Um Fio: História Dos Quilombos No Brasil*. São Paulo: Companhia das Letras.
Robles, W. 2018. Revisiting Agrarian Reform in Brazil, 1985–2016. *Journal of Developing Societies* 34 (1): 1–34.
Rosset, P.M., and M.E. Martínez-Torres. 2012. Rural Social Movements and Agroecology: Context, Theory, and Process. *Ecology and Society* 17 (3): 17.
Saad-Filho, A. 2015. Brazil: The Debacle of the PT. *Monthly Review Online*: 1–13.
Sabourin, E., C. Craviotti, and C. Milhorance. 2020a. The Dismantling of Family Farming Policies in Brazil and Argentina. *International Review of Public Policy* 2 (1): 45–67.
Sabourin, E., C. Grisa, P. Niederle, S.P. Leite, C. Milhorance, A.D. Ferreira, S. Sauer, and J.M. Andriguetto-Filho. 2020b. Le Démantèlement Des Politiques Publiques Rurales et Environnementales Au Brésil. *Cahiers Agricultures* 29 (31).
Sauer, S., K.O.A. Leite, and A. Shankland. 2019. *The Implications of Closing Civic Space for Sustainable Development in Brazil*. IDS and ACT Alliance.
Sauer, S., A.Z. Leite, and N.L.G. Tubino. 2020. Agenda Política Da Terra No Governo Bolsonaro. *Revista da ANPEGE* 16 (29): 285–318.
Sauer, S., and G. Mészáros. 2017. The Political Economy of Land Struggle in Brazil under Workers' Party Governments. *Journal of Agrarian Change* 17 (2): 397–414.
Schiavoni, C.M. 2017. The Contested Terrain of Food Sovereignty Construction: Toward a Historical, Relational and Interactive Approach. *The Journal of Peasant Studies* 44 (1): 1–32.
Schmink, M., J. Hoelle, C.V.A. Gomes, and G.M. Thaler. 2017. From Contested to 'Green' Frontiers in the Amazon? A Long-Term Analysis of São Félix Do Xingu, Brazil. *Journal of Peasant Studies* 46 (2): 377–399.
Schmitt, C., P. Niederle, M. Ávila, E. Sabourin, P. Petersen, L. Silveira, W. Assis, J. Palm, and G.B. Fernandes. 2017. A Experiência Brasileira de Construção de Políticas Públicas em Favor da Agroecologia. In *Políticas Públicas a favor de la Agroecología en América Latina y el Caribe*, 44–69. Brasília: Red Políticas Públicas en América Latina y el Caribe (PP-AL).
Schneider, S., S. Shiki, and W. Belik. 2010. Rural Development in Brazil: Overcoming Inequalities and Building New Markets. *Rivista Di Economia Agraria* 45 (2): 225–259.
Sencébé, Y., F. Pinton, and A.A. Cazella. 2020. On the Unequal Coexistence of Agrifood Systems in Brazil. *Review of Agricultural, Food and Environmental Studies* 101: 191–212.

Silva, M. K. S, and C.J. Schmitt. 2012. Politicas em rede: uma analise comparativa das interdependencias entre o Programa de Aquisicao de Alimentos e as redes associativas no Rio Grande do Sul e na Bahia. Anais do Encontro Anual da ANPOCS, 36, Aguas de Lindoia (SP).

Sonnino, R., C.L. Torres, and S. Schneider. 2014. Reflexive Governance for Food Security: The Example of School Feeding in Brazil. *Journal of Rural Studies* 36: 1–12.

Tarlau, R. 2014. *Occupying Land, Occupying Schools: Transforming Education in the Brazilian Countryside*. Berkeley: University of California Press.

Tarlau, R. 2015. Education of the Countryside at a Crossroads: Rural Social Movements and National Policy Reform in Brazil. *The Journal of Peasant Studies* 42 (6): 1157–1177.

Tranjan, J.F. 2016. *Participatory Democracy in Brazil: Socioeconomic and Political Origins*. Notre Dame: University of Notre Dame Press.

Val, V., P.M. Rosset, C.Z. Lomelí, O.F. Giraldo, and D. Rocheleau. 2019. Agroecology and La Via Campesina I: The Symbolic and Material Construction of Agroecology Through the Dispositive of 'Peasant-to-Peasant' Processes. *Agroecology and Sustainable Food Systems* 43 (7–8): 872–894.

Veras Soares, F., R. Nehring, R.B. Schwengber, C.G. Rodrigues, G. Lambais, D.S. Balaban, C. Jones, and A. Galante. 2013. *Structured Demand and Smallholder Farmers in Brazil: The Case of PAA and PNAE*. Brasília.

Wezel, A., S. Bellon, R. Doré, C. Francis, D. Vallod, and C. David. 2009. Agroecology as a Science, a Movement and a Practice: A Review. *Agronomy for Sustainable Development* 29 (4): 503–515.

Wittman, H. 2010. Agrarian Reform and the Environment: Fostering Ecological Citizenship in Mato Grosso, Brazil. *Canadian Journal of Development Studies* 29 (3–4): 281–298.

Wittman, H. 2011. Food Sovereignty: A New Rights Framework for Food and Nature? *Environment and Society: Advances in Research* 2 (1): 87–105.

Wittman, H. 2015. From Protest to Policy: The Challenges of Institutionalizing Food Sovereignty. *Canadian Food Studies* 2 (2): 174–182.

Wittman, H., and J. Blesh. 2015. Food Sovereignty and Fome Zero: Connecting Public Food Procurement Programmes to Sustainable Rural Development in Brazil. *Journal of Agrarian Change* 17 (1): 81–105.

Wolford, W. 2010a. Participatory Democracy by Default: Land Reform, Social Movements and the State in Brazil. *Journal of Peasant Studies* 37 (1): 91–109.

Wolford, W. 2010b. *This Land Is Ours Now: Social Mobilization and the Meanings of Land in Brazil*. Durham: Duke University Press.

Wright, E.O. 2010. *The Real Utopias Project: A General Overview*. Retrieved from https://www.ssc.wisc.edu/~wright/OVERVIEW.html.

15

Decolonial and Feminist Approaches to Critical Food Systems Education

David Meek and Rebecca Tarlau

Food systems education is a critical component of contemporary global struggles for more just economic, political, and ecological systems. It is also an expanding area of scholarship and encompasses a diversity of pedagogies and curricular offerings (Galt et al. 2012; Lieblein et al. 2000; Parr et al. 2007; Salomonsson et al. 2009). Food systems education takes place both within grassroots social movements (Meek 2020; Tarlau 2019) and throughout the traditional formal educational system. For example, in universities, food systems courses are found in both disciplinary and interdisciplinary departments, such as agronomy, anthropology, ecology, environmental studies and sciences, geography, and sociology (Galt et al.

D. Meek (✉)
University of Oregon, Eugene, OR, USA
e-mail: dmeek@uoregon.edu

R. Tarlau
The Pennsylvania State University, State College, PA, USA
e-mail: rzt70@psu.edu

A. A. R. Ioris and B. Mançano Fernandes (eds.), *Agriculture, Environment and Development*, https://doi.org/10.1007/978-3-031-10264-6_15

2013; Parr and Trexler 2011). Food systems education takes place in many primary and secondary schools through garden-based learning. In contrast to many food systems education programs in formal education systems, the food systems education initiatives led by grassroots social movements are most often linked to a radical political vision for more egalitarian societies.

We have previously advocated for a critical food systems education (CFSE) framework, which integrates the innovative pedagogical approaches in the field of food systems education, the insights in the literature on food justice, agroecology, and food sovereignty, and the pedagogical initiatives developed through decades of grassroots educational practices. In this previous work (Meek and Tarlau 2016), we highlighted the lessons we can take from contemporary formal food education practices while also lifting up the pedagogical innovations of grassroots movements. We aligned ourselves with scholars (Flowers and Swan 2011; Rosing 2012; Sumner 2013) who argue that certain elements of food systems education, such as garden-based learning and farm-to-school programs, are still open to food justice movement's long-standing critique that these practices have racial and class-based assumptions (cf. Allen and Guthman 2006; Guthman 2008). We also suggested that food systems educators need to incorporate political forms of agroecology (de Molina 2013) and food sovereignty as central aspects of their educational approach. We argued that popular educational practices, which are embedded within grassroots movements (McCune et al. 2014), can help us bridge the gap between food systems education programs and the actual movements that are attempting to transform the food systems themselves to these connections.

In this chapter, we revisit these arguments in the current political context, where we have seen the rise of a virulent right-wing fascism that has popularised through the vilification and dehumanisation of Black, indigenous, Muslim, immigrant, landless, lower caste, women, and transgender communities globally. Although the particularities of these racist and misogynistic strongmen vary in each country, they are linked through a global project of white supremacy that has direct implications for our food system. Simultaneously, we have also seen the rise of a new generation of social movements, including the movement for

Black Lives in the United States, which builds on a long history of the Black radical tradition, as well as the Indian farmers' movements whose revolts enveloped the subcontinent in 2020–2021. Given the current context, we believe it is essential to integrate a decolonial and critical feminist perspective into our previous critical food systems education (CFSE) framework. In this chapter, we revisit the four pillars of a critical food systems education approach—popular education, food justice, food sovereignty, and agroecology. Then we review the decolonial and critical feminisms literature, highlighting their pedagogical contributions and how they can inform critical food systems education.

Critical Food Systems Education: Integrative Theory, Pedagogy, and Policy Vision

Food systems education refers to the myriad ways that professors, teachers, activists, and communities are teaching, learning, and intervening in their food systems. We can group food systems education into university-level food systems courses, garden-based learning in primary and secondary schools, and most importantly for this article, the educational initiatives of food justice organisations and grassroots social movements. In these latter cases, we are not referring to food systems education as simply the action of the food justice organisation, but rather, the spaces where the action is combined with time for learning, reflection, and study. For example, U.S. grassroots organisations, such as Cooperation Jackson in Mississippi, Detroit Freedom Farmers, and Oakland's Phat Beets, among many others, are incorporating education as a component of their concrete food justice campaigns. These educational practices include post-farmers' markets workshops, speaker series at organic farms, courses on food sovereignty, or exchanges between different grassroots organisations. In Latin America, grassroots organisations linked to *La Vía Campesina* have developed extensive multi-year educational programs focused on agroecology, through national programs such as Brazil's Programme for Education in Areas of Agrarian Reform (PRONERA, see Martins and Rodrigues 2015; Tarlau 2019) and the Latin American Agroecology Institutes (McCune and Sánchez 2019;

McCune et al. 2017; Meek 2020). These educational spaces within grassroots organisations are both the central foundation of and the inspiration for a critical food systems approach (CFSE).

CFSE is as at once a theoretical framework, a set of pedagogies and pedagogical methods, and a vision for policy. Thus, through this tripartite contribution, CFSE is not simply a theoretical perspective for analysing the relationships between education and food systems, but also a set of practices that can transform how food systems courses are taught and a perspective that can suggest policies at the intersection of education and agricultural systems. This tripartite linkage between theory, pedagogy, and policy is a central element that distinguishes CFSE from traditional food systems education. Whereas many food systems education courses seek to help students develop an interdisciplinary understanding of the food system, the objective of CFSE is to leverage education and innovative pedagogical techniques so that students and educators can transform the food system, and develop transgressive subjectivities that promote grassroots agricultural alternatives (Sawyer 2004). Drawing on the lessons of university level and grassroots food systems education approaches, we propose integrating the theoretical contributions from critical pedagogy, agroecology, food sovereignty, and food justice into an overall food systems education approach. Our vision for CFSE is neither confined to the U.S. context, nor the university level; rather, we see CFSE as a platform for food systems education at diverse educational levels and across international contexts.

Popular Education and Critical Pedagogy: Merging Theory and Pedagogical Practice

First, in this CFSE framework, the link between education and social transformation is critical. This educational approach emanates from Brazilian educational theorist Paulo Freire's maxim of praxis (1973): the dialectical connection between understanding and learning to change the inequitable social, economic, political, and agricultural systems that shape our lives. Thus, our inspiration for developing a CFSE framework is based on the popular educational practices developed by social movements, in which the purpose of education is social transformation. These theories and pedagogical practices, often referred to as "critical pedagogy" in the U.S. academy, are a cornerstone of our CFSE framework.

Tarlau (2014) highlights two overarching trends in the field of critical pedagogy. First, as a pedagogical perspective, it is based on the acknowledgement that knowledge production functions as an "ideological apparatus" in schools (Althusser 1984), serving to produce school culture, and further processes of racialisation and gender identity creation (Apple 2004; Aronowitz and Giroux 1991; McLaren 2003). This focus in the critical pedagogy literature is directly connected to theories of social reproduction in education, which highlight how the educational system reproduces dominant societal structures, norms, and career pathways (Anyon 1981; Bowles and Gintis 1976; Willis 1977). Second, critical pedagogues seek to help students critically reflect on the nature of these realities and transform them by introducing concepts such as the teacher-intellectual, banking versus problem-solving education, and education as resistance (Tarlau 2014: 372). The emphasis is on how to turn education and schools into forces for liberation. Critical pedagogy is directly relevant to CFSE because it offers a way to both reflect on how education functions as a form of social reproduction and theorise schools' transformative potential. Although many educational programs at the university level engage with Deweyan theory and the importance of experiential learning, a more overtly Freirean framework necessitates that this action be connected to systematic transformation.

Popular education plays a pivotal role in influencing these critical educational approaches. While critical pedagogy is an academic field of literature that emerged in the early 1980s, we use the term "popular education" to refer to the concrete, often informal, educational practices that have been a central component of political mobilisation throughout the twentieth century. From the Highlander Center in Tennessee, offering educational workshops that influenced many leaders of the civil rights movement (Morris 1984), to Catholic priests espousing liberation theology (Berryman 1987), the Black Panther schools in Oakland (Payne and Strickland 2008), the literacy campaign in Nicaragua (Arnove 1986), and workers' colleges (Altenbaugh 1990), social movements have prioritised education as part of their broader social and political struggles. Thus, both the theoretical field of critical pedagogy and concrete popular education practices pose important questions for food systems education: Do courses, degree programs, school garden projects, and study abroad opportunities replicate existing social structures, class, racial and gender hierarchies, and career pathways? Or, do they explicitly employ education as a means of challenging these entrenched systems? Within CFSE, the objective is clear. Food systems educators need to rethink how they can help their students connect interdisciplinary knowledge to transformative systems changes. Drawing on Freire (1973), we see these types of educational practices as a form of food system *conscientização* (critical consciousness), where students both learn to analyse their world of food production and access, and take actions to change these systems.

Food Justice: The Pedagogies and Policies of CFSE

Second, we place food justice as another pillar of the CFSE framework. Food justice is a grassroots movement that is predicated upon a critique of the racial and class-based inequalities in the food system. The food justice movement emerged out of the U.S. civil rights and environmental justice movements of the 1960s and 1970s (Holt-Gimenez and Shattuck 2011; Mares and Alkon 2011). Food justice is directly relevant to questions of critical food systems education due to its critical attention to the interconnections of race, privilege, and the production of alternative food spaces—questions that need to be dealt with in advocating for more cooperative and emancipatory forms of food systems education. For instance, food justice's focus on the structural racial and class-based inequality in the food system should be directed at the school food system and its intersection with alternative agriculture. Farm-to-school programs, which seek to provide a portion of a school's food from local farms, and school garden projects, are largely located in upper class white school districts, unless they are funded by private foundations or heavily subsidised by the state (Allen and Guthman 2006).

Drawing upon the food justice perspective, we advocate turning critical attention to the ideals underpinning agricultural programming in schools. How do NGOs ground their garden-based learning projects in particular conceptions of the racialised bodies of their subjects, and understandings of food access, individual choice, and what constitutes "good food" (Guthman 2008)? Etmanski (2012) provides one example of how to fruitfully synthesise food justice and questions of critical education through a critical race and class analysis of learning in the organic food movement.

In summary, we have chosen to highlight food justice as a separate and critical cornerstone of CFSE, due to its attention to racial disparities in the food system. Much of the current food activism work in the United States is being conducted through this framework. We believe that attention to class and race dynamics should be a central goal of any food systems education program or theoretical approach. As we already mentioned, many food justice organisations have concrete educational practices that are aligned with the CFSE approach. We believe that these experiences should be the building blocks for developing more critical food systems educational programs at all schooling and informal educational levels.

Agroecology: Bringing Politics to the Front

Third, while agroecology is already a central part of food systems education (Francis et al. 2001; Jordan et al. 2005; Ostergaard et al. 2010), we make this field a separate cornerstone of the CFSE framework, because we believe food systems educators must engage directly with the political nature of agroecology. The conception of agroecology that is employed in some traditional university agricultural departments is that of agroecology as either a science or a set of practices (Wezel et al. 2009). Thinking about agroecology both as a science and as a set of practices is incredibly important, and to be clear, we would be bereft to not include lessons from these approaches. For example, critical food systems educators should continue to draw upon agroecology as a science in which innovative agriculture practices are developed, enabling the reduction of external inputs, and the maximisation of space-intensive production.

While drawing upon this accumulation of scientific agroecological knowledge, the CFSE approach aligns itself with the new and innovative food systems education programs at the university level, which are incorporating the political nature of agroecology into their theories, pedagogies, and courses. For example, Lieblein and Francis (2007) describe an action-education approach to agroecological education in Norway, where students use case studies and interviews with key stakeholders to chart the environmental and social impacts of the food system, as well as the potential impact of their proposed alternatives. Agroecology is an inherently political phenomenon, because agroecosystems are socioecological constructions that are mediated by power relations (de Molina 2013). For example, the proliferation of monocultural agroecosystems is partially a result of the powerful large farming lobbies. Agroecology seeks to contest these power relations and transform industrial agriculture by redesigning agriculture around ecological design principles. In addition to being political, agroecology is a broad and diverse movement (Wezel et al. 2009), advanced by various social movements, such as *La Vía Campesina* (Rosset and Martínez-Torres 2012). For students to understand the constraints facing agroecosystem management, there needs to be a move beyond agroecology as a science and set of practices, to agroecology as a political project. This requires focused recognition of the feedback between ecological dynamics, political economic processes, and agroecological movements. Many food systems educators, especially at the university level, already embody this approach; for example, Hillimire et al. (2014) describe how food systems students can simultaneously participate in a fair-trade movement and read critical analyses of market-based interventions. However, in many cases these movement-based experiential learning opportunities do not characterise the mainstream or majority of food systems education. In particular, the connection between agroecology and political intervention is often lost. In this light, we move to our final cornerstone of CFSE, food sovereignty.

Food Sovereignty: Pedagogies in Movement

Fourth and finally, a CFSE framework must be connected to the global movements mobilising around these issues, in this case, the international food sovereignty movement. Food sovereignty is a multifaceted and rapidly evolving perspective and political program with various definitions and principles (Rosset 2008). In the 2007 Nyéléni Declaration, *La Vía Campesina* describes food sovereignty as "the right of peoples to healthy and culturally appropriate food produced through ecologically sound and sustainable methods, and their right to define their own food and agriculture systems" (Patel 2009: 665).

Scholars increasingly recognise that the concept of food sovereignty can be a powerful corrective for food systems education. Davila and Dyball (2015) draw upon educational theories related to transformative learning and critical consciousness to explore how urban food production in Australia can help build control over food production practices, and collective knowledge of the ways food choices influence global food systems. Meyer (2014) describes how indigenous cosmologies are incorporated into an educational project that promulgates a native Hawaiian conception of food sovereignty as synonymous with land and sustenance. In the context of international development, Phillips and Roberts (2011) critically explore the history of farm schools in Tanzania, arguing that "school cultivation curricula should aim at education that does not simply target the knowledge and skill deficit of rural people, but should also address the political, social, and economic orders that construct and produce those deficits in the first place" (73).

While each of these previous examples explores independent initiatives, *La Vía Campesina* provides a concrete example of how food sovereignty education can be scaled up, drawing attention to the educational politics of scale (Meek 2015b). *La Vía Campesina* has a network of more than forty agroecological schools throughout the world, and offers extensive trainings, workshops, and exchanges in other locales (McCune et al. 2014). What makes the agroecological courses of *La Vía Campesina* distinct from the majority of agroecology courses and programs is that these initiatives are directly linked to political movements demanding to not only learn about, but also to implement these alternative agricultural models. Drawing upon these grassroots experiences, we suggest that the purpose of critical food systems education is to contribute to the global food sovereignty movement.

In summary, CFSE is a tripartite perspective consisting of a theoretical framework, set of pedagogies, and vision for policy. It is a perspective that recognises that educators have an explicitly political choice to make: they can use education to reproduce the current food system, raise awareness about the inequities of the food system, or utilise education as a means to form individuals who are determined to transform the food system. In a CFSE framework, critical food systems educators are committed to employing educational processes as part of a larger movement for transforming food systems. We now turn to one concrete example of CFSE in practice.

Integrating Decolonial and Critical Feminist Perspectives into CFSE

Building on this previous framework (Meek and Tarlau 2016), we argue that decolonisation is an important component of critical food systems education. As the following analysis will illustrate, decolonial literature contributes to CFSE by making visible the geopolitics of knowledge, decentring Western and White colonial systems of knowledge, and lifting up grassroots epistemologies and forms of knowing and being in the world. Meanwhile, critical feminism offers a pedagogical approach that acknowledges interlocking/intersectional oppressions, the relationship between experience and consciousness, the role of art, poetry, and performance in learning, the heterogenous collective, education of not for the oppressed, transnational rather than global versus local processes, and the importance of solidarity.

Decolonisation and Food Systems

Herein, we define decolonisation as a material and epistemological process through which historically marginalised groups deconstruct colonial systems of knowledge and land access and management and rebuild sovereignty over knowing, being, and engaging with the natural world. We base this definition upon our reading of the broader "decolonial turn", which is not a single theoretical school, social movement, or identity group but rather a constellation of views that understand coloniality as an ongoing problem in the modern era and decolonisation and decoloniality as unfinished processes (Maldonado-Torres 2011; Mignolo 2009; Ndlovu-Gatsheni 2015). More specifically, our analysis of decolonisation is explicitly grounded in an extensive corpus of scholarship that has and continues to emerge through the modernity/coloniality/decoloniality project (Escobar 2010; Mignolo and Walsh 2018; Pérez 1999). This largely Latin American initiative argues that non-Eurocentric forms of knowing and being, emanating from indigenous, peasant, and other groups on the periphery, are powerful alternatives to the ecologically destructive forces of colonial modernity.

Our usage of decolonisation is both similar to and different from many indigenous perspectives (Alfred 2009). For various indigenous groups and critical scholars, decolonisation centres upon deconstructing colonial ideologies, which hold that Western knowledges and understandings of history are unassailable. Decolonisation from this vantage requires dismantling the colonial social and psychological structures and discourses that are embedded in the mind (Fanon 1967), and continues to preclude indigenous sovereignty. Yet decolonisation from this perspective does much more than deconstruct; it is a culturally grounded process of valorising traditional knowledge systems, practices, and ways of being (Pirbhai-Illich et al. 2017). To decolonise necessitates we reexamine our basic understandings of history and knowledge, asking epistemological questions about how we arrived at such knowledge and doing the difficult work of changing the misconceptions, prejudice, and assumptions surrounding indigenous peoples. Our usage differs from that of Tuck and Yang (2012: 1): "Decolonization brings about the repatriation of indigenous land and life; it is not a metaphor for other things we want to do to improve our societies and schools". They argue that social justice advocates, despite their best intentions, mobilise the concept as a metaphor and that white, immigrant, postcolonial, and other oppressed peoples inadvertently advance settler colonialism by framing their struggles around decolonisation. From this perspective, decolonisation is an explicitly indigenous-led process directed towards the repatriation of native lands, the dissolution of colonial social and psychological structures of oppression, and the creation of spaces within which indigenous groups can prefigure sovereign futures.

Critical scholars and indigenous groups will likely argue—and perhaps correctly so—that it is an inherent contradiction to describe movements—such as those that comprise *La Vía Campesina*, and others that have adopted Marxist analyses, as advancing decolonisation, when it is engaged in the explicit colonisation of land. In this context, agrarian reform movements, regardless of their politics, may not be considered by some as engaged in decolonisation because their ultimate end is not the repatriation of indigenous land, but rather the furthering of settler colonialism (Snelgrove et al. 2014). These are undoubtedly messy contradictions, but ultimately we believe that they are productive ones for both critical scholars and movements of diverse orientations. Rather than "domesticating decolonization" (Tuck and Yang 2012: 3), we believe that our usage of the concept will help more fully actualise indigenous groups' transformative vision by helping restructure a discussion of the "politics of the possible", or the perceived repertoire of forms of political change individuals think of as actual possibilities (Guthman 2011). Analysing the interconnections between the decolonisation of land and knowledge within critical food systems education does the work of opening the doors within movements themselves to thorny debates concerning the radical potential of indigenous-peasant coalitions for structural agrarian reform (Gaztambide-Fernández 2012).

We can contribute to such a broader project of solidarity by recognising that decolonisation is a shared struggle among black, indigenous, and peasant movements. These diverse groups are united around the need to restructure material and subjective relations to land, to deconstruct the social and psychological structures of colonialism, and to bring new visions of modernity into existence. Land for many black and indigenous groups, as well as non-indigenous peasants, has ontological primacy. It is an assemblage of relations of obligations and identity; land mediates how individuals understand their subjectivities and engage with others and the natural world in non-dominating and non-exploitative relations. If decolonial struggles are united in their broad relation to land, then they are also unified in their commitment to transforming the "colonial relation", which consists of cumulative manifestations of interlocking axes of oppression, such as patriarchy, white supremacy, heteronormativity, and totalising understandings of state power (Coulthard 2014). The object of transformation for many black, indigenous, and peasant movements is the colonial relation, a scaling out that emphasises a background context of similar structural relations of oppression. Lastly, black, indigenous, and peasant movements share a drive to dismantle the legacy of colonialism on the mind (Fanon 1967). Colonialism created psychological structures of prejudice and self-negation, which colonised peoples throughout the world have long internalised. Decolonising the mind requires epistemically delinking from modernity and the assumed primacy of Eurocentric knowledge and simultaneously epistemically reconnecting with subaltern forms of knowledge as a process of developing alternative futures in the present (Mignolo and Walsh 2018; Quijano 2007).

Schools are a central location to explore how decolonisation can advance such a visionary political and ecological project. Educational institutions have historically served as a colonising force, controlling "what knowledge-making is allowed, disavowed, devalued or celebrated" (Mignolo 2009: 176). Schools and universities have functioned as instruments of colonialism because they train individuals who will be epistemologically obedient. Yet our research shows that schools and other spaces of learning can also serve as powerful decolonising instruments, challenging assumptions of agricultural modernisation and singular conceptions of what constitutes scientific knowledge. Various educational institutions that we have analysed in our respective research are all spaces of epistemic disobedience (Meek 2015a, b, 2016, 2020; Tarlau 2014, 2015, 2019). They valorise peasant systems of agrarian knowledge and practice, training students to delink from the "magic of the Western idea of modernity, ideals of humanity and promises of economic growth and financial prosperity" (Mignolo 2009: 161). We believe that teaching about agroecology, studying agroecology, discussing agroecology with one's family, and practising agroecology on one's land all constitute epistemic disobedience because these are processes of liberating what critical scholar and activist Vandana Shiva (1993) aptly terms the "monoculture of the mind". Shiva's phrasing is powerfully symbolic; it highlights how Western modernity is colonised by a homogenous epistemology in which only particular Western conceptions of empirical knowledge exist but also importantly that landscapes defined by monoculture are a product of this singular mentality. One of the major lessons we learned in our research is that decolonisation is not just a process of transforming the relations of knowledge production in schools but also a much broader project of transforming dominant conceptions of livelihoods and landscapes.

CFSE and Intersectional and Transnational Feminisms

In addition to this decolonial approach, we also believe it is important to incorporate the theoretical, practical, and pedagogical lessons from critical feminisms into a critical food systems education approach. Here we are defining critical feminisms as the intersectional theorising of both women of colour in the United States, including Black feminism, Chicana feminisms, and Indigenous feminisms, as well as the postcolonial and transnational feminisms developed primarily by women from the global South (Tambe and Thayer 2021).

Perhaps the most globally recognised concept emerging from U.S. Black feminist thought is the idea of intersectionality, a term coined by Crenshaw (1991) "to denote the various ways in which race and gender interact to shape the multiple dimensions of Black women's employment experiences" (1244). Previous Black feminist theorising had also suggested the need for an integrated framework for understanding class, race, and sexuality. For example, the Black feminist Combahee River Collective wrote in 1977 that, "The most general statement of our politics at the present time would be that we are actively committed to struggling against racial, sexual, heterosexual, and class oppression and see our particular task the development of integrated analysis and practice based upon the fact that the major systems of oppression are interlocking" (271). This approach, distinct from the forms of identity politics that silo different struggles and ignore class analysis, promotes the recognition of difference within anti-capitalist struggles.

Black feminist thought has developed in relationship to the histories of black women's activism and black women's intellectual contributions, the latter of which has often appeared through poetry and other forms of art. Hill Collins (1991) has described six components of this theoretical and activist approach, including a dialectical relationship between theorising and activism or experience and consciousness, a recognition of diversity within similarities, the importance of group knowledge, prioritising the contributions of Black women intellectuals and activists, taking into consideration changing social conditions, and the direct relationship to other projects of social justice (Hill Collins 1991). Chicana and Latinx feminisms, which have developed in conversation with and alongside Black feminisms (Moraga and Anzaldúa 2021), have theorised from the space of the borderlands, or the liminal space between false binaries. Anzaldúa (1987) argues that rather than a location of powerlessness, this liminal space produces a mestiza consciousness, which has the potential for a "massive uprooting of dualistic thinking in the individual and collective consciousness is the beginning of a long struggle, but one that could, in our best hopes, bring us to the end of rape, of violence, or war" (102). Thus, it is the marginalisation of women, immigrants, Black, and indigenous communities that they hold the potential for developing a new consciousness that radically breaks with the current paradigms.

Finally, postcolonial and transnational feminisms are deeply aligned with these intersectional feminisms, while also taking distinct global processes as their focus of analysis. Postcolonial feminisms have interrogated the colonial project from the perspective of the third world women, critiquing discourses and strategies of control in which "white men saving brown women from brown men" (Spivak 1988: 297). Furthermore, postcolonial feminisms have questioned the representation of third world women within the academy (Minh-ha 1987) and the attempts to speak for or on behalf of marginalised violence, which Spivak (1988) argues is a form of "epistemic violence". Transnational feminisms have critiqued theories of globalisation that position the "global" as an all-powerful force that imposes itself on "local" communities, and instead, advocated for studies that examine the "the real and concrete effects of global restructuring on raced, classed, national, sexual bodies of women in the academy, in workplaces, streets, households, cyberspaces, neighbourhoods, prisons, and social movements" (Mohanty 2003: 516). This method of inquiry and activism examines interconnections between diverse geographical locations and how transnational processes both shape and are reshaped by local communities in order to promote ethical collaboration and solidarity (Collective and Nagar 2006; Tambe and Thayer 2021).

These critical feminisms have important implications for our approach to food systems education. First, in contrast to liberal and white feminist approaches, intersectional and transnational feminisms centre the question of class while also examining how race, gender, and sexuality shape those class experiences and communities' position within the global capitalist system. This approach has been central to U.S. food justice movements, which have been led by the communities of colour that are also facing the most critical food needs. This intersectional approach is also embodied in the rallying call of grassroots agrarian social movements: "There is not socialism without feminism!" A critical food systems education will necessarily include an examination of a multiplicity of oppressions within an overall anti-capitalist framework. Second, Black and Latinx feminists' emphasis on the relationship between experience and consciousness, and the importance of theorising from liminal spaces, is directly relevant to CFSE and the Freirean idea of praxis. It is not from the academy that the accumulation of knowledge about alternatives to the food systems emerge, but rather, through the theorising of the practices of actual existing food justice and food sovereignty movements, who are able to develop these alternatives because they exist in the liminal space between the dominant agribusiness model and the urban economy. In addition, women of colour's experience and intellectual contributions are often documented through poetry and artwork, suggested the need to centre art and culture within critical food systems education. The already central role of art and performance in many food systems education programs, known in Latin

America as *mística*, is a testament to the importance of the educational value of cultural production.

Third, the focus on group knowledge as well as the heterogenous collective allows for the development of distinct food systems educations as well under the banner of one global project for food sovereignty, an approach epitomised by *La Vía Campesina* and its more than a hundred members organisations. Fourth, the critique in all these feminisms of speaking on behalf of women of colour in the global North and women throughout the global South, and the characterisation of this representation as a form of epistemic violence, suggests that grassroots movements should be the central protagonists in developed food systems education. This is reminiscent of the importance of developing a Pedagogy *of* the Oppressed (Freire 1973), not for or on behalf of the oppressed, or similarly an Education *of* the Countryside, not simply in the or for the countryside (Pinheiro Barbosa 2017; Pinheiro Barbosa and Rosset 2017).

Fifth, transnational feminists' rejection of global capitalism as an all-powerful force is an important contribution, especially when examining the implications of global agribusiness and the neoliberal food regime (Rosset 2006). Transnational feminism suggests that we should examine these global processes as constituting but simultaneously constituted by agrarian social movements, who have appropriate transnational discourses for their own ends. These movements and their food systems education programs have developed a transnational agrarian "counter-public", or "an oppositional space in which networks, organizations and individuals who share certain values or identities engage with one another around a core theme.... articulate collective identities, debate discourses and strategies and construct alternative forms of relations with one another" (Thayer 2010: 26). Finally, the role of solidarity in all these critical feminisms is a central goal of food systems education, which includes solidarity across intersecting oppressions and globally across agrarian movements. Embodying this solidarity are the many agroecology programs in Latin America, which bring together organisations from diverse countries to study and learn together. During the current global pandemic, the Brazilian Landless Workers' movement were leaders in this praxis of solidarity, drawing on their vast social movement infrastructure to donate 6 thousand tons of food to poor communities throughout Brazil. In this example, donating food grown through agroecological methods is in itself an educational process, teaching communities that their ability to survive a pandemic will depend not on the state but the struggle for food sovereignty.

Conclusion: Educating for Food Sovereignty

Food systems education is rapidly transforming, informed by diverse popular movements throughout the world. Throughout this chapter, we have advocated for a critical approach to food systems education, taking as our starting point Apple and Aasen's reminder that "education, every aspect of it one can imagine, is political" (Apple and Aasen 2003: 1).

The first section of this chapter reviewed the four basic pillars of critical food systems education, and explored its tripartite on theory, pedagogy, and policy. Notably, these four cornerstones of CFSE are not simply bodies of academic literature, but are also actual practices that social movements have been incorporating into their grassroots struggle for decades. We then drew out the lessons from each of the four cornerstones of the CFSE framework. First, from the practices of popular education, we take Freire's ideas of education as a tool for conscientisation and liberation; the importance of praxis, connecting theory to practice; the non-apologetic acknowledgement of education as a political act; and specific pedagogical practices such as problem-posing education, generative themes, and community-based research. Second, from the experiences of food justice struggles, we take the importance of analysing race- and class-based divisions in the food system, and the need to confront head-on questions of race and class in all food systems programs. Just as importantly, we integrate the idea that marginalised groups have the right to define their own food systems and become active participations in the construction of any educational program concerning this system.

The third cornerstone we highlight is agroecology, which is a common practice within many food systems education programs. However, we choose to highlight not only the scientific and practical contributions of agroecology, but also the political nature of this agroecological approach.

Valuing traditional ecological knowledge and practice is a political act, not simply a scientifically more efficient form of agricultural production. In other words, our focus on incorporating the practices of agroecology into our CFSE framework, such as the reduction of external inputs and the creation of feedback loops, is about promoting an alternative political and economic project that contests the currently dominant agribusiness agricultural model. The fourth cornerstone of CFSE is its connection to the global food sovereignty movement. Both the practice and theory of food sovereignty pay attention to the influence of politics and economic systems on food, and furthermore, the role that power, scale, and history play in determining agricultural production. However, most important for our analysis, food sovereignty is also about creating a global movement of thousands of groups struggling to implement a fairer, more just food system. Food systems education programs should begin to think about themselves in relationship to its contribution to this global movement.

Our chapter then made two novel theoretical additions to the critical food systems education framework. First, we sought to integrate a decolonial perspective. Acknowledging the very real political differences between disparate understandings of decolonisation, we drew upon the modernity–coloniality–decoloniality project, in advocating a conception of decolonisation that can be relevant to any of the black, indigenous, peasant, and other subaltern movements that are seeking to undo the long histories of colonialism on the mind, land, and education system. Second, we lifted up the voices and insights of black, Chicana, and transnational feminists, and integrating central elements of their political vision into the CFSE framework. Through developing concepts, such as intersectionality, these movements have made key theoretical inroads in visibilising structures of oppression and opportunities for social transformation. Drawing upon these movements, we advocate taking seriously the importance of solidarity, transnationalism, and group knowledge, and integrating these visions into future critical food systems education initiatives both within and outside the academy. More than a set of simply theoretical innovations, this chapter has sought to lift up and synthesise the diverse examples of global social movements who remain on the front lines of critical food systems education.

References

Alfred, T. 2009. Restitution Is the Real Pathway to Justice for Indigenous Peoples. In *Response, Responsibility, and Renewal*, ed. J. Dewar, G. Younging, and M. DeGagne, 179–187. Ottawa: Aboriginal Healing Foundation.

Allen, P., and J. Guthman. 2006. From "Old School" to "Farm-to-School": Neoliberalization from the Ground Up. *Agriculture and Human Values* 23: 401–415.

Altenbaugh, R.J. 1990. *Education for Struggle: The American Labor Colleges of the 1920s and 1930s*. Philadelphia, PA: Temple University Press.

Althusser, L. 1984. *Ideology and Ideological State Apparatuses: Essays on Ideology*, trans. B. Brewster. London: Verso.

Anyon, J. 1981. Social Class and School Knowledge. *Curriculum Inquiry* 11: 3–42.

Anzaldúa, G. 1987. *Borderlands/ La Frontera: The New Mestiza*. Spinsters/Aunt Lute.

Apple, M.W. 2004. *Ideology and Curriculum*. New York: Routledge Falmer.

Apple, M.W., and P. Aasen. 2003. *The State and the Politics of Knowledge*. New York: Routledge.

Arnove, R.F. 1986. *Education and Revolution in Nicaragua*. New York: Praeger.

Aronowitz, S., and H.A. Giroux. 1991. *Postmodern Education: Politics, Culture, and Social Criticism*. Minneapolis, MN: University of Minnesota Press.

Berryman, P. 1987. *Liberation Theology: Education at Empire's End*. New York: Pantheon Books.

Bowles, S., and H. Gintis. 1976. *Schooling in Capitalist America*. New York: Basic Books.

Collective, S.W., and R. Nagar. 2006. *Playing with Fire: Feminist Thought and Activism Through Seven Lives in India*. Minneapolis, MN: University of Minnesota Press.

Combahee River Collective. 1977. *The Combahee River Collective Statement*. Boston.

Coulthard, G.S. 2014. *Red Skin, White Masks: Rejecting the Colonial Politics of Recognition*. Minneapolis: University of Minnesota Press.

Crenshaw, K. 1991. Mapping the Margins: Intersectionality, Identity Politics, and Violence Against Women of Color. *Stanford Law Review* 43 (6): 1241–1299.

Davila, F., and R. Dyball. 2015. Transforming Food Systems Through Food Sovereignty: An Australian Urban Context. *Australian Journal of Environmental Education* 31 (1): 34–45.

de Molina, M.G. 2013. Agroecology and Politics: How to Get to Sustainability? About the Necessity for a Political Agroecology. *Agroecology and Sustainable Food Systems* 37: 45–59.

Escobar, A. 2010. Latin America at a Crossroads: Alternative Modernizations, Post-Liberalism, or Post-Development? *Cultural Studies* 24 (1): 1–65.

Etmanski, C. 2012. A Critical Race and Class Analysis of Learning in the Organic Farming Movement. *Australian Journal of Adult Learning* 52 (3): 484–506.

Fanon, F. 1967. *Black Skin, White Masks*, trans. C.L. Markmann. New York: Grove.

Flowers, R., and E. Swan. 2011. "Eating at Us": Representations of Knowledge in the Activist Documentary Film Food, Inc. *Studies in the Education of Adults* 43 (2): 234–250.

Francis, C.A., G. Lieblein, J. Helenius, L. Salomonsson, H. Olsen, and J. Porter. 2001. Challenges in Designing Ecological Agriculture Education: A Nordic Perspective on Change. *American Journal of Alternative Agriculture* 16 (2): 89–95.

Freire, P. 1973. *Pedagogy of the Oppressed*. New York: Continuum International.

Galt, R., D. Parr, and J. Jagannath. 2012. Facilitating Competency Development in Sustainable Agriculture and Food Systems Education: A Self-Assessment Approach. *International Journal of Agricultural Sustainability* 11 (1): 69–88.

Galt, R.E., D. Parr, J. Van Soelen Kim, J. Beckett, M. Lickter, and H. Ballard. 2013. Transformative Food Systems Education in a Land-Grant College of Agriculture: The Importance of Learner-Centered Inquiries. *Agriculture and Human Values* 30 (1): 129–142.

Gaztambide-Fernández, R.A. 2012. Decolonization and the Pedagogy of Solidarity. *Decolonization: Indigeneity Education and Society* 1: 41–67.

Guthman, J. 2008. Bringing Good Food to Others: Investigating the Subjects of Alternative Food Practice. *Cultural Geographies* 15: 431–447.

Guthman, Julie. 2011. Weighing In: Obesity, Food Justice, and the Limits of Capitalism. Vol. 32. University of California Press.

Hill Collins, P. 1991. *Black Feminist Thought: Part I: The Social Construction of Black Feminist Thought*. New York: Routledge.

Hillimire, K. et al. (2014). Food for Thought: Developing Curricula for Sustainable Food Systems Education Programs. *Agroecology and Sustainable Food Systems* 722–743.

Holt-Gimenez, E., and A. Shattuck. 2011. Food Crises, Food Regimes and Food Movements: Rumblings of Reform or Rides of Transformation? *Journal of Peasant Studies* 38 (1): 109–144.

Jordan, N., D. Andow, and K. Mercer. 2005. Ecology of Agricultural Systems: A Service-Learning Course in Agroecology. *Journal of Natural Resources and Life Sciences Education* 34: 83–89.

Lieblein, G., and C. Francis. 2007. Towards Responsible Action Through Agroecological Education. *Italian Journal of Agronomy* 2 (2): 83–90.

Lieblein, G., C. Francis, W. Barth-Eide, H. Torjusen, S. Solberg, and L. Salomonsson. 2000. Future Education in Ecological Agriculture and Food Systems: A Student–Faculty Evaluation and Planning Process. *Journal of Sustainable Agriculture* 16 (4): 49–69.

Maldonado-Torres, N. 2011. Thinking Through the Decolonial Turn: Post-Continental Interventions in Theory, Philosophy, and Critique—An Introduction. *Transmodernity: Journal of Peripheral Cultural Production of the Luso-Hispanic World* 1 (2): 1–15.

Mares, T.M., and A.H. Alkon. 2011. Mapping the Food Movement: Addressing Inequality and Neoliberalism. *Environment and Society: Advances in Research* 2 (1): 68–86.

Martins, M.F.A., and S.S. Rodrigues (eds.). 2015. *PRONERA: Experiências de Gestão de Uma Política Pública*. São Paulo: Compacta Gráfica Editora.

McCune, Nils, and Marlen Sánchez. 2019. Teaching the territory: Agroecological pedagogy and popular movements. *Agriculture and Human Values* 36 (3): 595–610.

McCune, N., J. Reardon, and P.M. Rosset. 2014. Agroecological Formación in Rural Social Movements. *Radical Teacher* 98: 31–38.

McCune, N. et al. (2017). Mediated territoriality: Rural workers and the efforts to scale out agroecology in Nicaragua. *The Journal of Peasant Studies* 44(2): 354–376. http://www.tandfonline.com/doi/abs/310.1080/03066150.03062016.01233868.

McLaren, P. 2003. *Life in Schools: An Introduction to Critical Pedagogy in the Foundations of Education*. Boston: Pearson Education.

Meek, D. (2016). Cultural Politics of the Agroecological Transition. *Agriculture & Human Values* 33(2): 275–290.

Meek, D. 2015a. Learning as Territoriality: The Political Ecology of Education in the Brazilian Landless Workers' Movement. *Journal of Peasant Studies* 42: 1179–1200.
Meek, D. 2015b. Towards a Political Ecology of Education: The Educational Politics of Scale in Southern Pará, Brazil. *Environmental Education Research* 21: 447–459.
Meek, D., and R. Tarlau. 2016. Critical Food Systems Education (CFSE): Educating for Food Sovereignty. *Agroecology and Sustainable Food Systems* 40 (3): 237–260.
Meyer, M.A. 2014. Hoea ea: Land Education and Food Sovereignty in Hawaii. *Environmental Education Research* 20 (1): 98–101.
Meek, David. 2020. *The Political Ecology of Education: Brazil's Landless Workers' Movement and the Politics of Knowledge*. West Virginia University Press.
Minh-Ha, Trinh T. 1987. Difference: 'A Special Third World Women issue'. *Feminist Review* 25 (1): 5–22.
Mignolo, W.D. 2009. Epistemic Disobedience, Independent Thought and Decolonial Freedom. *Theory, Culture and Society* 26 (7–8): 159–181.
Mignolo, W., and C.E. Walsh. 2018. *On Decoloniality: Concepts, Analytics, Praxis*. Durham, NC: Duke University Press.
Mohanty, C.T. 2003. "Under Western Eyes" Revisited: Feminist Solidarity Through Anticapitalist Struggles. *Signs: Journal of Women in culture and Society* 28 (2): 499–535.
Moraga, C., and G. Anzaldúa. 2021. *This Bridge Called My Back, Fortieth Anniversary Edition: Writings by Radical Women of Color*. Albany, NY: State University of New York Press.
Morris, A.D. 1984. *The Origins of the Civil Rights Movement: Black Communities Organizing for Change*. New York: Free Press.
Ndlovu-Gatsheni, S.J. 2015. Decoloniality as the Future of Africa. *History Compass* 13 (10): 485–496.
Ostergaard, E., G. Lieblein, T.A. Breland, and C. Francis. 2010. Students Learning Agroecology: Phenomenon-Based Education for Responsible Action. *Journal of Agricultural Education and Extension* 16: 23–27.
Parr, D. M., et al. (2007). Designing sustainable agriculture education: Academics' suggestions for an undergraduate curriculum at a land grant university. *Agriculture and Human Values* 24(4): 523–533. http://link.springer.com/article/510.1007/s10460-10007-19084-y.
Parr, D., and C.J. Trexler. 2011. Students' Experiential Learning and Use of Student Farms in Sustainable Agriculture Education. *Journal of Natural Resources and Life Sciences* 40: 172–180.

Patel, R. 2009. What Does Food Sovereignty Look Like? *Journal of Peasant Studies* 36 (3): 663–706.

Payne, C., and C.S. Strickland, eds. 2008. *Teach Freedom: Education for Liberation in the African-American Tradition*. New York: Teachers College Press.

Pérez, E. 1999. *The Decolonial Imaginary: Writing Chicanas into History*. Bloomington: Indiana University Press.

Phillips, K.D., and D. Roberts. 2011. Cultivating Schools for Rural Development: Labor, Learning, and the Challenge of Food Sovereignty in Tanzania. In *School Food Politics: The Complex Ecology of Hunger and Feeding in Schools Around the World*, ed. S.A. Robert and M.B. Weaver-Hightower, 71–93. New York: Peter Lang.

Pirbhai-Illich, F., P. Shauneen, and F. Martin, eds. 2017. *Culturally Responsive Pedagogy: Working Towards Decolonization, Indigeneity and Interculturalism*. New York: Springer.

Pinheiro Barbosa, L. 2017. Educação do Campo [Education for and by the Countryside] as a Political Project in the Context of the Struggle for Land in Brazil. *The Journal of Peasant Studies* 44 (1): 118–143.

Pinheiro Barbosa, L., and P.M. Rosset. 2017. Educação do Campo e Pedagogia Camponesa Agroecológica na América Latina: Aportes da La Vía Campesina e da Cloc. *Educação e Sociedade* 38 (140): 705–724.

Quijano, A. 2007. Coloniality and Modernity/Rationality. *Cultural Studies* 21 (2–3): 168–178.

Rosing, H. 2012. Demystifying the Local: Considerations for Higher Education Engagement with Community Food Systems. *Journal of Agriculture, Food Systems, and Community Development* 2 (4): 79–84.

Rosset, P.M. 2006. *Food Is Different: Why We Must Get Out of Agriculture*. London: Zed Books.

Rosset, P.M. 2008. Food Sovereignty and the Contemporary Food Crisis. *Development* 51 (4): 460–463.

Rosset, P.M., and M.E. Martínez-Torres. 2012. Rural Social Movements and Agroecology: Context, Theory, and Process. *Ecology and Society* 17: 3.

Salomonsson, L., A. Nilsson, S. Palmer, A. Roigart, and C. Francis. 2009. Farming Systems Education: Case Study of Swedish Test Pilots. *Renewable Agriculture and Food Systems* 24: 48–59.

Shiva, V. 1993. *Monocultures of the Mind: Perspectives on Biodiversity and Biotechnology*. New York: Palgrave Macmillan.

Snelgrove, C., R.K. Dhamoon, and J. Corntassel. 2014. Unsettling Settler Colonialism: The Discourse and Politics of Settlers, and Solidarity with

Indigenous Nations. *Decolonization: Indigeneity, Education and Society* 3 (2): 1–32.

Spivak, G. 1988. Can the Subaltern Speak? In *Marxism and Interpretation of Culture*, 271–313. Urbana: University of Illinois Press.

Sawyer, S. 2004. *Crude Chronicles: Indigenous Politics, Multinational Oil, and Neoliberalism in Ecuador*. Chapel Hill, NC: Duke University Press.

Sumner, J. 2013. Food Literacy and Adult Education: Learning to Read the World by Eating. *Canadian Journal for the Study of Adult Education* 25 (2): 79–92.

Tambe, A., and M. Thayer. 2021. *Transnational Feminist Itineraries: Situating Theory and Activist Practice*. Chapel Hill, NC: Duke University Press.

Tarlau, R. 2014. From a Language to a Theory of Resistance: Critical Pedagogy, the Limits of "Framing", and Social Change. *Educational Theory* 64 (4): 369–392.

Tarlau, R. 2015. How Do New Critical Pedagogies Develop? Educational Innovation, Social Change, and Landless Workers in Brazil. *Teachers College Record* 117: 110304.

Tarlau, R. 2019. *Occupying Schools, Occupying Land: How the Landless Workers' Movement Transformed Brazilian Education*. Oxford: Oxford University Press.

Tuck, E., and K.W. Yang. 2012. Decolonization Is Not a Metaphor. *Decolonization: Indigeneity, Education and Society* 1 (1): 1–40.

Thayer, M. 2010. *Making Transnational Feminism: Rural Women, NGO Activists and Northern Donors in Brazil*. New York: Routledge.

Wezel, A., S. Bellon, T. Dore, C. Francis, D. Vallod, and C. David. 2009. Agroecology as a Science, a Movement, and a Practice. *Agronomy for Sustainable Development* 29 (4): 503–515.

Willis, L. 1977. *Learning to Labor: How Working Class Kids Get Working Class Jobs*. New York: Teachers College Press.

16

Territorial Resistance and Peasant Food Systems in Brazil

Camila Ferracini Origuéla

Introduction

This chapter is aimed at analysing territorial resistance and peasant food systems built by socio-territorial movements, such as the Landless Rural Workers Movement (*Movimento dos Trabalhadores Rurais Sem-Terra*—MST). Territorial resistance can be understood as struggles, actions, practices, organisations, and autonomous and emancipatory techniques and technologies created and applied by peasant movements in a given space for food production, processing, and commercialisation, especially agroecological food. Depending on its level of organisation, scope of action and acquired political, economic, environmental, and cultural features, territorial resistance can give rise to peasant food systems (Origuéla 2019), as it brings peasants closer to nature, traditional knowledge, regional cultures, and to other peasants, by means of organising

C. F. Origuéla (✉)
São Paulo State University (Unesp), São Paulo, Brazil
e-mail: camila.ferracini@unesp.br

A. A. R. Ioris and B. Mançano Fernandes (eds.), *Agriculture, Environment and Development*, https://doi.org/10.1007/978-3-031-10264-6_16

steering groups, farmers' associations, agricultural cooperatives, family or collective agro-processing, and gathering both rural and urban workers together around farmers' markets.

Understanding the scope, characteristics, potential, and challenges of territorial resistance and how fundamental this is for the construction of peasant food systems are the main contributions of this article to those interested in studies on peasantry in the twenty-first century. To understand territorial resistance and peasant food systems, we analyse experiences of food production, processing, and commercialisation developed by the landless movement in rural settlements located in the state of Rio Grande do Sul (RS), more specifically in the Metropolitan Region of Porto Alegre (RMPA).[1] In this administrative area, since the 1990s, peasants settled in agrarian reform territories have been producing agroecological foods, such as vegetables and rice. The state has a history of struggle for land that contributed to the formation of the MST in the 1970s and to its consolidation in 1984. Besides, the RMPA stands out in the production of agroecological foods, and the initiative has both national and international relevance in terms of peasant resistance.

In order to understand territorial resistance and peasant food systems, the methodological approach included: (a) literature review; (b) survey and systematisation of primary and secondary data; (c) participant observation; (d) semi-structured interviews with peasants settled in the RMPA, members of MST's cooperatives and agro-processing facilities and members of public institutions, such as the National Institute of Colonisation and Land Reform (*Instituto Nacional de Colonização e Reforma Agrária*—INCRA), the Technical Assistance and Rural Extension Services Association—Rio Grande do Sul branch (*Associação Riograndense de Empreendimentos de Assistência Técnica e Extensão Rural*—EMATER- RS/Ascar) and the State Department for Rural Development, Fisheries and Cooperatives. All interviewees were involved in the organisation and development of the experience carried out in RMPA. The number of interviews was defined according to parameters of diversification—interviews include different key informants—and

[1] The RMPA comprises 33 municipalities. There are rural settlements in eight of them, namely: Capela de Santana, Charqueadas, Eldorado do Sul, Guaíba, Montenegro, Nova Santa Rita, São Jerônimo, and Viamão. Altogether, there are 17 rural settlements and 1172 settled families.

saturation—when arguments emerging in the interviews become repetitive.

The chapter is organised around three topics. In the first topic, territory, territorialities, and peasant territorial resistance are addressed through the peasant approach of Agrarian Question Paradigm (AQP). The theoretical and methodological pathways of AQP are fundamental for conducting a geographical, that is, territorial analysis of the peasantry. In the second topic, aiming at understanding the actions, practices, and struggles for autonomy and emancipation in food production, the analysis is focused on the territorialisation of MST in Rio Grande do Sul, especially in the RMPA. Finally, as the third topic, the development of peasant food systems is addressed, with a focus on their social, political, economic, and environmental prospects and challenges in the face of the agrarian question.

Territory, Territorialities, and Territorial Resistance

The expansion of capitalist agriculture, as well as its consequences for peasant farmers, can be approached from two paradigms, the Agrarian Question Paradigm (AQP) and the Agrarian Capitalism Paradigm (ACP). Both paradigms constitute theoretical and methodological pathways that comprise theories, methods, concepts, and intentionality (Fernandes 2008a, b). In brief, the AQP sees the agrarian question as a structural problem of capitalism. The ACP, in turn, sees the peasant as the problem—a person unable to integrate into a capitalist mode of production, who, therefore, becomes a family farmer—and understands agribusiness as the only model for agricultural development. The AQP, taken here as a theoretical and methodological framework, consists of two trends: the *proletarianist* trend, for which the capitalist mode of production will either lead to the disappearance of the peasantry (Kautsky 1998) or to its proletarianisation (Lenin 1982); and the *peasantist* trend, which argues that peasantry tends to be (re)created by the uneven and combined development of the capitalist mode of production in agriculture (Oliveira 1991) or by reaction in the form of land occupation (Fernandes 2000).

By being (re)created through land occupation—an action organised by socio-territorial movements that struggle for land—the peasant is reintroduced into production of non-capitalist relations of production (Martins 1981). Although peasants hold a piece of land for the production of goods, this does not make them capitalists (Chayanov 1984). The purpose of peasant work is to satisfy the needs of the family, instead of making an average rate of profit, as is the case with capitalists. Capital enters the peasant economic unit through the subjection of land income by means of subordinating it through the action of monopolies (Paulino 2012). This can bring about the territorialisation of monopolistic capital or the monopolisation of the territory by capital (Oliveira 1991). This last process is also addressed as territoriality of capital in peasant territory (Ross 2015). Territorialisation of capital may lead to peasant deterritorialisation in a certain place followed by reterritorialisation somewhere else (as expressions of the politicised production of space). Under monopolisation of territory by capital, however, the condition for peasant reproduction is subordination.

Territorialisation of capital establishes both capitalist relations of production, such as wage labour, and non-capitalist relations of production, which contradict it, such as family work. The territorialisation of capital allows capitalists to directly control the social relations that constitute the territory. Monopolisation of territory by capital, on the other hand, allows indirect control. Such control is exerted through the commodity, the product of peasant work, and, in most cases, through the way it is produced, that is, which techniques, technologies, and agricultural inputs are used in the production process—all of it provided by agribusiness companies. Peasants are, thus, integrated into the market through the work contained in their product. The appropriation of land income by capitalists is guaranteed by the monopoly this social class exerts not only over land ownership, but also over the circulation of goods and production of agricultural inputs. In the processes of territorialisation, deterritorialisation, and reterritorialisation of either capital or peasants, territories are constructed, destroyed, and disputed, producing conflicts (Fernandes 2008a, b).

The territorialisation of capital in the countryside produces, on the one hand, the deterritorialisation of peasants and, on the other hand, their reterritorialisation. That is, while peasants are expropriated in some places, they are (re)created elsewhere. Those who are not (re)created by the uneven and combined development of the capitalist mode of production organise into socio-territorial movements, occupy land, set up camps, hold demonstrations and, thus, struggle for land and agrarian reform. Peasants reclaim and occupy unproductive properties setting up camps and demanding the expropriation of large estates. Land occupations and encampments compel the Brazilian State to create rural settlements. Rural settlements are peasant territories. In this case, the territorialisation of the peasantry occurs through struggles, resistances, confrontations, and demonstrations in the countryside and in the city. Once land is secured, peasants struggle to stay on it, resist by devising different political and economic strategies.

Peasant territory is distinct from capitalist territory, although they are related under the wider, hegemonic logic of a capitalist society. This relationship is antagonistic since it produces, simultaneously, subordinate peasant territorialities and autonomous peasant territorialities. To understand them, we can resort to the theorisation of Shanin (1979), according to which, it is not possible to understand the peasantry without analysing the society within which peasants are inserted. Accordingly, the peasant territory must be analysed both as such, considering its particularities—that is, the social relations that are typical of the peasant individual and family—and as part of the societal structure within which it exists and resists. Ploeg (2016), in turn, understands that peasant economic units are governed by balances, or attempts to achieve them. Such balances are both internal and external. The internal ones involve balances between people and nature, production and reproduction, internal and external resources, autonomy and dependence, scale and intensity. The external ones are between city and countryside, agriculture, processing, and commercialisation; State and peasant; agrarian growth and demographic growth.

Even presenting social and economic characteristics common to any society, the peasant family unit develops relationships with the hegemonic mode of production, the State, and other social and economic agents by producing balances. In peasant territory, and in its relationship with capitalist territories, different territorialities are produced. The territory can be understood as a social, historical, and relational construction, constituted in the process of appropriation of space by individuals (Raffestin 1993). Territory is the result of the interactive agency of individuals or agents in space-time. Territorialisation produces territories and, consequently, territorialities. These, in turn, are systems of political, economic, and cultural relations between humans and between humans and nature. The social relationships that produce peasant territories are family-based, based on family work, on coevolution relationship between humans and nature, on traditional knowledge. Such relationships aim at the reproduction of the family on the land based on food production.

In the peasant territory, depending on its relationship with the capitalist mode of production, new territorialities are built; in other words, new socio-territorial actions, relationships, and practices. These territorialities are categorised here as subordinate territorialities and autonomous territorialities. Such analytical framework allows for a deeper understanding of peasant territories. Subordinate territorialities occur when peasants move away from nature, traditional knowledge, socio-territorial movements, modes of peasantry organisation, and farmers' markets and approach the capitalist mode of production, particularly, agribusiness companies. When this happens, peasants begin to produce commodities using the technological package of agribusiness, participating in global chains. Many establish connections with large companies and may produce directly for these.

Peasants' approach to capital, although seemingly advantageous, deepens their subordination, since agribusiness companies define what should be produced and how, constantly pressuring peasants to adapt to their norms and production methods. Companies determine all stages of the production and distribution process. They define seeds, food, knowledge, techniques and technologies, financing, and markets. Even production span of time is dictated by them, no longer by nature, not even by traditional knowledge, but by capital. It means a total techno-economic subordination. The peasant family does not decide, control, or determine anything. They just work and the results of this work, the commodity, belong to the company. Capital subordinates the peasant territory and subjugates its labour process. Besides appropriating land income, agribusiness companies determine the prices of food produced and all the agricultural inputs necessary for its cultivation. Every time peasants interact with the forces of capital, they become more subordinate.

The peasant's relationship with nature is fundamental in defining subordination, because the further away from nature and, consequently, from its knowledge, the closer it is to capital. This is why agroecology has become so important for peasant social reproduction, providing peasants with relatively more autonomy in recent decades. Although subordination seems to be a path of no return, not all peasants are able to adapt to the norms and the production methods of agribusiness companies. Most of them have no access to financial resources for this. Others, who manage to integrate, soon realise how subordinate they are and the consequences of this for the family. In this context of marginalisation, dependence, and subordination, resistance is built. Such resistances are based on peasant territories and territorialities, that is, on a coevolution relationship between peasant and nature, on recovering traditional knowledge and on the use of peasant techniques and technologies.

When territorial resistances take place, autonomous territorialities are constructed. Such territorialities consist of political, economic, and cultural relationships that aim at reducing or eliminating subordination of the peasant to capital. These are agroecological and cooperative actions, struggles, and practices that often go beyond peasant territories. In these cases, the peasant has autonomy in the production process, produces the seed and the food in a coevolution relationship with nature. By organising into collectives, groups, associations, and cooperatives, peasants can more easily afford agricultural inputs, without depending on companies for this. They can even process food, thus diversifying marketing opportunities, and build farmers' markets, getting closer to rural and urban workers.

During their life, peasants develop different strategies for social reproduction, and for individual and collective resistances (Shanin 2008).

Being based on peasants' territories and territorialities, such resistances can be classified as territorial resistances, which highlights the importance of both the peasant territory and peasant use of the territory, allowing for their resistance in the capitalist mode of production. Therefore, although having land is fundamental for the peasants' existence, its use, that is, the socio-territorial relations produced on it, is vital for their resistance in the face of the perversities of capital. Depending on the political, economic, and cultural relations built in their territories, peasants' social reproduction becomes more autonomous. Therefore, peasants produce both subordinate and autonomous territorialities. A subordinate peasant can become autonomous and vice versa.

Territorialisation of MST in the Metropolitan Region of Porto Alegre: Autonomy in the Production of Agroecological Foods

The consolidation of the Landless Workers' Movement (MST) as a socio-territorial movement comprised three distinct periods, according to Fernandes (2013). The first one (1978–1985) corresponds to the processes of conception and formation of the movement in the South and Southeast regions, respectively. The second period (1985–1990) refers to the spread of the movement to other regions of the country. In this period the MST organised, establishing a structure with coordination, national board, national secretariat, and sectors. In 1990, the third period began with the institutionalisation of the movement. The 1990s and early 2000s were marked by the intensification and massification of land occupations across the country, which resulted in the creation of several rural settlements. A fourth period can be added, starting in the 2000s, with the organisation of autonomous and emancipatory socio-territorial actions and practices, such as production of agroecological foods, rise of peasant cooperatives, construction of food sovereignty, and creation of popular markets.

The organisation of MST in Rio Grande do Sul began in 1978. In that year, Kaingang indigenous people, from the Nonoai Indigenous Reserve, started a parallel movement to recover their territory and managed to expel 1800 settler families who were living in their land. Upon being deterritorialised, some settlers migrated to the North of Brazil in the scope of colonisation projects, others migrated to the nearest cities. Several peasant families, however, did not accept migrating to another region or cities, deciding to fight for land in RS. Some groups, then, occupied the Macali and Brilhante plots of land, others occupied the Forest Reserve of *Fazenda Sarandi* (Sarandi Farm), in the municipality of Rondinha. The plots Macali and Brilhante were public land but were leased for the cultivation of soybeans. The RS state government registered the encamped families, committing to settling them. Some of them were settled in the municipality of Bagé, the others in a colonisation project in the state of Mato Grosso in the centre of the country. However, the number of encamped families continued to rise.

In 1980, the Encruzilhada Natalino camp was created between the municipalities of Passo Fundo, Ronda Alta and Sarandi. The camped landless rural workers were evicted several times by the police, but they resisted. Three years later they were settled. In 1982, the landless organised a regional meeting, convening people from various states of centre-south regions of the country. Two years later, they held the First Meeting of Landless Rural Workers, in Cascavel, Paraná State. The meeting succeeded in unifying struggles for land, giving rise to the MST. Following the consolidation of MST, several land occupations were carried out in Rio Grande do Sul. Between 1988 and 1989 there were 12 land occupations involving 2435 families, mainly in the northwest of the state. In the 1990s, there were 75 land occupations by 28,066 families, and in the 2000s, 127 land occupations by 29,634 families, mainly concentrated in the Metropolitan Region of Porto Alegre (RMPA) and in the centre-south of the state of Rio Grande do Sul (Dataluta 2017). In the 2010s, these figures decreased significantly, reflecting the agrarian reform policy adopted by recent governments.

To the extent that land occupations, encampments, and public demonstrations were carried out aimed at putting pressure on state and federal governments, rural settlements were gradually created in Rio Grande do Sul. Once these territories were established, other demands, imperatives, resistances, and struggles began to emerge, comprising the daily life of settled peasants. The main demand of the families concerned the economic viability of their territories. To this end, they started growing different foods and organising producer associations and agricultural cooperatives. The first MST cooperatives were regarded with suspicion by most peasants, as they were not used to producing collectively. Furthermore, to a certain extent, the cooperatives reproduced the model of agricultural production implemented by the Green Revolution, based on capital-intensive agriculture and the consumption of industrial inputs. High production costs led some cooperatives to default. Even peasants who were not members of cooperatives, but adopted this production model, ended up in debt.

The MST increasingly invested in technical training of settlers. Many of them have even achieved academic degrees. This was an important step towards the economic organisation of the settlers. Moreover, the movement allied with other peasant organisations in Brazil and other countries, when it joined International Via Campesina in the 1990s. This approach led to a dialogue of knowledge, creation of agendas of struggle, and the development of collective coping and resistance strategies. The economic organisation, technical and/or academic training, and dialogue between different socio-territorial movements allowed peasants to improve food production. Among such improvements was the rise of agroecology that allowed settlers to achieve greater autonomy from industrial inputs, and also contributed to the reorganisation of cooperatives, with the creation of agro-processing facilities and consolidation of farmers' markets.

Dependence on capital and industrial inputs, produced by agribusiness companies, distances peasants from nature, from their way of life and their way of producing food, thus moving them apart from what is essential for their existence and resistance in the capitalist mode of production. Agroecology, conversely, allows peasants to grow food through relations of exchange with nature, regional culture, and the humanity. The first agroecological experiences in rural settlements in the RMPA began in the late 1980s with the cultivation of vegetables. Then, they were referred to as alternative agriculture as opposed to the agricultural model implemented by the Green Revolution. The word agroecology was not yet used by settled peasants. Many families could not afford the costs of production using industrial inputs. They lacked capital, access to credit, infrastructure, and technical assistance to adopt this technological package. Vegetable cultivation expanded in rural settlements of the RMPA without the proper attention of MST in the early years. However, this experience inspired the agroecological transition of rice in the same region in the late 1990s. Not only in the RMPA, though in this same context, other agroecological experiences emerged in the state. An example is the creation, in 1997, of Bionatur, the first MST agroecological seed cooperative in the country. These mentioned experiences were fundamental for MST, during its territorialisation process, to start encouraging agroecological production in peasant territories, making them viable not only from an economic point of view, but also from social, political, and environmental perspectives. The production of agroecological rice is part of this context. To understand it, other important elements should be analysed, such as the economic crisis that affected rice cultivation in the RMPA in the 1990s.

In the late 1980s, peasant farmers from the state of Santa Catarina began leasing land in the RMPA for growing conventional rice. The *catarinas*, as they were called (because of their state of origin, Santa Catarina), had capital and machinery, both financed by rice mills of Santa Catarina. While *catarinas* were producing rice, the landless peasants occupied unproductive rural properties in this region, putting pressure on the state to create rural settlements. Once being settled, the peasants started growing grains on their plots, especially soybeans, maize, and beans. As they had experience with these crops before being deterritorialised in the north and northeast of the state, they believed it would be easier to develop them. However, the RMPA has some natural aspects, such as being a floodplain area, which is favourable to some types of agricultural crops and unfavourable to others.

For rice cultivation the conditions are more propitious. This is a crop that was adapted to the *banhado*, that is, the floodplain area, characteristic of the RMPA. The *catarinas* even used pre-germinated seeds, a technique used today in agroecological production. MST's settlers could not afford to invest in agricultural production and the rural settlements lacked basic infrastructure. Faced with this, many peasants began to lease part of their plots to the *catarinas*. Land leasing weakened MST's political action in the region for a while (Martins 2017), as the movement was opposed to this practice. Other settlers also started to cultivate conventional rice, organising into cooperatives. In 1998, rice cultivation went through a crisis due to a drop in the price of this item in the market. Those who leased their plots lost an important source of income, as many *catarinas* went bankrupt, and those who farmed on their own were in debt as well, since they could not maintain production in view of high costs and low prices. One of the cooperatives involved in the rice production also faced financial insolvency.

It is in this context of bankruptcy and indebtedness that peasants, some of them already members of cooperatives, began to rethink their productive practices, resorting to agroecology. In 1999, settled peasants started producing pre-germinated rice without the use of chemical inputs. In 2002, they organised the First Ecological Rice Seminar. At this event, they analysed the principles of agroecology, highlighting the advances and challenges of rice cultivation. They decided to restructure the Cooperative of Settler Workers of Porto Alegre Region (*Cooperativa dos Trabalhadores Assentados da Região de Porto Alegre*—COOTAP) so that it could act in the processes of drying, storing, and marketing agroecological rice. In 2004, the Ecological Rice Steering Group (*Grupo Gestor do Arroz Ecológico*—GGAE) was created with the aim of organising production, which also contributed to its expansion not only in the RMPA, but all over the state. Later, for the same purpose, the Vegetables Steering Group (*Grupo Gestor de Hortaliças*—GGH) was created.

Both steering groups defined that peasant families should keep their autonomy during the whole production process, either fully or partially freeing themselves from the trammels of capital, having control over seeds, certification, machinery, storage, processing, and marketing. The steering groups are formed by various organisational structures, including production group, formal association group, cooperatives of agricultural production, irrigation district, and the collectives of seeds, drying and storage, and, finally, marketing. These structures are important because, besides planning and operating agroecological practices, they democratise decision-making. Thus, peasants participate in the whole production process, deciding what should or should not be done. Steering groups are arenas for the construction of territorial resistance, which, in many cases, go beyond peasant territories.

Peasant Food Systems: Advances, Possibilities, and Challenges for the MST

Peasant food systems, developed in the scope of cultivation, processing, and commercialisation of agroecological foods such as rice and vegetables in the RMPA, began to be built during the spread and territorialisation of MST in the state. This means that, without the struggles for land, land occupations, encampments, and demonstrations in the countryside, and without agrarian reform policies, many of the studied rural settlements would not exist. This does not mean that existent settlements fully succeeded in deconcentrating land ownership in the country; they are, however, territories of resistance and confrontation with capital. The territorialisation of MST is not limited to creating rural settlements, on the contrary, it continues in the struggles for housing, infrastructure, credit, public policies, among others. During this process, different territorialities are built by peasants while producing food.

In the RMPA, subordinated territorialities were initially predominant, and still are present in peasant territories. However, after experiencing marginalisation and dependence in a situation of subordinated production bound to the dictates of, above all, industrial capital, settled families realised that, to remain on the land, reproduce socially, develop their territories economically, they should have more autonomy, not only regarding production but also in processing and commercialising their produce. Agroecology made this autonomy possible, since more than knowledge and an agricultural practice it is a social movement that requires building a closer relationship with nature, local knowledge, regional cultures, and the society. Agroecology can be understood as a territorial resistance that has allowed the construction of autonomous peasant territorialities in rural settlements. And these resistances and territorialities are the foundations of peasant food systems. Only through them these systems may exist.

In recent decades, peasants settled in the RMPA have created different forms of social, political, and economic organisation, which were developed in the context of struggles, disputes, resistance, and conflicts. Besides agroecology, steering groups and cooperatives also stand out in this regard. Steering groups are arenas for peasant organisation and socialisation, which are essential for building autonomous territorialities. These groups presuppose the democratisation of knowledge and decisions. All those who are involved in agroecological production, in some way, participate in these arenas, contributing to their organisation. They comprise horizontal social relationships, built 'from the grassroots' and distinguished from relationships of dependence and subordination. Autonomy consists, among other things, in power of choice, of deciding about what and how to do things. In the current agrarian conjuncture, attaining such power is fundamental for the resistance of peasant territories. Thus, through these groups, peasants determine and obtain the knowledge required for agroecological practices, the essential seeds for crops, the irrigation districts that are fundamental in rice production, the cooperatives and their respective functions, in addition to agro-processing facilities and their operations.

Once the cooperatives were restructured, they were able to refocus on agroecological production, settled families started to obtain the necessary knowledge and infrastructure. Thus, peasants have access to machinery, cars and trucks, irrigation equipment, drying and storage facilities, agro-processing facilities, and peasant markets. The cooperative, in this case, operates as a kind of extension of the peasant farm (Chayanov 2017). Resources that a peasant alone could not afford, for being too expensive, are available through the cooperative, which is responsible for organising the stages of production process that are often unfeasible for a peasant alone. This also applies to agro-processing and markets. Cooperatives, together with producer associations, build processing units and manage peasants' access to public policies aimed at food purchasing from peasant farming, such as the National School Feeding Programme (PNAE) and the Food Acquisition Programme (PAA), and to farmers' markets, some even created by peasants themselves.

After growing agroecological foods, some items, such as rice, vegetables, and fruits, are taken to agro-processing units. The rice goes to a drying and storage unit located in the rural settlement Lanceiros Negros, in the municipality of Eldorado do Sul, which is managed by the cooperatives. Having control of this production process has substantially increased peasants' autonomy. Before that, agroecological rice was dried and stored together with conventional rice in structures rented by the cooperatives, which hindered agroecological certification by the settled families, let alone marketing the rice as a differentiated product. Vegetables and fruits are processed in the Terra Livre processing unit, located in the rural settlement Itapuí, in the municipality of Nova Santa Rita. With the processing of these foods, peasants have access to various markets, such as public procurement and farmers' markets, thus diversifying their strategies for social and economic reproduction.

Peasants settled in the RMPA market their food in different farmers' markets, such as those that take place in Porto Alegre and other municipalities of the RMPAs. Some of these farmers' markets have been around since the early 1990s, others were created more recently aiming at bringing those who produce food closer to those who consume it. There are about twenty-five such markets in the RMPA. Another important peasant market is MST's agrarian reform store, located in the Porto Alegre Municipal Market. There, besides rice and vegetables, other items produced in rural settlements are sold. There are similar stores of MST in other cities in the country, such as São Paulo and Belo Horizonte. MST intends to expand such stores across the country. These markets, in addition to bringing peasants closer to consumers, dialogue with urban workers about the importance of agroecology in the production of healthy foods. And, without peasant territories, there is no agroecology.

The existence of the peasant in the capitalist mode of production also depends on those who are not peasants, such as governments (Shanin 2008). Agrarian reform policies and public policies aimed at peasant agriculture are fundamental. These policies were created in Brazil, among other factors, as a result of political-geographical actions performed by MST in recent decades. The construction of peasant food systems always begins with land occupations, encampments, and demonstrations. It is an extensive, arduous, and often contradictory political process for those who are part of the landless movement. Besides providing access to land, some public policies have added techniques, instruments, and fundamental infrastructure for food production to peasant territories. Thus, these persons do not need to resort to industrial or financial capital to obtain basic items, such as tools, machinery, and agro-processing. Having control over objects and relationships, that is (im)materialities, is fundamental in struggles for autonomy.

The food systems being built by the landless movement face some challenges, such as internal conflicts within steering groups and cooperatives, and the fact that they depend, to a certain extent, on public policies. To exist and expand in peasant territories, food systems depend on peasants' multidimensional and multi-scale organisation, especially regarding steering groups and cooperatives. This means organising in the political, economic, and environmental dimensions, and acting on the local, regional, national, and international levels. Such dimensions and levels require well-defined organisational structures that include those who grow food and those who perform other activities, such as processing and commercialisation. Although steering groups and cooperatives value democracy in decision-making, peasants do not always agree with decisions made, which may lead sometimes to conflicts and internal disputes.

A certain degree of dependence of peasant food systems on the State, especially regarding food purchase, also persists. These policies have been instrumental in building peasant food systems. Without them, the landless movement would hardly be able to reach the current level of economic organisation. However, dependence on these policies has a perverse side, because they can be changed or have fewer financial resources available as governments alternate in power. If the government understands agriculture through the perspectives of the ACP, especially agribusiness, those public policies may lose resources. That's why peasants are already investing in other markets, the popular ones. Nevertheless, even having available several farmers' markets and agrarian reform stores, public purchases are still the main channel for selling peasant food produce. Despite the advances and potential of peasant food systems, the challenges analysed here can harm the organisation of steering groups and cooperatives, as well as the development of agroecological production in rural settlements.

Conclusion

In this piece we examined territorial resistance and peasant food systems in the RMPA. Territorial resistances are struggles, actions, practices, and organisations that aim at greater autonomy and emancipation in food production. In the case of the RMPA, territorial resistance was born in the first rural settlements created in the 1980s, with the production of vegetables without using industrial inputs. After that, in the 1990s, in a context of indebtedness and bankruptcy of peasants and agricultural cooperatives, the landless movement decided to invest in agroecological rice cultivation. Over the years, these experiences have become more complex, with the organisation of steering groups, the creation of agro-processing facilities, and the organisation of farmers' markets. These struggles, actions, and agroecological practices allowed for construction of autonomous territorialities. These, in turn, are the foundation of peasant food systems; systems in which peasants, in their territories, control the production, industrialisation, and marketing of food, freeing themselves from the trammels of industrial and financial capital, agribusiness, and global food chains.

References

Chayanov, A. 1984. *La Organización de la Unidad Económica Campesina*. Buenos Aires: Nueva Visión.

Chayanov, A. 2017. *Teoria das Cooperativas Camponesas*. Porto Alegre: Editora UFRGS.

Dataluta. 2017. *Relatório DATALUTA Brasil*. Presidente Prudente: Unesp.

Feliciano, C.A. 2003. *O Movimento Camponês Rebelde e a Geografia da Reforma Agrária*. MSc dissertation. São Paulo: USP.

Fernandes, B.M. 1996. *MST: Movimento dos Trabalhadores Rurais Sem-terra: Formação e Territorialização em São Paulo*. São Paulo: Hucitec.

Fernandes, B.M. 2000. *A Formação do MST no Brasil*. Petrópolis: Editora Vozes.

Fernandes, B.M. 2008a. Entrando nos Territórios do Território. In *Campesinato e Territórios em Disputa*, ed. E.T. Paulino and J.E. Fabrini, 273–301. São Paulo: Expressão Popular.
Fernandes, B.M. 2008b. Questão Agrária: Conflitualidade e Desenvolvimento Territorial. In *Luta pela Terra, Reforma Agrária e Gestão de Conflitos no Brasil*, ed. A.M. Buainain, 173–224. Campinas: Editora da Unicamp.
Fernandes, B.M. 2013. Re-peasantization, Resistance and Subordination: The Struggle for Land and Agrarian Reform in Brazil. *Agrarian South: Journal of Political Economy* 2 (3): 1–21.
Kautsky, K. 1998. *A Questão Agrária*. Brasília: Linha Gráfica Editora.
Lenin, V. 1982. *O Desenvolvimento do Capitalismo na Rússia*. São Paulo: Abril Cultural.
Martins, J.S. 1981. *Os Camponeses e a Política no Brasil: As Lutas Sociais no Campo e seu Lugar no Processo Político*. Petrópolis: Vozes.
Martins, A. 2017. *A Produção Ecológica do Arroz nos Assentamentos da Região Metropolitana de Porto Alegre: Territórios de Resistência Ativa e Emancipação*. PhD thesis. Porto Alegre: UFRGS.
Oliveira, A.U. 1991. *A Agricultura Camponesa no Brasil*. São Paulo: Contexto.
Origuéla, C.F. 2019. *Território e Territorialidades em Disputa: Subordinação, Autonomia e Emancipação do Campesinato em Assentamentos Rurais no Rio Grande do Sul*. PhD thesis. Presidente Prudente: Unesp.
Paulino, E.T. 2012. *Por uma Geografia dos Camponeses*, 2nd ed. São Paulo: Unesp.
Ploeg, J.D.V. 2008. *Camponeses e Impérios Alimentares: Lutas por Autonomia e Sustentabilidade na Era da Globalização*. Porto Alegre: UFRGS.
Ploeg, J.D.V. 2016. *Camponeses e a Arte da Agricultura: Um Manifesto Chayanoviano*. São Paulo and Porto Alegre: Unesp/UFRGS.
Raffestin, C. 1993. *Por uma Geografia do Poder*. São Paulo: Ática.
Ross, D. 2015. *Contradições na Construção dos Territórios Camponeses no Centro-Sul Paranaense: Territorialidades do Agronegócio, Subordinação e Resistências*. PhD thesis. Presidente Prudente: Unesp.
Simula, G. 2015. *Milking Money: Exploring the Struggle for Autonomy from Theory to Practice*. MSc dissertation. The Hague: International Institute of Social Studies.
Shanin, T. 1979. *Campesinos y Sociedades Campesinas*. México: FCE.
Shanin, T. 2008. Lições Camponesas. In *Campesinato e Territórios em Disputa*, ed. E.T. Paulino and J.E. Fabrini, 23–47. São Paulo: Expressão Popular.

17

The Difficult but Not Impossible Defeating of Right-Wing Populism and the Exploration of a Socialist Future

Saturnino M. Borras Jr.

Introduction[1]

Both in the past and at present, pro-capitalist, right-wing or even fascist movements and political parties have often found support from rural communities. Yet there is nothing inherently conservative in rural politics: history has also shown how working people in the countryside—peasants, landless labourers, and others—joined the proletariat and other social forces at the barricades and in the trenches during anti-feudal or socialist revolutions, and anti-colonial national liberation struggles. Key questions that scholars have grappled with during the past century

[1] Adapted from Borras, SM. 2020. Agrarian Social Movements: The Absurdly Difficult but not Impossible Agenda of Defeating Right-wing Populism and Exploring a Socialist Future. *Journal of Agrarian Change*, 20(1), 3–36. Reproduced under permission.

S. M. Borras Jr. (✉)
International Institute of Social Studies (ISS), Den Haag, The Netherlands
e-mail: borras@iss.nl

A. A. R. Ioris and B. Mançano Fernandes (eds.), *Agriculture, Environment and Development*, https://doi.org/10.1007/978-3-031-10264-6_17

include how and why such radical transformations happened from one society to another, and with what implications, and which strata of the working people in the countryside were the most open to such revolutionary projects. Some of these scholars come from the same broad theoretical tradition but have competing interpretations, such as Wolf (1969) and Paige (1975); others represent opposing worldviews, such as Scott (1976) and Popkin (1979). But most of them were concerned with the 'big picture' meanings and implications of the consciousness and exercise of political agency of working people in the countryside, as was Barrington Moore Jr. (1967). Their basic question was similar to that posed by Huizer (1975): how do peasants become revolutionary? Or, indeed—the flip side—how do peasants become reactionary?

Although the era of peasant-based revolutions has passed, Huizer's question remains relevant, and could be broadened and updated for the current global conjuncture, when there are parallels, resemblances, and interconnections between agrarian populism and the populism of broader political agitations that are regressive in nature, and when the challenge is how to transform such interconnections into a political opportunity that can contribute to eroding right-wing populism. The question can be *broadened* by asking: how do 'working people' in the countryside—peasants, landless labourers, indigenous communities, fishers, pastoralists, lower middle class, the vast number of people in the informal sector including those who live and work in nearby small towns—become revolutionary? And it can be *updated* by asking: how could they join other social forces in the fight against neoliberal capitalism and contemporary right-wing populism, and rally to support a reconceptualised notion of anti-capitalist left-wing populism that is broadly anchored on socialist principles? Socialism here is broadly defined, following Erik Olin Wright, as '*a deeply democratic and egalitarian organization of power relations within an economy*' (Wright 2016: 102).

In my view, the current conjuncture calls, and could provide a fertile ground, for contemporary anti-capitalist agrarian social movements to propel a broad-based socialist agenda. At the same time, I would argue: (i) that the future of contemporary agrarian social movements, including the broader food sovereignty movement, depends not only on how consistently anti-capitalist they are, but even more importantly, on their willingness and ability to build their movements and frame their agendas within a socialist framework and perspective (along Wright's broad definition); or at least on their being not anti-socialist, and thus their willingness and their ability to actively co-construct a pluralist platform for a broad-based, common political project; and (ii) it is essential that any anti-right-wing populist political project centred on left-wing populism has a significant component which comes from and relates to rural communities. Historically, successful reforms or revolutions were direct outcomes of the dialectically linked weakening of a ruling elite and the strengthening of challengers. Political work in defeating contemporary right-wing populism thus cannot rely solely on either weakening right-wing populists or strengthening democratic challengers: both elements will be needed. It is in this context that the agrarian world could make important contributions to challenging resurging right-wing populism because it has the material basis and political potential to split the ranks of right-wing populists, at the same time that it can contribute to broadening a counter-current of class-conscious left-wing populism.

This chapter addresses the twin challenges of *splitting the ranks* of right-wing populists while *expanding the united front* of democratic challengers. It aims to explore an agenda for political conversation, primarily, and academic research, secondarily, on possible contributions to overall efforts at defeating contemporary right-wing populism. It builds on initial and often rough assumptions and tentative hunches, and generates questions to which I do not have the answers. For instance, I do not know what foundational theoretical logic and practical political scaffolding are necessary and possible for building a class-conscious left-wing populism, beyond proposing a vision that is outside capitalism and which is framed here as a socialist alternative. Finally, I am writing this chapter with practical political, and not academic, concerns as a principal starting point, coming from an activist worldview. This is a continuation of the conversation with comrades inside the radical emancipatory movements in the Philippines during the late 1980s and the first half of the 1990s.[2]

[2] An intense debate inside the Communist Party of the Philippines (CPP) and fraternal movements erupted in the late 1980s, leading to multiple breakaway groups during the first half of

In this chapter, I use the term 'populism' to mean *the deliberate political act of aggregating disparate and even competing and contradictory class and group interests and demands into a relatively homogenised voice, i.e. 'we, the people', against an 'adversarial them' for tactical or strategic political purposes.* As such, populism is inherently relational. It tends to be a means towards an end rather than an end itself, giving it a very generic character that is open-ended and flexible, facilitating easy adaptation by various ideological camps, even competing ones. Two types of populism are the main subjects of this analysis. The first is right-wing populism, i.e. a regressive, conservative, or reactionary type of populism that promotes or defends capitalism in the name of 'the people'; in its current manifestation, it is also xenophobic, nationalist, racist, and/or misogynistic. The second is agrarian populism, i.e. the political bundling of various rural-based or rural-oriented social groups and class interests and issues into a homogenised category, 'the people of the land'; many variants of agrarian populism are anti-capitalist and try to advance a 'peasant way' of alternative development.

I use the term 'right-wing populism' for lack of a better term. The logic of the notion of 'populism as a matter of degree' (see discussion in the next section) extends to the notion of 'right-wing populism', i.e. some populist currents lean further towards the right than others, even when they are all fundamentally right-wing, given that: (i) they are champions of contemporary capitalism, although the latter may take a variety of forms; (ii) they are generally anti-socialist; (iii) they have disdain for basic democratic institutions, especially human rights; (iv) they share a tendency towards authoritarianism and militarism; and (v) they are xenophobic or racist, and many are misogynistic. Actually existing populisms will not find a perfect fit in the term 'right-wing populism', but they

the 1990s. A significant part of that debate involved the National Peasant Secretariat of the CPP that started to experiment with newer forms and goals of peasant mass movement struggles relative to the Maoist orthodoxy of 'people's protracted war'. The leadership of the CPP dismissed such experiments as 'impetuous petty bourgeois agrarian populism' and 'infantilism', which were to be shunned and suppressed. Later, the broader party debate deteriorated into a 'revolutionary–counterrevolutionary' framework in which any comradely discussion became impossible. I have been deeply involved in radical agrarian movements in the Philippines since the early 1980s, and have thus been engaged in this debate not in the classrooms but in the trenches. For further background, see Franco and Borras (2009).

have no better fit in other terms either, such as 'authoritarian populism' (Akram-Lodhi 2018), 'populist authoritarianism' (Docena 2018), neoliberal authoritarianism/authoritarian neoliberalism (Saad-Filho 2018), or 'populist nationalism' (Vanaik 2017a, b; Win 2018), as there will always be significant outliers. Most of these populist currents have a strong tendency towards authoritarianism, but again, it is a matter of degree, and a regime may dynamically oscillate towards and away from populism and authoritarianism (as in the case of Cambodia; see Schoenberger et al. 2018). There will always be exceptions once we deploy a defining term, which is by nature a delimiting term. But resorting to using the unqualified term 'populism' loses the distinct character of the political moment, which is partly marked by some kind and degree of 'right-wing-ness' and 'authoritarianism'.

Moreover, the term 'right-wing populism' dovetails with our subsequent discussion of a potential counter-current, namely, a 'reformulated left-wing populism'. That discussion will clarify the fundamental differences between what is a right-wing or a left-wing populism, and demonstrate why such clarity is important. It will help to illustrate the absurdity of some casual commentaries which put the United States's Trump and Venezuela's Maduro in one basket, for instance. It will also show that the term 'left-wing populism' suffers from a similar semantic problem: for example, Bolivia's Morales is a left-wing populist, but arguably employs some capital-accumulation strategies with features that are more to be expected from a contemporary right-wing regime than a left-wing regime, such as neoliberal (neo)extractivism (McKay 2017), just as the Workers' Party (PT) in Brazil under Lula and Dilma constructed a similar type of populism (Andrade 2019). Finally, 'right-wing' and 'left-wing' populisms are used here rather like bookends, i.e. as ideal-types, or as heuristic tools. In reality, only rarely will any populist current fit neatly into either ideal-type. The bookends will, however, allow us to see a *dynamic continuum* rather than fixed categories in between, with populist currents and regimes consolidating features of one of the ideal-types but regularly borrowing features from the other. We will see a shifting, fluid situation in which populists straddle various points between these bookends, *constantly morphing* away from ideal-types. Perry Anderson's analysis of the brand of populism of Brazil's Bolsonaro, and how it is similar to but at the same time significantly different from its contemporary counterparts in Europe or Trump is a good illustration (Anderson 2019). This is an important feature of populism today, and one which renders the terms 'right-wing' or 'left-wing' populism imperfect and imprecise, but analytically useful.

In its attempt at homogenising disparate, often competing interests of various classes and groups, each of the contemporary right-wing populisms and progressive agrarian populisms is marked and defined by internal *contradictions* and, at times, *antagonisms* (based on class relations, ideological positions, political calculations, among others), even while the two ideologically opposed populist groups target broadly similar issues and adversaries.[3] Scoones et al. (2018) offer a closer, if preliminary, look at the possible connection between 'authoritarian populism' and the rural world, trying to frame new ways of asking questions in order to understand such a relationship. The current chapter builds on this.

Despite big claims that the world is now urban, the fact remains that nearly half the world's population, that is more than three billion people, can be categorised as rural. Rural political tendencies have become swing factors in many settings and political moments, including electoral politics and democratisation more generally.[4] Where rural voters represent a sizeable proportion of voters, if not the majority, the influence of rural issues on populist discourses and agitation is significant. We saw this in the 2014 elections in India[5] and in the current political situation in

[3] Brass (1997) offers a critical examination of the relationship between the 'new' right and what he clusters together and labels as 'new' populism in the 1960s through the 1990s, in which agrarian themes form (Brass argues), a common bond for the two. This speaks to the themes explored in the present chapter, but with different categorisations of objects of analysis.

[4] For the latter, see Fox's edited volume on rural democratisation with perspectives from Latin America and the Philippines (Fox 1990).

[5] BJP and BJP Allies (National Democratic Alliance—NDA) won a combined total of 212 seats of the total 342 rural seats (out of the total 543 seats), as compared to Congress Party and Allies (United Progressive Alliance—UPA) total of 40 rural seats. In terms of vote share, the NDA secured 36.9% of the total rural vote, compared to 23.7% for the UPA (NES 2014: 130). Thanks to Amod Shah for alerting me to these data, and for helping me interpret them.

the Philippines,[6] for instance. Even in societies where the rural population is no longer significant compared to that of cities and mega cities, votes from rural communities—which usually include small towns—can nevertheless be critical. This was true of the 2016 elections in the United States,[7] the 2017 elections in Germany,[8] and the 2018 elections in Russia,[9] and in Turkey.[10] In some countries, there is no clear urban–rural divide in electoral voting behaviour, but rather an upsurge of support in both urban and rural spaces for right-wing candidates, such as in the 2018 elections in Brazil.[11] Moreover, recent transformations in

[6] According to the December 2018 Social Weather Station (SWS) survey: '74% of Filipino adults said they were satisfied with Duterte's performance, while 15% were dissatisfied, for a net satisfaction rating of +60, correctly rounded. This translates to a "very good" rating, said SWS. The remaining 11% were undecided. … The survey also showed that his net satisfaction rating… in urban areas [was] +64, [and] in rural areas: +57 [note: +57 for rural areas is just 3 points below the overall net satisfaction rating of + 60]' (Bueza 2018: n.p.).

[7] 'Between 2008 and 2016, Republicans' share of the urban vote barely changed, and Democrats' share fell by four points. In the suburbs, Republicans likewise didn't change much, and Democrats lost five points. The shifts were larger in rural areas, where Republicans gained by nine points, and Democrats lost 11 points. Of course, some places are more "rural" or "urban" than others —there's a continuum. Strikingly, as you move up that continuum, from the largest metro areas to the most isolated and sparsely populated rural areas, the vote steadily grows more Republican' (Kurtzleben 2016: n.p.).

[8] 'German right-wing party AfD finds more support in rural areas with aging populations: The AfD [Alternative für Deutschland] received 12.6 percent of all votes in the parliamentary election on September 24, 2017, making them the third most powerful party in the German parliament. … Generally, the AfD performs better in rural areas with negative demographic trends – a phenomenon that occurs more frequently in eastern German districts than in western districts. This allows for the conclusion that perspective is lacking among those living in rural areas with negative demographic developments' (Franz et al. 2018: 70). Thanks to Sofia Monsalve for bringing this to my attention.

[9] On the eve of the Presidential Election (18 February 2018), Russian Public Opinion Research Centre VTsIOM conducted a survey: 'If the presidential elections were held next Sunday and the list of candidates would look like this, then which of these candidates would you most likely vote for?' The results were as follows: 76.40% of rural residents would vote for Putin. The official results of the election (18 March 2018): 76.69% of Russian people [not just rural] voted for Putin (VTsIOM 2018). With thanks to Natalia Mamonova.

[10] 'The AKP's vote share has been consistently higher in the countryside than in the cities. … The approval of the constitutional amendment in the referendum on 16 April 2017 established a super presidential system without checks and balances. The amendment was passed with a slight majority (51.4%), but the share of the yes vote in the rural areas was much higher, estimated to be between 56 percent (Konda 2017) and 62 percent (Yetkin 2017)' (Gurel et al. 2019: 458).

[11] I thank Daniela Calmon for sharing with me comprehensive and fascinating data on the 2018 elections in Brazil, and how these can be interpreted through the current paper's lens.

the global political economy require a more nuanced understanding of intermediate geographic spaces, namely, small and medium towns that have internalised some features of both urban and rural social life.[12]

Furthermore, the issues that help condition the rise of populism in one geographic region may originate from or be linked to a distant place. For example, the rise of the populous and wealthier industrial belts in southeast China is linked to the massive rural-to-urban migration from other parts of China, the phenomenon of the left-behind population in the countryside, and the widening gap between and rural and urban worlds that forced the national government to adopt a populist programme, the New Socialist Countryside.[13] The rise of these Chinese industrial belts is linked in turn to the decay of many rural and urban communities in the United States that used to host factories, many of which shut down as corporate operations migrated to southeastern China, among other destinations. Thus, the populist impulses in multiple settings—rural China, urban/industrial China and de-industrialised, abandoned and neglected rural and urban communities in the United States—are concretely connected. It is not surprising that, despite the differences between them, right-wing populists worldwide are increasingly supporting or encouraging each other. This has prompted Edelman to raise a question that needs serious reflection: 'To what extent are the world's autocrats – Trump, Duterte, Erdoğan, Modi, Orbán, Putin, among others – simply a mutually reinforcing collection

[12] In this context, the 2016 national elections in France merit a close and careful look. 'Nièvre, the poorest *département* in Burgundy, is a traditional heartland of the French left. For 40 years it was the rural power base of the former Socialist president François Mitterrand, who was mayor of the small town of Chäteau-Chinon for 20 years. "This place has been leftwing since the French revolution", one local Socialist politician boasted, adding that Nièvre was a focal point for the French resistance during the second world war. And yet, the Front National more than doubled its vote here in the previous regional elections, and it is here in Burgundy that Le Pen is hoping for some of her highest scores. … Le Pen's rural target is not just farmers, who are shrinking in number in France and represent about 1% of the electorate. Her base comprises people living in modest towns and country villages far away from big cities, who have felt the sharp edge of France's decades of mass unemployment, who have seen factories close and local shops and services disappear, in places where the population is ageing, young people are leaving and those who stay have to drive long distances to see a doctor, or sometimes even to post a letter' (Chrisafis 2017).

[13] See Ye et al. (2013).

of erratic rulers? Or are they taking shape as a *global authoritarian populist axis*?' (Edelman 2018: 1, emphasis added).

All these factors have resurrected the issue of agrarian populism in broad new ways, requiring us to revisit and critically examine it against varying contemporary populisms, especially right-wing populism. It is not only that contemporary right-wing populism has concrete links with the rural world, but that there are awkward parallels, resemblances, and interconnections between right-wing populism and agrarian movements, and these are *not* random accidents. The political economy upon which such populisms emerged over long periods of time (Akram-Lodhi and Kay 2010a, b) partly shapes the kind of broader politics that characterise the present conjuncture (Saad-Filho 2018). The boundaries between right-wing populist currents and their social base in the countryside on the one hand, and the populism of agrarian movements on the other, are porous, blurred, and malleable.

It is in the context of this perspective on agrarian populism—partly in light of its possible interconnection with the rise of right-wing populism worldwide today—that we take up the (unexpected) suggestion of a leading sceptic of contemporary agrarian movements and food sovereignty, Henry Bernstein, to go 'beyond the comfort zone of class purism' and not to dismiss today's agrarian populism. Revisiting the Russian revolution, Bernstein (2018: 1146) notes that the challenge for adherents of Marxist political economy, whose strength is in socioeconomic analysis, is to have a better grasp of agrarian politics:

> The route from the former to the latter entails many additional determinations and complexities, as well as capacity to confront the contingent, the indeterminate and unanticipated, and to change positions, that goes far beyond the comfort zone of class purism and other illusions. ... This points towards a paradox ... namely that while the best of Marxism retains its analytical superiority in addressing the class dynamics of agrarian change, for a variety of reasons agrarian populism appears a more vital ideological and political force. ... In my view, the challenges facing any Marxist agrarian politics would be helped by critical engagement with the most progressive (anti-capitalist) of today's agrarian populism, and the diverse rural struggles it embraces, rather than dismissing *a priori* all agrarian populism as necessarily and equally 'wrong' and 'reactionary'.

In relation to Bernstein's proposition, it is worth stating again that by *left-wing* agrarian movements and the broad food sovereignty movements, I mean those social movements that are anti-capitalist in their ideological orientation, whether this manifests systematically or not, explicitly or otherwise. Many, but not all, of these movements have socialist ideologies, influences, and leanings, or are, at least, not anti-socialist (especially when seen from Wright's broad definition of socialism, cited above). These are widely diverse social movements in terms of ideological and political provenance, and are heterogeneous in their class origins and intersecting social identities. Their anti-capitalist stance is not uniformly, evenly, and consistently found across and within movements, and does not emerge from a unified ideology. These are movements of small farmers from the industrial North, poor peasants from the Global South, landless labourers, migrant farmworkers, indigenous peoples, pastoralist organisations, fishers' movements, women's movements, environmental advocacy groups, climate justice associations, and various radical food activist groups—bound, to varying degrees, by some principles of food sovereignty—whose class formations are further complicated by their intersecting and messy social identities around gender, generation, race, ethnicity, religion, and/or nationality. La Via Campesina and the IPC for Food Sovereignty (with their regional and national affiliates) are the most politically significant and publicly known global hubs of these broadly left-wing agrarian social movements (Edelman and Borras 2016). These are the movements that are often lumped together, pejoratively, as populists and dismissed as such.

What Is Populism?

What we are keen to examine, following Rancière (2016: 102), are the 'diverse or even antagonistic figures of the people, figures constructed by privileging modes of assembling certain distinctive traits, certain capacities or incapacities: an ethnic people defined by the community of land or blood; … an ignorant people that the oligarchs keep at a distance'. Rancière elaborates that the 'notion of populism itself constructs a people characterised by the formidable alloy of a capacity – the brute force of great number – and an incapacity – the ignorance attributed to that same great number' (ibid.). This connects with Laclau's unit of analysis which is 'not to be the *group*, as a referent, but the socio-political *demand*' of particular groups (Laclau 2005: 224, original emphasis). It is in this political process that *a section* of the community/people gets projected as 'the people', and the people is reduced to mean that particular section (ibid.: 214). With these basic concepts as reference points, we can construct a set of seven characteristics which apply to contemporary populism: (i) a matter of degree, (ii) 'politics of appearances', (iii) a matter of variation, (iv) oscillation between rhetoric and reality, (v) differentiated and layered in its composition, (vi) politically volatile and capricious, and (vii) relevant whether in or out of state power. These will be elaborated below.

First, populism is not an 'either/or' question; rather, it is *a matter of degree*. It is better understood not as a thing but as a relationship, not in black and white, but in shades of grey, as we compare, for instance, the varying strands and degrees of populism, and tendencies towards militarism, authoritarianism, or democratisation of Zimbabwe's Mugabe, Thailand's Thaksin, Philippines' Duterte, Cambodia's Hun Sen, Myanmar's Aung San Suu Kyi, India's Modi, the United States's Trump, France's Le Pen, Turkey's Erdogan, Brazil's Bolsonaro, or Russia's Putin. As Laclau (2005: 45, original emphasis) puts it: 'To ask oneself if a movement *is* or *is not* populist, is actually to start with a wrong question. The question that we should, instead, ask ourselves, is the following: *to what extent* is a movement populist?' The logic of this argument is derived from some of the key tasks of political actions to homogenise diverse interests (complementary, competing, or contradictory) of social groups and their political positions into a singular stand or voice, aimed at achieving greater salience partly by trying to *deliberately* blur or cloak the sharp contradictions and differences between social groups and classes, highlighting selective unified features that are usually deliberately distorted if not largely imagined. Often invoked is a homogeneous 'people'.

Second, to a large extent, populism is about *politics of appearances*, having some parallels with Tsing's notion of 'economy of appearances', i.e. 'the self-conscious making of a spectacle [that] is a necessary aid to gathering investment funds' (Tsing 2000: 118). She elaborates: 'In speculative enterprises, profit must be imagined before it can be extracted; the possibility of economic performance must be conjured like a spirit to draw an audience of potential investors. The more spectacular the conjuring, the more possible an investment frenzy' (ibid.). We can call the right-wing populist political version of Tsing's 'economy of appearances' the *politics of appearances*: the self-conscious making of a *spectacle* that is a necessary mechanism in gathering political support. The possibility of political performance must be conjured like a spirit to draw an audience of potential voters, supporters and investors; the more spectacular the conjuring, the more possible a frenzy of political support. Of course, all 'politics' engages in 'politics of appearances'; the difference with contemporary right-wing populism is the deliberate attempt at making a spectacle. All right-wing populist agitations engage in spectacles (where spectacular activities and leaders feed into each other; see also Saad-Filho 2018, in the context of Brazil), while their core group, supporters, and sympathisers are investing political support as a form of speculation for rewards or benefits in the shape of social reforms or for rent-seeking opportunities.

Third, there are *varied types of populism* in relation to democracy and authoritarianism. There are right-wing and left-wing authoritarian populisms, and in between them lies a diversity of possible combinations. Authoritarianism, seen as a dynamic political process, is inherently uneven and replete with contradictions, and a regime is seldom completely democratic or totally authoritarian. Populist currents malign the institutional establishment with pejorative labels, such as 'establishment insiders', for very tactical reasons and moments. They eschew or try to subvert conventional institutions, at least selectively and tactically. Furthermore, competing variants of populism do co-exist at the same time and in the same political-administrative territory, and do clash head-on: Trump versus Sanders in the United States in 2016, and Le Pen versus Melenchon in France in 2017 (and again in 2022) are examples. It is within this wide-ranging terrain that one has to examine populists' attitude towards social structures and institutions, or towards the character of the state (e.g. class basis, ideology), and, following Scoones et al. (2018), debates on key concepts like 'authoritarian populism' (Hall 1985).

Fourth, populism *oscillates between rhetoric and reality*, that is, 'populism in word' and 'populism in deed'. Many of the neo-extractivist left-wing regimes in the Latin American 'pink tide' governments veered towards 'populism in deed', at least partially. One of the defining features of these regimes has been the adaptation of extractivist neoliberal capitalism with accompanying distributive social reforms, such as cash transfer schemes and food distribution programmes carried out by governments under the leadership of Lula and Dilma in Brazil, Morales in Bolivia, and Correa in Ecuador—at times under creative populist banners such as *buen vivir*, or indeed, 'food sovereignty' (see Arsel et al. 2016; Chappell 2018; Gudynas 2011; Veltmeyer and Petras 2014; Vergara-Camus and Kay 2017). Many of these reforms are now under threat from a resurgent right in the region. Meanwhile, several right-wing populist groups got into power on the promise of sweeping social reforms, but there remains a huge gap between what was promised and what was delivered. Whether this will continue to be the case is something to watch closely.

Fifth, a populist current (right-wing populists or progressive agrarian movements) is *inherently internally differentiated and layered* in terms of groups and political tendencies. The layers include leaders, a core group, and a social base of supporters and sympathisers. These sets of actors have varying agendas, roles, and degrees of commitment to the framing of the populist agenda and agitation which are not necessarily unified, with each one trying to use the other. A core group is usually a mixture of diverse sub-groups: ideologues committed to particular worldviews, oligarchs, racists, and sub-layers of brokers, speculators, scammers, swindlers, and often circles of organised crime. The willingness of ordinary people to let populist leaders claim them, act in their name, speak on their behalf, and bundle them together as 'the people' (with the whole group often rebundled in an even narrower manner, to be known simply by the name of the populist leader) may not reflect their belief in and commitment to the populist rhetoric or trust in the populist leader. It may only indicate that their distrust in the old establishment or traditional elitist system is so deep that they prefer to take a gamble on something different and unconventional. Moreover, a core group, or sub-groups within a core group, do not emerge from nowhere.

A broad right-wing populist agitation (broad, in the sense that it includes diverse currents of conservative and right-wing groups) mobilises and emboldens moribund or marginalised fringe groups engaged in hate politics, such as white supremacists and other racist groups, and religious extremists whether of Buddhist, Christian, Hindu (Vanaik 2017a, b), or Islamic (Hadiz 2016) variants. Sixth, each layer of actors within a populist group (leader, core group, sympathisers, or supporters) is *politically volatile and capricious* in an ever-fluid situation, where the leadership, core group, and mass base may behave differently over time, often in self-contradictory ways. They can change their discourse at any time, as they are situational and tactical but at the same time strategic in their political calculation. Duterte in the Philippines is an example: engaging in left-wing rhetoric one day, and right-wing the next day (Curato 2017). This capriciousness is largely determined by the constant need to conjure a spectacle based on what the ringleaders think the people want or do not want.

Seventh, a populist group is *relevant whether in or out of state power*. The right-wing agitations that we are interested in may or may not be in power, but their significance stands, regardless. Their location vis-à-vis state power has an influence on most of the characteristics discussed here, such as how they frame their discourse or forge alliances. Right-wing populists that gain state power do not necessarily change the character of the state (class basis, ideology); rather, they tend to focus on changing the regime of political rule and determining who is in and out in government among the factions of the ruling classes and social groups. At the same time, right-wing populists that are outside state power should not be dismissed or taken for granted because they can significantly influence the character and trajectory of state power. Even where other political parties promise never to include them in a coalition, the emergence of parties like Geert Wilders' Party for Freedom in the Netherlands has forced ruling coalitions to adopt some of the right-wing populists' rhetoric and policies. Some groups long considered to be fringe groupings and politically irrelevant and unpopular could, in the event of a sudden change of political conjuncture, reinsert themselves into a broader and more significant right-wing populist current, such as the white supremacists and alt-right in the United States, in

Austria, Germany, and Sweden, the region of Andalucía in Spain (with the rise of the Vox party in the 2018 elections),[14] or even in Indonesia, as seen in the rising influence of Islamist groups. Right-wing populism, viewed from an 'inside/outside the state' perspective, is better seen as a continuum, and from a long historical sweep. Indeed, understanding Trump's right-wing populism and the progressive populism of La Via Campesina and its US affiliates requires us to understand the long history and moments and instances in different historical conditions of American populist agitations (see Chrisman 2016; Gaventa 1982; Hobsbawm 1987; Taggart 2000).

The issue of populists being in or out of state power is partly related to the discussion on the episodic character of populism. Mouffe (2005: 70) concludes that: 'It is no doubt encouraging to see that the appeal of [right-wing political parties] diminishes once they become part of the government, and that they seem able to thrive only when in opposition'. Taggart (2000: 106) similarly argues: 'The episodic nature of populism as a political phenomenon owes much to its highly ambivalent relationship to institutions. This makes it necessarily short-lived'. I agree that populism tends to be episodic, but I am not convinced by the explanations of Mouffe and Taggart. One defining character of the political moment is precisely that many right-wing populists are in power, and utilising the very institutions they once selectively attacked. In my view, the reason for the episodic character of populism is *not* due to populists' ambivalence towards formal liberal institutions, but rather it is a result

[14] 'The Vox party in Spain has become the first far-right group to triumph at the ballot box since the country's return to democracy after the death of Francisco Franco in 1975. Vox … exceeded all predictions to take 12 seats in the Andalucían regional election on Sunday. Although the ruling Spanish socialist party (PSOE) won the elections, taking 33 of the 109 seats in the regional parliament, its support collapsed in the heartland it has ruled since 1982. Even with the support of the Podemos-led Adelante Andalucía coalition—which won 17 seats—the PSOE would still be short of the 55 seats needed for a majority in the regional parliament. The conservative People's party (PP) took 26 seats, while the centre-right Ciudadanos party won 21. Were the two rightwing parties to join forces with Vox, they would jointly command a majority, with 59 seats' (Jones 2018). In January 2019, the three right-wing parties brokered an agreement that guarantees that the PP's candidate will become the next regional president. This will be the first time since the end of Francoism that Andalucía has a right-wing government.

of the cyclical nature of the crises of capitalism, and by extension, the crises of political rule (see also the analysis on the Brazilian situation by Saad-Filho 2018).

Important Debates on Agrarian Populism

There is nothing inherently progressive in agrarian populist movements—despite the often celebratory claims by such social movements and their supporters in contemporary agrarian politics. Historical cases include instances when agrarian movements took a conservative or reactionary political stance. One objective of this essay resonates with Paxton's reason for studying the 1920s and 1930s French countryside in order to understand fascism in France. Paxton (1997: 6) laments: '[I]t was in the countryside that both Mussolini and Hitler won their first mass following, and it was angry farmers who provided their first mass constituency. Yet, so far, every student of fascism in France has ignored the countryside'. He concludes: 'Given the salience of angry farmers in the success of fascism elsewhere and the importance of the peasantry in the French society, that is a crippling omission' (ibid.). Today, nearly all of the right-wing populists, both those in power and those not (yet) in power, have very significant electoral and political support from the rural population. How are we to understand this situation, and what is to be done?

Populist agitation is always antagonistic to an 'other'. The 'us' cannot be constructed without conjuring up a 'them'. But to the extent that the 'us' is defined or constructed, this is always framed as 'the anti-status quo', 'the anti-establishment', 'the subaltern', 'the underdog', 'the wronged', 'the violated', 'the looked-down', 'the pushed aside', 'the left-behind', or 'the challenger'.[15] In critical agrarian studies, populism has a broad meaning similar to that discussed so far. The provenance of contemporary agrarian populism is the left-wing Russian *narodniks* during the second half of the nineteenth century who aimed to overthrow tsarist rule and to rescue the surviving Russian peasant communes (*obshchina*) and

[15] See further elaboration in Panizza (2005).

their organisational structure (*mir*), which they believed could contain the seed for a possible socialist future. *Narodnism* ('*narod*' broadly means 'people') was a 'restorative struggle' with a tendency to romanticise communities where capitalist relations had not yet fully taken hold. Thus, the peasantry was seen as a route to socialism without having to pass through the capitalist phase of development.

It was estimated that about 2000–3000 urban students went to the Russian countryside in 1874, with a certain degree of spontaneity, without any written programme or organisation. These young intellectuals did not know much about peasant life or the practicality of political work. 'Moving from village to village, they distributed revolutionary pamphlets and talked indiscriminately to the peasants who crossed their path about the need to radically redistribute land and engage in revolution' (Taggart 2000: 50). The *narodniks* would soon be frustrated by what they discovered about peasants' politics: the peasantry did not have the appetite for revolution. The urban intellectuals imagined and expected peasants 'to be oppressed, idealistic and ripe for revolution. In practice they found the peasants to be acquisitive, conservative and profoundly suspicious of the students' (ibid.: 52). Many of these peasants would tip off the authorities about the presence of the *narodniks*. By 1877, most of the students, about 1611, had been arrested. In Taggart's words (ibid.: 52): 'The summer of 1874 showed what a group of activists could do. More than that, it showed what the peasantry would not do'. The *narodniks* shifted strategy from educating the peasantry to engaging in armed struggle, in the form of assassination attempts at the tsarist authorities, especially the tsar, some successful, mostly not. Two organisational groupings came about, 'Land and Freedom' (*Zemlya i Volya*) and the 'People's Will' (*Narodnaya Volya*); the latter successfully assassinated Tsar Alexander II in 1881 (see also Bernstein 2018).

The intellectuals in the People's Will read *Capital* and got into direct contact with Karl Marx. Vera Zasulich wrote to Marx: '[W]e often hear it said that the rural commune is an archaic form condemned to perish by history, scientific socialism and, in short, everything above debate. Those who preach such a view call themselves your disciples *par excellence*: "Marksists Marxists"'. She continued: 'Their strongest argument is often: "Marx said so". You would be doing us a very great favour if you were to set forth your ideas on the possible fate of our rural commune, and on the theory that it is historically necessary for every country in the world to pass through all the phases of capitalist production' (Zasulich 1983: 98–99). To which Marx responded, after several lengthy draft versions of his reply: 'The analysis in *Capital* … provides no reasons either for or against the vitality of the Russian commune. But the special study I have made of it, including a search for original source material, has convinced me that the commune is the fulcrum for social regeneration in Russia' (Marx 1993: 124). The exchange between Zasulich and Marx has been the subject of much controversy and debate in the literature of Marxist agrarian studies about populism (see Bernstein 2018; Shanin 1983a).

While the original Russian populism was short-lived, its legacy and influence would continue on, partly because of its key element, namely, its principal commitment to socialism—albeit trying to take a route via the peasantry. As Hobsbawm (1977: 199, cited in Bernstein 2018: 1131) puts it, *narodnism* '"is not significant for what it achieved, which was hardly anything, nor for the numbers it mobilised, which hardly exceeded a few thousand … [but that it] … formed, as it were, the chemical laboratory in which all the major revolutionary ideas of the nineteenth century were tested, combined and developed into those of the twentieth century"'. These ideas were inextricably linked to parallel and subsequent debates in Marxism, from Engels' formulation of the peasant question and Kautsky's formulation of the 'agrarian question' (Engels 1894; Kautsky 1988), to Russian revolutionary ideas and practices from Leninism and the Chayanovian socioeconomic logic of the peasant economy (Chayanov 1966; Lenin 2004),[16] and even to contemporary Marxism (see Akram-Lodhi and Kay 2010a, b; Levien et al. 2018)—all of which are relevant in the current, early twenty-first-century conjuncture. *Narodnism*, decimated after 1881, was reincarnated 'in the

[16] Chayanov's theories of the peasant economy would become a key influence in subsequent agrarian discourse and among towering agrarian scholars such as Shanin (1972), Scott (1976), and van der Ploeg (2013). The extent to which the original *narodnism* and Chayanov have informed contemporary agrarian populism is something that, in my view, is generally assumed or theoretically extrapolated rather than demonstrated. This is relevant, especially because most of the important agrarian movements do not make explicit the theoretical provenance of their political frameworks, and the few that do refer explicitly to their theoretical inspirations actually invoke Marx, and sometimes even Lenin—but never Herzen, Chernyshekvskii or Chayanov.

form of a "Social Revolutionary" party in the early 1900s', and it would 'become the major rural party of the left' (Hobsbawm 1987: 295). For Shanin (1983a: 271):

> The crux of the originality and illumination of the Russian revolutionary populist lies ... in the posing of a number of fundamental questions concerning capitalist society, its 'peripheries' and the socialist project. The attempts to disqualify those questions as belonging to the past only, i.e. representing the Russian social backwardness in the 1880s or the petty bourgeoise nature of its peasantry, have proved wrong by historical experience. The decline of peasant Russia did not make those questions disappear; quite on the contrary, most of them became increasingly global and pertinent also in super-industrial environments. Such questions left unanswered come back to haunt socialists time and time again, and will proceed to do so until faced, theoretically and politically. They can be avoided only at socialism's peril.

It is important to clarify a central point: how did the politically loaded term 'populism' originate, evolve, and come to have such a negative meaning in the Marxist academic and political tradition? In the history of some communist parties, '(neo)populism' was viewed from a 'revolutionary–counterrevolutionary' ('R-CR') framework that could—and did—lead to recurring bitter factional purges, a position impacted by a small, but at some point influential section of sectarian Marxists. We turn to Shanin once again for his interpretation of the history of this term that is central in the current text. He explains it in the context of Marxism and the *narodniks*, demonstrating that the history of this term was intertwined with right-wing populism:

> The label 'populist', like that of 'marxist', is badly lacking in precision; the heterogeneity of both camps was considerable. In Russian speech a populist (*narodnik*) could have meant anything from a revolutionary terrorist to a philanthropic squire. What makes it worse is the fact that there are today no political heirs to claim and defend the heritage of Russian populism – political losers have few loyal kinsmen, while the victors monopolise press, cash and imagination. Lenin's major work ... from which generations of socialists learned their Russian terminology,

> used 'populism' as a label for a couple of writers who stood at that time on the extreme right wing of the populists. ... This made Lenin's anti-populist argument of 1898 easier, while increasing the obscurity of the populist creed to his readers of today. (Shanin, 1983a: 8)

Agrarian populism is plural and diverse. Terence J. Byres, in his 1979 classic critique of the populism of Michael Lipton (1977), identified three types of agrarian populism: *classical populism*, *neo-populism*, and *liberal populism* (Byres 1979). He would later, in 2004, advance the notion of '*neoclassical neo-populism*' to categorise the body of work of Griffin et al. (2002), as well as Lipton.[17] Neo-populism is essentially identified with Chayanov (1966). Byres' basis for categorisation includes the position taken on social differentiation among the peasantry, the role of rich peasants, industrialisation, revolution, private property, and socialism. It is a useful heuristic tool that can help improve our understanding of the so-called agrarian populist movements today, within and across right-wing and left-wing agrarian movements. This is particularly helpful when problematic narratives tend to assume that the neoclassical economics version of populism is progressive, while Marxist advocacy for socialism is outdated and dogmatic.

Following Byres' categories, and looking at the examples of La Via Campesina and IPC for Food Sovereignty, one can conclude that the

[17] In his 1979 critique of Lipton, Byres argued that Lipton embraces classical populism in as much as he has 'an almost mystical faith in the mass of the people (who happen to be rural "countryfolk") – not some of the people, but all of them who are capable ... of uniting against their urban oppressors and establishing egalitarian Utopia' (Byres 1979: 238). He further characterised Lipton as a classical populist because of his belief that 'the small farmer is more efficient ... than the large', as well as 'a distaste for industry and a conviction that industrialisation ... is undesirable; an anti-capitalist stance; a determination to confront and reject Marxism, allied to a curious fascination with Marxist ideas' (ibid.). Byres then argued that Lipton is a neo-populist because of his 'defence ... of rich peasants ... in his claim that he actually accepts the need for industrialisation, but in the distant future, and not if an efficient agriculture is possible; and in his aversion to revolution' (ibid.) Byres tagged Chayanov (1966) as the father of neo-populism. Finally, Byres argued that Lipton is a liberal populist because of his 'aversion to revolution' and 'accompanying professed faith in reformist solutions and in the power of reason and argument to secure social justice (even from dictators)' (ibid.). Twenty-five years later, Byres (2004) criticised the work of Griffin et al. (2002) on land reform, implicating Lipton, and put forward an argument that Griffin et al. and Lipton are in fact 'neo-classical neo-populist', with their fundamentals anchored in neoclassical economics. For the latest relevant discussions, see: Bernstein (2018), Bernstein et al. (2018), van der Ploeg (2013, 2018), White (2018).

majority of the affiliated movements have the following characteristics: (a) anti-capitalist; (b) not anti-revolution; (c) not anti-industrialisation (although industry and progress have to be redefined); (d) believe in the power of small farmer production—but only under certain conditions of democratic property regimes and technological advancement; (e) do not consider people in the countryside to be an undifferentiated mass; and (f) do not support the polarised view of the 'oppressed rural mass versus urban oppressors'. Are there *tendencies* within these movements to gravitate around each of Byres' ideal-type populisms? Definitely. Some members of La Via Campesina from Central and Eastern Europe do not want to hear anything about a socialism especially of the type that existed in their region in the past (although they may be open to the kind of socialism defined by Wright). In addition, key La Via Campesina movements in India are champions of Byres' neo-populist pole. Tendencies to pick up some elements of classical populism, especially those combining an anti-capitalist and socialist discourse, with agrarian societies as spearhead, are palpable in some discourses of influential individuals and movements within some sections of La Via Campesina. Yet, on the opposite pole, one will see a commitment to Marxism, or even disciplined adherence to classic Leninist principles, from other member movements of La Via Campesina. What appears at first sight as a unified position within La Via Campesina as a global movement is in fact better seen as a series of temporary outcomes of conjunctural political struggles internal to the movements that are in turn underpinned by various ideological currents. Understanding these ideological currents is key, and Byres' typology of agrarian populism is a useful reference point.

Following this discussion around Byres' categories of agrarian populism, it becomes relevant to see Bernstein's (2018) emphasis on not dismissing a priori agrarian populists as, partly and importantly, a *methodological* question that calls for concrete analysis of concrete conditions. Heeding this call, and taking a closer look at contemporary agrarian movements, we will see highly differentiated national and transnational agrarian movements based on class, ideology, and politics (Borras and Franco 2009; Edelman and Borras 2016). However, there is a persistent tendency in the debates to see and label them as a unified and homogeneous movement. They are not. The various social groups that comprise this catch-all category of 'agrarian (neo-) populism'—and the movements that compete to (re)present these, separately and collectively—are linked through class relations, and their internal politics are, at times, marked by antagonism. The breadth and diversity increase as these (sub-)national movements link horizontally across classes, and unite vertically as they forge transnational coalitions. It is therefore unfortunate that such movements are often lumped together, understood, and labelled, a priori and pejoratively, as amorphous agrarian populist movements, and dismissed on that basis. Such tendencies can inadvertently undermine the potency of the Marxist framework in advancing urgent and necessary critiques of contemporary agrarian movements on many fundamental issues, such as: uneven and inconsistent anti-capitalist stance and campaigns; reluctance to take an explicitly socialist platform; inconsistent position and actions in explicitly building class-conscious broader alliances among working people, especially among landless labourers in the countryside; and other class-oriented issues for which a Marxist framework is best positioned.

Contrary to the caricaturised depiction of agrarian neo-populism, many of the key movements within La Via Campesina represent entirely different stories. I will illustrate this point by looking at three of the founding members of La Via Campesina, namely, Brazil's MST (*Movimento dos Trabalhadores Sem Terra*) which is a broadly Marxist agrarian movement (Wolford 2010), Philippine Peasant Movement (KMP, *Kilusang Magbubukid ng Pilipinas*) which is within the close orbit of a Marxist-Leninist-Maoist left movement (Borras 2007; Franco 2011; Putzel 1995), and Andalucía's SOC (*Sindicato Obrero del Campo*) which comes from a broad left-wing tradition with significant anarcho-syndicalist influences. All three movements are anti-capitalist, with broad socialist perspectives, and all have played key leadership roles in La Via Campesina at various stages of the history of this transnational movement. These three movements *do not* have a perfect fit in the description of classical agrarian populism, or in any of the subsequent formulations of 'neo-populism'; none of the three movements is nostalgic-romantic, conservative, reactionary, or utopian; none of the three is class-blind in their political work; none of the three is anti-industrialisation; and they bear little resemblance to any of the prominent Indian rich peasant movements in terms of class base, ideology, and politics. These movements take agrarian landed classes as the principal class adversary, with demand for land as central—even when they take agribusiness based outside the countryside as *equally important* targets. This is fundamentally different from rich peasant agitations, such as those in India, that deliberately avoid landed classes as targets and land redistribution and labour issues as central demands, and instead target 'urban elites' and/or foreign corporations as the principal adversaries. My hunch is that it is most likely that the overwhelming majority of the movements affiliated with La Via Campesina are closer, in varying degrees and extents, to MST, KMP, or SOC in terms of social base, ideology, and politics than to the caricature of an agrarian populist, whether classical or neo. Interpretation, presentation, or claims about these social movements by observers—allies, advocates, admirers, sympathisers of agrarian movements, whether academic researchers or NGOs—may not always be a precise reflection of what actually exists, and cannot replace the need for more careful *empirical* investigation of such questions.

There are also two international farmer federations which are worthy of note here, namely, the International Federation of Agricultural Producers (IFAP, established in 1946 and self-liquidated in 2010) (Desmarais 2007), and, arguably, its reincarnation in the form of the World Farmers Organisation (WFO) (Edelman and Borras 2016). These organisations are movements of rich commercial farmers, or are politically led or influenced by the latter; they are largely based in industrialised countries with a few members in developing countries that represent medium-scale and rich agricultural producers and farmers, such as the Zambia National Farmers' Union. Their sets of demands are fundamentally different from La Via Campesina's. Generally, they are defenders of capitalism, and opposed to socialism. They endorse the World Trade Organisation (WTO), with reforms. Most of them can easily fit in Byres' categories of 'liberal populism' and/or 'neoclassical neo-populism'.

It will be interesting and important to examine whether and how, and to map the extent to which, their mass bases are linked to contemporary right-wing populism, and compare whether progressive agrarian populists acted differently as compared to their conservative counterparts in interacting with or confronting right-wing populism. It is, for example, important to investigate empirically—rather than assume theoretically—how members of European Coordination Via Campesina (ECVC) and the European affiliate of WFO (and IFAP previously), namely, COPA-COGECA (Committee of Professional Agricultural Organisations/General Confederation of Agricultural Cooperatives) are voting in the contemporary electoral contests across Europe. It is legitimate to ask because we know that the rise of right-wing populist groups in Europe today, from France to Sweden, and from Austria to Germany, has been electorally supported by rural voters. There is a material basis here for possible contradictions, including the fact that many of the farmers associated with these organisations are likely to be engaged in hiring (sometimes illegal) migrant farmworkers from eastern Europe or north Africa, an issue that is at the heart of the right-wing upsurge. In a similar vein, it would be relevant to examine more closely the politics of their American counterparts, such as the Farm Bureau, especially in relation to their attitude towards Trump's position on immigration, on the one hand, and the issue of widespread reliance on (sometimes illegal) farmworkers from Latin America, on the other hand. Indeed, it is important to interrogate the two key gatekeepers of La Via Campesina in South Asia, namely, the Karnataka State Farmers Association (KRRS) and Bhartiya Kisan Union (BKU), not just in terms of the Modi regime, but especially on the broader issue of rising Hindu populist nationalism (Vanaik 2017a, b), and the demands they are putting forward to the current national government more broadly. These are empirical questions.

The justification for the lengthy discussion above is to demonstrate that agrarian populism is, in reality, far more differentiated than its homogenised and caricaturised depiction by some sections of the Marxist intellectual community (academics or party cadres). Deliberately lumping groups together or failing to discern this differentiated nature can lead to a flawed deployment even of objectively rigorous theoretical ideas, and can result in unfortunate or even disastrous political miscalculations in terms of practical politics; or worse, as is often the case, it can lead to such movements—all movements—being dismissed altogether. Conservative rich peasant movements exist, but these are not included in the broader category of 'progressive agrarian populism/populists' that I use in this chapter, the minimum defining character of the latter being that they are radically anti-capitalist.

Contemporary progressive agrarian movements are relatively vibrant, as even sceptics like Bernstein (2018) admit. Based on our definition of populism discussed above, I would argue that their *political actions are populist* centrally because they are attempts at re-bundling socioeconomically differentiated class and group interests and issues into a more homogenised voice—'people of the land'—in relation to a constructed 'other'. The rise of transnational agrarian movements (TAMs) during the past couple of decades, particularly La Via Campesina, and the subsequent platform for action, that is food sovereignty, are perhaps the most significant political processes in the social justice movements' global front since neoliberalism took a hold in the early 1980s and subsequently debilitated conventional workers' trade unions and movements (Desmarais 2007; Edelman and Borras 2016; Edelman et al. 2014; Martinez-Torres and Rosset 2010; McMichael 2008; Shattuck et al. 2015; Wittman et al. 2010). La Via Campesina is, in practice, a populist movement, based on our definition of the term (which is *not necessarily* pejorative). As we have emphasised earlier, this brand of agrarian populism is not necessarily class-blind, as is often insinuated in polemics.

More generally, the assumption that *populism and class blindness necessarily and always go together* should be interrogated empirically. In my view, it is precisely the class-consciousness within some of these TAMs and food sovereignty movements that has led them to aggregate disparate interests and demands among differentiated social classes and groups in political projects that are *deliberately framed as multi-class initiatives*. This does not mean they were able to resolve the inherent contradictions, or even antagonism, in some of the social relations that underpin potential movement participants, e.g. farmers and (migrant) farmworkers, or rich farmer movement versus poor peasant movement members. It only means that these class dynamics are actually flagged, and are being addressed even if *unevenly and inconsistently* within and between many of these movements across societies, across political spaces, and over time. This is a necessarily tension-filled and conflict-ridden *process*; an arena of struggle within and among movements. This issue remains one of the most under-explored themes in the study of contemporary agrarian movements: one possible reason for this is because it is an extremely sensitive issue and thus quite difficult to research.

Moreover, and relatedly, many serious scholars studying agrarian movements and politics in and related to the countryside who do not identify their scholarship solely, fully, and strictly with orthodox Marxism are not necessarily class-blind or uninformed by relevant Marxist concepts about questions of agency and politics—as a quick, random glance at some of the influential work of numerous important scholars from across disciplines and generations would remind us.[18] A simplistic, polarised division of progressive researchers in critical agrarian studies between strict orthodox Marxism on the one hand, and an ideal-type Chayanovian neo-populism on the other hand, does not fully capture the diversity of intellectual and political currents that actually exist, and is, politically, not very productive. Meanwhile, it is equally important to emphasise that even class-based politics is, like agrarian populism, diverse and plural. Dogmatic interpretations of Marx and Lenin, or of Mao, is one strand, but is not the only possible route to the interpretation of class-based politics, as the contrasting approaches to understanding agrarian class agency and agrarian class politics by Eric Wolf (1969) and Jeffrey Paige (1975), for instance, remind us.[19]

Populism, Class Politics, and Crisis

In the era of contemporary right-wing populism, class and class politics have become even more relevant. E.P. Thompson (2013: 9) explains that, 'the class experience is largely determined by the productive relations into which men [*sic*] are born – or enter involuntarily. Class-consciousness is the way in which these experiences are handled in cultural terms:

---

[18] For example, Baud and Rutten (2004), Baviskar (1999), Clapp and Isakson (2018), Edelman (1999), Fairbairn (2014), Hall (2011), Holt-Giménez (2017), Isakson (2014), Li (2007, 2014), Martinez-Alier (2014), McMichael (2008), Moore (1967), Newell and Wheeler (2006), Patel (2009), Peluso (1992), Peluso and Lund (2011), Ribot and Peluso (2003), Scoones (2015), Scott (1976, 1985), Shanin (1972), Tsikata and Yaro (2014), Weis (2007), Wolford (2010).

[19] And as some more recent studies point out, such as Bernstein (2018), Bernstein et al. (2018), Cousins et al. (2018), Lerche and Shah (2018), Levien et al. (2018), White (2018).

embodied in traditions, value-systems, ideas and institutional forms'. He elaborates further: 'If the experience appears as determined, class-consciousness does not. We can see a logic in the responses of similar occupational groups undergoing similar experiences, but we cannot predicate any law'. 'Consciousness of class', he concludes, 'arises in the same way in different times and places, but never in just the same way' (ibid.).

Right-wing and left-wing populists are two ideologically opposed groups *targeting broadly similar issues and antagonists*. Thus, even though each takes an adversarial and antagonistic stance towards the other, their discourse, actions, and shared elements of their anti-establishment narratives can be objectively mutually reinforcing. In the 2016 US elections, Bernie Sanders' attack against 'the establishment' for having caused the closure of American factories was cast firmly within an anti-capitalism/anti-neoliberalism frame, while Donald Trump, also speaking out against factory closures, did not blame capitalism/neoliberalism but unpatriotic American corporations. Sanders and Trump both gained popularity in the US rust belt communities. It is thus not surprising that the social base of an insurgent right-wing populist group can sometimes also be the mass base of a challenger left-wing/progressive initiative, *or at least the boundaries between them can be porous, blurred, or malleable*. The partially overlapping mass (class) bases usually comprise those who are *not* part of the ideological core of the right- and left-wing populist groups, and can constitute the swing vote that can be lured from left-wing issues to right-wing populism, and vice versa. Again, the American rust belt politics during the 2016 US elections is a good example, with traditionally Democrat states voting for Trump when Sanders failed to become the Democratic Party presidential candidate.[20] Sanders and Trump are ideological opposites for some of their core groups—'alt-right' for Trump, socialists for Sanders—but they have common supporters and sympathisers, and the basis of the latter, at least in the beginning, is not their ideological stances, but the immediate, concrete issues addressed, such as factory closures amid corporate migration outside the country.

[20] For elaboration and insiders' accounts, see Sanders (2016) and Bond and Exley (2016).

We see this dynamic play out across various themes and regions of the United States, such as coal mining and the Appalachian communities.[21]

Right-wing populist groups and anti-capitalist agrarian movements are both insurgent anti-establishment challengers, although they tend to operate on different scales (with the former often in the spotlight of big politics). Transnational agrarian movements like La Via Campesina and food sovereignty movements are populist movements in the sense discussed in this essay. They juxtapose 'the people' or the 'community' as the 'us', and the big corporations (agro-chemical corporations, food empires, banks, and so on) and landed classes and oligarchs as the 'them'. Similarly, the progressive left-wing slogan, 'the 1% versus the 99%', is a populist but definitely class-conscious formulation in relative terms, compared to Trump's populism that deliberately lumps together those in the 1% and the 99%. But still, the formulation of '99%' conceals important class conflict and tension, and is problematic in many ways. The way food and agrarian movements construct powerful punchy slogans reflects such consciousness in the constant invocation of the 'us' and the 'other': 'industrial agriculture heats up the planet, small-scale agriculture cools down the earth', 'not about us without us', 'people before profit', 'no to agriculture without farmers', 'small-scale farmers feed the world', or 'WTO kills'. Their ability to condense complex conditions of their disparate base in short, graphic slogans that become key reference points and mobilising narratives is a brilliant populist strategy of constructing the 'us' and its 'other'.

Many contemporary right-wing populist groups have evolved in a similar manner by aggregating disparate social classes and groups in the rural world, forging a homogenising discourse (i.e. deliberately class-blinding narratives) and projecting a future in ways that converge somewhat with agrarian movements and food sovereignty movements, even when the basis, reason, and implications are fundamentally different. Le Pen's projection of 'the forgotten French countryside'[22] resonates with

[21] This is well captured in the video 'Why the Poorest County in West Virginia has Faith in Donald Trump: Anywhere but Washington', at https://www.youtube.com/watch?v=eqceHviNBC4.

[22] As depicted for instance in the video 'Marine Le Pen's Rise in "Forgotten France"', at https://www.youtube.com/watch?v=N-ooZ96nA8g.

Confederacion Paysanne's long-standing framing of these issues—but the two perspectives are ideological opposites as to why this happens and what is to be done. The feeling of extreme frustration is common to both the right-wing populist mass bases and agrarian movements. In many contemporary settings, this mounting frustration derives from problems around cheap food provisioning, social security, and health care, jobs, neglect of public services, social decay including the menace of drug addiction, absence of redistributive reforms like land reform, persistent violence, and criminality, rural–urban migration flows, and massive indebtedness[23] that, separately or together, generate much angst and anger among people in (the cities and) the countryside. These issues are magnified during socioeconomic and political crises. It is the *crisis* situation that provides the fertile ground for the populist seed to grow: the fact or threat of socioeconomic losses among ordinary people—not only the working class, or 'sub-proletariat', but also the middle class, as argued by Perry Anderson in the context of the rise of Bolsonaro in Brazil (Anderson 2019) and Bello in the context of Chile, Thailand, and the Philippines historically (2018)—*together with the state's inability or unwillingness to act* according to what people perceive to be its moral obligation to respond in times of crisis. To make matters worse, in many instances it is the state that takes the lead in assaulting the remaining potential capability of poor people for social reproduction, namely, dispossessing poor people of their access to land, often without offering alternative livelihood or employment (Levien 2011; Li 2010, 2011; Wolford et al. 2013).

Crisis revives the collective memory of past cycles of broken promises or broken systems. It is the simultaneous *convergence* of multiple crises—including crisis in political rule (see, e.g. Anderson's [2019] analysis of the rise of Bolsonaro in Brazil)—that has triggered the current right-wing populist upsurge. Historically, the rise of right-wing populism has been a response to, or accompanied by, crisis. The French 'Greenshirts' peasant movement of the 1920s and 1930s emerged in response to a triple crisis: *economically* ('an ever-deepening decline of farm prices that lasted so

[23] Gerber (2014) offers a powerful analysis of the role of debt in capitalism, with reference to the rural world, which is extremely relevant in deeper analyses of the politics of right-wing populism and the countryside but cannot be elaborated further here for reasons of space.

long and plunged so low that even the most diligent efforts could barely keep a family alive'), *culturally* ('the low esteem for peasants' life, values, and needs'), and *politically* (a 'crisis of representation') (Paxton 1997: 11). These past cycles of broken promises of rural reforms and reform of the global food system have led to the accumulation of frustration and anger among rural and urban social groups and the rise of relatively vibrant left-wing transnational agrarian, environmental, climate justice, and food sovereignty movements—all with strong populist impulses and tendencies. They are increasingly linking their movements together.[24]

Yet the same manifestations of capitalism in crisis have also spurred the current right-wing populist agitations, although the latter frame such crises differently, as for example, the climate change denial of right-wing US populists, Trump's electoral promise to revive the dying coal industry, and the Brazilian Bolsonaro's similar spin on these issues (Saad-Filho 2018). Whether and how the Trump administration will fundamentally differ from past administrations in terms of its treatment of American agriculture and the countryside remains to be seen, although current indications suggest that it will continue with similar policies. As Chrisman (2016: 1) explains '[a]s of 2012, just four percent of farms produced two-thirds of agricultural value – that's a lot of wealth concentrated in just a few hands. Today, a small number of farmers are "efficiently" producing more grain than ever, while metropolitan and suburban populations have ballooned'. She continues, 'But what of those who remain? 46 million Americans still live in the countryside, with many hollowed out towns, few job prospects, and the near impossibility of making a living off the land for all but the biggest farm operators' (see also Edelman 2018; Ulrich-Schad and Duncan 2018).

How the converging multiple crises are experienced, and what impacts they have on the dynamic reconfiguration of material class interests and social identity formation across social classes and groups in society, provide part of the basis for an *objective alliance* between right-wing populism and agrarian populism to emerge. In other words, although their actions are independent from each other, and they may even take an

[24] See Borras et al. (2018), Brent et al. (2015), Claeys and Delgado Pugley (2017), Edelman and Borras (2016), Mills (2018), Tramel (2016, 2018).

adversarial stance against each other, they actually reinforce one another. The danger will come if and when this objective alliance evolves into a *subjective alliance*, that is, the conscious construction of an organised coalition of forces and actions. If this were to happen, with disgruntled rural populations voting en masse for right-wing candidates, the forces of reaction could gain much ground and political momentum. One challenge, therefore, is how to prevent any transformation of the existing objective alliance into a subjective alliance between right-wing populism and agrarian populism, or even how to undermine the objective alliance between them. Holt-Giménez and Shattuck (2011) offer powerful analytical and political reference points about some of these key issues in relation to food movements.[25]

Social media is a crucial instrument in the dynamic and fluid construction of populist identities (see Soriano 2015) for both types of populisms discussed here. Social media enabled, facilitated, and expanded right-wing populists' tactical rejection of conventional and institutionalised channels of contact with ordinary people. This was most effective in penetrating the countryside in the fastest way possible, and without filters from traditional elite brokers such as churches or small town *caciques* and other elites. Smart phones, Facebook, and Twitter have, under certain conditions and in various ways, democratised access to information, whether fake news or otherwise, and provided mass access to photo, audio, and video technology and resources with ease and minimal cost. The traditional role of elite allies of geographically and socially marginalised rural population—rural school teachers, church leaders, educated young intellectuals, *caciques*, and others (a key subject of research in classic agrarian studies)—has been radically transformed and increasingly replaced by more diverse in situ and distant information brokers and populist agitators. For example, a blogger and social media agitator who is part of the core group of a right-wing populist leader can have millions of followers on Twitter or Facebook, including those who live in the countryside. Clearly, this can be a double-edged sword: the same technologies enabled the volunteers for Sanders to directly, quickly, and frequently connect with millions of people who would become

[25] See also Roman-Alcalá (2018) for a related argument.

supporters of the electoral insurgency; but the right-wing upsurge led by Trump was also largely enabled by the same technologies.

While recognising the dangers inherent in the expansion—in numbers and political influence—of core groups linked to right-wing populism, there are also dangers in fighting them. The way that some groups oppose right-wing populism by addressing populist leaders, core groups, and supporters can have an impact on whether the right-wing group is undermined, or consolidated and expanded. Some popular ways of confronting right-wing populist groups are deeply problematic: for instance, considering a populist group to be ideologically homogeneous; perceiving a right-wing group to be homogeneous in terms of social class, race, and ethnicity; and assuming that there are no inherent contradictions within right-wing populist groups between leaders, core group, supporters, and sympathisers.

Moreover, dismissive, insulting, or condescending remarks addressed to leaders, core groups, or supporters and sympathisers can, under certain conditions, have a counter-productive effect, consolidating the group's perception of the antagonism between 'them' (as 'the underdog') and the 'others' (the elite). Fake news is generated and routinised in this process. For instance, insulting Red Shirt movement participants as stupid, ignorant peasants might help to strengthen and expand the populist core group among Thailand's northeast rural communities. Mocking Duterte supporters for their imperfect and unpolished Tagalog (the minority language of the national capital and the traditional elite) might only consolidate Duterte's base and maintain his popularity. Addressing mainstream parties and media, Santiago Abascal, leader of the far-right Vox party that made major electoral gains in the Andalucían regional elections in 2018, put it succinctly: 'You haven't understood anything. … Every time you insult us, you are insulting the millions of Spanish people who listen to us and identify with our message' (Jones 2018: n.p.). When any of these things happen, there is a danger that the core group grows as disgruntled supporters and sympathisers join them, thus strengthening the previously tenuous relationship between populist leaders and their mass sympathisers. This inadvertently supports the deliberate efforts by right-wing populists to downplay or dismiss class differentiation among their mass base, cementing the feeling that the latter is a homogeneous mass that has been uniformly wronged. This volatile scenario is even more pronounced in the era of social media where the back-and-forth exchanges are often public and where passions flare up so quickly.

Fundamental Differences Between Right-Wing Populism and Progressive Agrarian Populism Today

It is important not to forget that, historically, rural social classes and groups have a chequered record in terms of supporting right-wing, even fascist political projects. The contemporary rural support for Trump in the United States (Ulrich-Schad and Duncan 2018), Putin in Russia (Mamonova 2016, 2018, 2019), Le Pen in France, Erdogan in Turkey (Adaman et al. 2019; Gurel et al. 2019), Modi in India, the Red Shirt movement in Thailand supporting Thaksin (Nishizaki 2014), and the Greenshirts of 1920s and 1930s French led by Henry Dorgeres (Paxton 1997) all remind us of past and present rural support to right-wing political ideas and initiatives. Bello (2018) offers a critical reflection on the relationship between the rise of fascism, the peasantry, and the middle class in Chile, Indonesia, Italy, the Philippines, and Thailand, seen from the contemporary context. While not forgetting and dismissing the problematic ties between right-wing populism and agrarian populism, it is also important to identify and clarify the *fundamental differences* and the bases for these. We can see this in a number of ways, two of which are as follows.

On the one hand, contemporary right-wing populists and progressive agrarian populists essentially differ in their take on the nature and causes of the current crises and how to address them, as mentioned earlier. While both populisms have elements of nostalgia, the right-wing current is straightforwardly reactionary: defending or promoting capitalism, or at least particular types of capitalism, such as Trump's 'Make America Great Again' agitation, or Bolsonaro's nostalgia of past Brazilian military rule and hope for a neoliberal capitalist future projected via his Trumpian slogan, 'Make Brazil Great'. Many, though not all, progressive agrarian and food sovereignty movements may have strong tendencies towards restorative and nostalgic narratives ('everything was fine before the agribusiness came to our villages' sort of formulation), but in contrast to right-wing populism these are usually set within broader and longer perspectives of emancipatory struggles towards alternatives to, or alternatives within, capitalism (Holt-Giménez and Shattuck 2011; Wright 2016).

On the other hand, the strategies and forms of collective action of right-wing populists and progressive agrarian populists tend to be generally antagonistic to each other. While it is not inherent in right-wing populism to be militarist or fascist (it can be at home with liberal democracy, as arguably demonstrated by the populist Corazon Aquino government, 1986–1992, in the Philippines[26]), recent and current right-wing populist variants are susceptible to or inclined towards militarist and fascist tendencies. In contrast, contemporary progressive agrarian movements, especially those allied in La Via Campesina, are deeply and fundamentally opposed to militarist and fascist methods of political rule. Right-wing populists have disdain for human rights principles, laws, institutions, and activists, while progressive agrarian movements have made human rights central in framing their political struggles (Monsalve 2013). Food sovereignty itself is founded on the broad framework of human rights (Claeys 2012), just as its core component of agroecology has been framed by a section of the human rights community (De Schutter 2014).

As discussed above, the rise of right-wing populism is partly due to the multiple crises caused by capitalism in a situation of unprecedented levels of wealth and affluence amid unprecedented degrees of inequality and precarity, in an ecologically fragile world. Right-wing populist agitations are anchored in a rhetoric in which salvation, ironically, is thought to depend on capitalism—the very cause of the crisis

[26] Where this type occurred, the ruling classes did not hesitate to mobilise the coercive power of the state to suppress democratic dissent, as exemplified for instance in the Mendiola Massacre of 22 January 1987 when the military and police guarding the Presidential Palace in Manila open fired on a 20,000 strong peasant march demanding land reform, killing 13 and wounding dozens more (Borras 2007). See also: 'Mendiola Massacre newsreel from AsiaVisions' at https://www.youtube.com/watch?v=wsXocipaJKA.

which the populist agitation purports to address. Unfortunately, the contradiction in this is not always obvious. This is where agrarian movements and food sovereignty become a potent social force that, together with other progressive social forces, can make a modest but significant contribution to countering the rise of right-wing populism. They are able to do this for a number of material reasons.

First, shared narratives about those who were/are in state power causing much of people's misery—whether because they are too soft or too hard on corporate giants, among others—easily resonate across sympathisers of both right-wing populist groups and agrarian movements. *Second*, as already mentioned, right-wing populist groups and agrarian movements may share a far larger common mass base in the countryside (and beyond) than perhaps we imagine. The general feeling of neglect, lack of wealth and power redistribution, joblessness and social decay in the countryside can easily become key issues for both political groups. *Third*, agrarian movements and food sovereignty movements are often marked by their multi-class character, and they have, on many occasions, been able to cross class lines to forge broader alliances. For example, rural constituencies for food sovereignty have forged alliances with urban sectors: workers, urban poor communities, middle-class urban consumers, and so on. There is an interesting difference or disconnect between the conditions of urban and rural classes and groups and the populisms that emerge from their narratives or movements. In general, urban working class and precarious middle class demand jobs, which gives the narrative of right-wing populism considerable currency in these communities, while progressive agrarian populists' principal demand is land (and to some extent, jobs).[27] If they can be cast in terms of a broader anti-capitalist agitation, these two sets of contemporary urban and rural demands (which make Kay's 2009 revisiting of the classic debate on 'urban bias' between Byres 1979, and Lipton 1977, both useful and timely) could dismantle the pro-poor façade of

[27] This also differentiates them from the conservative version of agrarian populism that is usually anchored in rich farmers' demands, namely, productivist and remunerative issues.

right-wing populism or that of conservative agrarian populism (especially agrarian neo-populism led by rich peasants) and put them on the defensive.

Fourth, despite their encompassing agenda, agrarian movements rarely aspire to replace an established faction of the elite in state rule, in the way that most 'big politics' right-wing populists do. This is a double-edged sword, depending on specific context, because the agendas of agrarian movements can be, and have been, vulnerable to co-optation by the establishment they detest, or by other populist groups, left-wing or right-wing. There are ample examples of this in Latin America's 'pink tide' countries where food sovereignty has been enshrined in the constitution or in national governmental policies, but interpreted in ways that contradict social movements' understanding, such as the incorporation of food sovereignty into the broader *buen vivir* platforms, funded, ironically, through neo-extractivism (see Andrade 2019; Clark 2017; Giunta 2014; Henderson 2018; McKay et al. 2014; Vergara-Camus and Kay, 2017). The Zero Budget Natural Farming (ZBNF) movement founded by Subhash Palekar in India has gained much momentum, praise, and support from prime minister Modi, but its awkward or even troubling overlap with the right-wing Hindu nationalist current poses difficult challenges and dilemmas even for its supporters from among the organised progressive agrarian populists. In a critical reflection, Khadse et al. (2018: 214) admit that, 'There is sometimes an uncomfortable closeness of some elements of Palekar's discourse to that of Hindutva's cultural chauvinism', in addition to their observation that the majority of ZBNF members are from the 'dominant caste/middle class'.

Fifth, there is one institutional connection that links current agrarian and food sovereignty movements to right-wing agitation, even if they generally despise each other, namely, NGO funding. However, the link operates in a competing, contradictory way.[28] The rise of the global NGO complex and its multi-billion dollar annual funding portfolio has had a strategic impact on social justice movements because the latter have been among the recipients of such funding. A significant

[28] The discussion on this issue draws on Edelman and Borras (2016), especially the chapter on the global NGO complex and its implications for agrarian movements.

chunk of this funding is government money from donor countries that is distributed through the international aid infrastructure, following the logic of the neoliberalisation of the development aid sector. Many NGOs have bought into the framework that promises to deliver concrete reforms that are measurable and based on 'inclusive multi-stakeholder' events. This has resulted, among other things, in advocacy for specifically identifiable and countable individual or household 'beneficiaries' of specific projects. Often, these projects conveniently and deliberately omit or completely dismiss class-oriented demands and actions as something unfeasible (it might be nearer the truth to admit that they are un-fundable). For instance, society-wide land reform was dropped in favour of targeted participatory community-based formal land titling; struggles for just systems of production and trade were replaced with initiatives for farmers to become inserted into the global value chain. This has transformed many social movements and NGOs by making them more politically moderate, stripping them of their original defining features: a radical social justice agenda for systems change, irreverence, and subversiveness. Despite the problematic character of the NGO fund distribution infrastructure, it has been a critical pillar supporting the rise of social justice movements, especially agrarian and environmental movements, and enabling them to maintain a significant presence. However, the rise of right-wing populism in many OECD countries, and the accompanying demands to end such funding, have rendered this NGO complex politically vulnerable. Some funding facilities have already partially collapsed amid right-wing pressure to get rid of them, or to redirect the funds to promote the interests of national corporations (e.g. funds to promote 'trade not aid', and directly related to national corporations). It is plausible that the weakened global donor complex has correspondingly weakened agrarian movements and their allies worldwide. One possible unintended—but positive—outcome of such a shift could be greater autonomy for agrarian movements. The challenge is how to extricate NGOs from their current entanglement with government funds, reclaim their subversive roots, and strengthen their deep commitment to autonomy, social justice, and help in building mass movements.

The *final* fundamental material basis that I want to note here for the opposite worldviews of right-wing populism and progressive agrarian movements is their competing notions of property or claims to fruits of labour. This is powerfully illustrated in the contrasting ideas about property around land, as argued by Wolford (2005) in the context of Brazil: MST's notion of property is based on the principle of 'land for those who work it', in contrast to the landed elite's principle of land for those who have a stronger legal claim in terms of productivity and the ability to buy land in a free market. Such contrasting views were played out in the late 1990s in the context of the World Bank's market-assisted land reform, and are now being resurrected in the Bolsonaro era. Following Wolford (2005), it is likely that social relations around property and labour will remain among the most important defining elements that differentiate, or create cleavages between progressive agrarian populism and right-wing populism.

What Is to Be Done? Ask Big and Act Insurgent, with a Socialist Perspective

The 'consensus at the centre' has dominated mainstream responses to addressing contentious issues confronting the people. This consensus, Mouffe (2016: 64, emphasis added) argues, 'removes from politics one of its constitutive elements – its *partisan* nature'. The consensus seeks reforms within the framework of 'what is do-able', leading to its inability—or unwillingness to even try—to go beyond the limits imposed by the status quo. Perhaps what is needed is a strategy of 'what is possible'. The difference between the two is that the former works within the limits and possibilities of what is do-able *within a given balance of social forces*, while the latter takes an insurgent approach *to disrupt a given balance of social forces* in order to pursue transformative deep social reforms. It pertains to what is possible in terms of disrupting the pre-existing balance of social forces and power in order to effect radical reforms that are otherwise unthinkable, and daring to tackle political agendas that are *absurdly difficult* but not impossible.[29] As Eric Olin Wright (2016: 102, original emphasis) reminds us:

> Pessimism is intellectually easy, perhaps even intellectually lazy. It often reflects a simple extrapolation of past experience into the future. Our theories of the future, however, are far too weak to really make confident claims that we know what *can't* happen. … The appropriate orientation towards strategies of social transformation, therefore, is to do things now which put us in the best position to do more later by working to create those institutions and structures that increase, rather than decrease, the prospects of taking advantage of whatever historical opportunities emerge.

The problems confronting the rural world are huge, persistent, and entrenched. Reviewing how capitalism has, over time, condemned humanity and ecology to their current dire situation, Patel and Moore (2017: 41) suggest that there is an urgent 'need to dream for more radical change than contemporary politics offer'.[30] The working class and the middle class worldwide were angry and were demanding radical reforms. Right-wing populists answered their call—with programmes that were *big and insurgent*—with relative success (in right-wing terms), and in many places in the world, at least for now.

Is it politically unfeasible to ask big and act in an insurgent way in a radically progressive context? Becky Bond and Zack Exley (2017) of the Bernie Sanders 'political revolution' platform and campaign don't think so: in fact, they think that it is the only way to go forward. They have shown, from the trenches, that it is *absurdly difficult but not impossible*. Bond and Exley's book's title, *Rules for Revolutionaries*, is not randomly catchy. It is in direct conversation with Saul Alinsky's classic book, *Rules*

[29] The concept of 'difficult but *not impossibly* difficult; meaning, agendas that are *absurdly difficult* but not impossible' is inspired by and draws from the debate between the dwarf Balin and the grey wizard, Gandalf, on the latter's proposed strategy of stealth to tackle the dragon Smaug in the Lonely Mountain by recruiting a hobbit, namely, Bilbo, in J.R.R. Tolkien's *Unfinished Tales* (1980: 430–431). See Ferguson (2015) for a framing of the enormity of such a challenge (of the politics of distribution) and why we should not back away from it.

[30] For a related discussion, but with more elaboration in the context of agrarian struggles, see Borras and Franco (2018).

for Radicals (Alinsky 1971). Bond and Exley recognised that the contribution made by Alinsky was his central focus on the value of community organising, anchored on immediate, concrete issues that could deliver palpable gains to community members, all done using organising and mobilising methods that are subversive and irreverent. Alinsky's approach found wide-ranging adherents worldwide, including, not surprisingly, among those who would become actors in middle-of-the-road politics.

Ultimately, however, Alinsky's approach only aspires to alleviate the condition of poor and socially marginalised local communities. For the Sanders team, the nature and extent of problems go far beyond a local community and are so enormous that real solutions require agents of social change to think big, to demand system change, and to use and pursue appropriate creative and insurgent methods of work and scale of mobilisation. This was subsequently framed in an open call for a 'political revolution' by the 99%, i.e. the people—the working class and the middle class—with a socialist perspective, as the Sanders campaign put it, generating political, financial, and logistical support from millions of ordinary people, especially from among millennials. It was a populist electoral insurgency, of a kind that was unheard of in the annals of American history. Sanders did not win the Democratic Party primary, but the campaign and the movement left us with a political treasure trove to mine for our own political projects wherever and whatever these might be. The principles of *ask big*, *act insurgent*, and dare to boldly advance a *democratic socialist perspective* in a context of a *political revolution* may actually gain traction in far more places than many might have conservatively assumed. 'Audacity is crucial … and if the democratic and progressive forces do not adopt a bold stance, we can be sure that the extreme right will do so', as Inigo Errejón (2016: 67), political secretary of *Podemos* in Spain, reminds us (see also Franquesa 2019).[31]

[31] The outcome of the late 2018 election in Andalucia offers an important insight in understanding the rise of right-wing populists, and the resistance to it. According to Jaume Franquesa: 'In terms of geographical and class spread, Vox's results are also interesting. Vox (and the whole right, in general) has done better in cities than in the countryside, and within those cities it has done particularly well in wealthy neighborhoods. But Vox has obtained very high results in several municipalities with highly marketised, migrant labor-based agriculture, the most eye-catching example being El Ejido—a place where there was a terrible anti-migrant pogrom at the turn of the century—which was the only Andalusian municipality where Vox won the

Before going into some propositions on what is to be done, I will look briefly at three existing strategies used in struggles against right-wing populism, which are popular but may actually be problematic. The first is engaging in and mobilising around *restorative narratives and campaigns*, insinuating that the period *prior* to right-wing populism was good for 'the people'. This is quite a popular tactic worldwide, especially among those who were in power prior to the current populist upsurge, such as liberal democrats and social democrats. There is thus a compelling need to take a two-pronged approach to social struggles, striking simultaneously against the old regime and against the current right-wing upsurge.

The second is the contribution of *small-scale and 'too polite and neat' campaigns and mobilisations* to subverting right-wing populism: however well-intentioned, these may prove to be insignificant. One aspect of the 'consensus at the centre' approach that is partly responsible for the current right-wing populist swing is the dogmatic imposition of 'multi-stakeholderism'—the codification of multi-stakeholder approaches (McKeon 2018) to resolving class and identity conflicts which completely fail to acknowledge the presence of power imbalances along class and other social hierarchies. Grand mismatches abound: for huge problems around the despotism of local rulers and chieftains, the answer is community-based conflict resolution mechanisms; for problems caused by extreme greed and bullying, practised with impunity by transnational corporations, the answer is multi-stakeholder platforms around voluntary self-regulation; for the problem of land grabbing, the answer is transparent and consultative administrative processes in land deals. There are serious class and power imbalances in these spaces where dominant actors actively ignore or deliberately misunderstand politicised notions of 'accountability' (Fox 2007; Gaventa 2006; Ribot 1999), or deliberately avoid radical interpretation and the use of governance instruments such as Free, Prior, Informed Consent (FPIC) (Franco 2014).

election (with about 30% of the vote)'. Personal email communication with Jaume Franquesa, 4 December 2018. See also Franquesa (2019) for a broader discussion in the context of the Spanish experience more generally.

The third strategy is sectoral struggle (e.g. trade union or peasant, or urban-centred political work) against right-wing populism. This is a necessary but *insufficient* banner for social justice framing and for a spearheading social movement that can unite the working class and other sectors and social groups that are struggling, or have the potential to struggle, against right-wing populism. In many societies today, the scale of operations and exposure of agrarian movements and food sovereignty movements is limited and small relative to that of right-wing populism. In fact, the ideas and organisations of agrarian and food sovereignty movements in these societies shrink dramatically when viewed from the perspective of big politics which right-wing populist groups steer. Today, the 'rural' cannot be automatically equated with agricultural; many rural-based small and medium sized towns have emerged worldwide, and these are becoming hotspots for right-wing mass base building.

In the era of converging global crises and right-wing populist agitations, the need for radical, deep social reforms has only grown more urgent. Five goals of deep social reforms and ways to pursue them are relevant in the current conditions of the rural world, namely, *redistribution, recognition, restitution, regeneration,* and *resistance*—or in short, *revolution.*[32] These are deep reforms that are directed at the heart of social structures and institutions that reflect and maintain class formation and class rule, and could, or should, ratchet up towards structural transformation that could undermine capitalism and pave the way for a socialist alternative. Such a set of radical reforms does not recoil from class politics; in fact, class politics is at the heart of the 5Rs.

[32] The discussion on the 5Rs draws from a long history of my work on land politics in collaboration with several colleagues and comrades, but most especially with Jennifer Franco. This work started with a focus on redistribution and recognition, which are basically direct and logical extensions of struggles for land and territory; more recently, especially informed by our work on land grabs and expulsions, the inseparable issue of 'restoration' of resource access, or 'restitution' has been added—thus, 3Rs. In the last few years, our research work has been devoted to the intersection between climate change politics (mitigation and adaptation) and land grabbing, leading to the addition of the 'regeneration' angle. The discussion and inclusion here of 'resistance' is in dialogue with Nancy Fraser's work on 'redistribution', 'recognition', and 'representation' (Fraser and Honneth 2003). For the progression of our work on this, see Franco et al. (2015) and Borras and Franco (2018). Patel and Moore (2017: 207–212) put forward a similar proposition for five Rs—recognition, reparation, redistribution, reimagination, and recreation—where individually and collectively some of these overlap with the 5Rs in the current chapter.

The proposed agenda for political conversation, namely, deep social reforms, is fundamentally different from the dominant 'third way' or social-democratic reforms because the latter actually legitimise and strengthen capitalism through incremental reforms that are ameliorative at best,[33] because they shy away from addressing existing class antagonisms and thus class politics. If taken seriously, the 5Rs proposed here constitute a political revolution, and not just a reform programme. These 5Rs require multi-class struggles far broader than agrarian or rural struggles. Nevertheless, agrarian movements may be able to make strategic contributions to these social struggles. We now turn to an abbreviated discussion of each of these spheres.

First, where wealth and the means of production to create wealth are monopolised by a few (veering towards the obscene 1–99% ratio), wealth and power *redistribution* becomes urgent and fundamental. In the context of agrarian societies, this includes redistribution of access to and control of the key means of production—land, water, seas, forests—thus contesting the very essence of capitalism. Second, where social exclusion, marginalisation, and discrimination by one dominant social group over other groups constitute an oppressive social complex layered with xenophobia, racism, misogyny, and so on, social struggles for *recognition* become an important struggle that can expose the fundamentally regressive nature of right-wing populism. In agrarian societies, this can entail recognition of the rights of indigenous peoples or ethnic minorities over their territory, or women's rights to their distinct access and control of the means of production: land, water, forests. Third, where people have lost their land, territory, houses, savings, pensions, and other important means of production and reproduction because of corporate resource grabs, or have lost healthcare because of scams by financial swindlers,[34] social struggles for *restitution* become a key front. In many rural areas, this means restitution of access to land, territory, water, forests, especially in light of global land grabbing during the past decade. Fourth, ecological and climate crisis is one of the hallmarks of the contemporary period

[33] Similarly, some groups may be opposed not to the fundamentals of capitalism, but only to some versions of capitalism, such as neoliberal capitalism.

[34] Harvey's broadly cast 'accumulation by dispossession' comes to mind (Harvey 2003).

in world history; it has deep roots in the history of capitalism, captured by the term Capitalocene (Moore 2017). Social struggles for ecological *regeneration* and environmental justice have become an integral part of broader social justice struggles, as Martinez-Alier et al. (2016) argue. This connects with what Patel and Moore (2017: 43) call 'reparation ecology', with struggles for agroecological farming systems (Rosset and Altieri 2018), or with struggles for 'agrarian climate justice' (Borras and Franco 2018). Fifth, these four goals of deep social reforms can only be accomplished through fierce, relentless, and disruptive *resistance* within and/or against capitalism—which really means political *revolution*. This puts agrarian movements directly in confrontation with right-wing populist groups, the latter being generally believers, champions, defenders, or apologists of capitalism.

Political actions around these five social reform agendas could expose the contradictory ideological positions between social justice movements and right-wing populism, and could contribute to progressively undermining the latter. Right-wing populists are unlikely to meet the social justice-oriented demands of 'the people' because, ultimately, their basic interests fundamentally clash with the people's demands. Vanaik (2018) reminds us, for example, that basic agrarian issues (land, food, rural jobs) render Modi politically vulnerable. The 5Rs discussed above are not to be treated like a checklist from which one can cherry-pick. The 5Rs are linked in their logic because they are responding to inherently interconnected social processes. These processes can be highlighted in two abbreviated illustrations.

On the one hand, contemporary capitalism has seized the opportunity to make profit out of responses to climate change via market-based transactions such as carbon sequestration and trading, biofuels, and neoliberal nature conservation. In essence, this is *regeneration by dispossession*—defined here as the capitalist project of operating within and regenerating nature by dispossessing ordinary villagers to facilitate continuous capital accumulation. Many of the conservation initiatives associated with climate change politics are associated, in varying degrees, with this strategy. Alternative ways to confront regeneration by dispossession are through *regeneration by restitution*, or *regeneration by recognition*, or *regeneration by redistribution*, or a combination of all of these, in the context explained above (see also Borras and Franco 2018).

On the other hand, *redistribution without transformation* represents the pre-populist agitation period, the elitist 'business as usual' era. The various forms of redistribution that did occur—usually concessions resulting from social struggles, such as partial land redistribution—were ultimately carried out within a hostile neoliberal system, so that any initial gains were easily cancelled out. Many land reforms in the past (both prior to and during neoliberalism) suffered the same fate: small-scale farmer beneficiaries lost out within a broadly neoliberal agricultural policy climate. Some redistributive reforms provoked tensions around issues related to identity politics, such as the recognition claims by indigenous peoples or women that were undermined by (or completely absent from) conventional land reforms. All these have contributed to the accumulation of frustration among ordinary people, many of whom have turned to support right-wing populists. The struggle against right-wing populism can be partly fought by pursuing struggles for redistribution that dovetail with further reforms: thus, *redistribution with transformation, redistribution with recognition*, or a combination of these (see Borras and Franco 2018).

If we take a closer look at the 5Rs, we can begin to understand how their logics are intertwined, how their failure (separately and collectively) under past conditions fed into the right-wing populist 'politics of appearances', but at the same time how they contain the potential to subvert right-wing populism. Middle-of-the-road, social-democratic type reformers have shied away from big and insurgent reforms, justifying themselves by arguing that these are un-doable, i.e. impossible. We must not forget that most of the rights and freedoms that many of us enjoy today, such as the eight-hour working day, maternity leave, minimum wage laws, voting rights, civil rights, the formal end of apartheid in South Africa, and so on, were big and insurgent issues in their own settings and historical moment, and seemed impossible—until the political momentum in their favour overruled any conservatism, scepticism, or pessimism.

Concluding Discussion: Towards a Class-Conscious Left-Wing Populism

Errejón (2016: 67) argues that right-wing populism 'has been able to recuperate the powerful idea "community" – that we must build a spirit of community at a time when there is more insecurity, more anxiety and fear, more uncertainty about tomorrow'. What is being emphasised here is that confronting right-wing populism should start with a good grasp of how and why right-wing populists have been effective in what they are doing. In the context of contemporary right-wing populism, Errejón (2016: 68) laments that there is a mistaken assumption that populist methods of work 'could be paving the way for [right-wing populist] ideas'. He disagrees: 'it helps [right-wing populism] much more if we abandon all forms of collective affects [i.e. "passion"], and thus cede this space to them'. Mouffe (2016: 124) insists on 're-appropriating the term populism'. She goes on: 'That it is being used in a derogatory way by parties who defend the status quo is no reason to abandon it. It's necessary to reclaim it … and to give it a positive meaning with the notion of "*left-wing populism*"'.

A priori dismissal of agrarian populism as class-blind, naively and romantically allocating everyone a seat at the table is a classic case of a straw person argument because such a populism does not exist. Populism as defined and discussed so far inherently includes groups of people across classes—and by the same token, and dialectically, it inherently excludes groups of people. The inclusion–exclusion process of 'othering' does not give a seat to everyone at the table. Progressive agrarian populism, in generic terms, is similar to Leninist or Maoist notions of a 'broad united front', or indeed, Mao's 'mass line' which is, arguably, essentially a form of populism; significantly different but fundamentally similar in the sense that both are attempts at aggregating varying, and even competing, social class and group interests (see Mao 1975). As Shanin (1983b: 270) reminds us: 'Russian populists did produce a class analysis, if a different one from that of Engels, i.e. concluding that unlike France of 1848, or of 1871, the main forces due to face each other in Russia are the state and a state-bred squiredom and capitalists versus the "labouring class", i.e. a plebeian front of peasants, workers and intelligentsia, allied with the radical soldiers'. Finally, perhaps most of the peasant movements studied by Eric Wolf, with special reference to the agency of the middle peasantry, may, to some degree, qualify in the category of agrarian populism (Wolf 1969). The flip side of this is that only the agrarian movements under the tight command of dogmatic strands of 'workerist' Leninist parties could not be accused of populism in the latter's pejorative meaning. Theoretically, this is questionable; politically, it is problematic because this type of agrarian movement does not constitute a significant global force today.

The use of the term 'working people' seemed to be generally acceptable to most Marxists, sometimes to denote the basic 'worker–peasant alliance', but as Shivji (2017) argues, it is poorly elaborated. Shivji himself elaborates on the concrete basis underlying the concept of working people, especially in the context of neoliberalism—an explanation that largely overlaps with Bernstein's (2010) notion of 'classes of labour':

> Commodification and privatisation of health care, education, water, sanitation and removal of subsidies from essential foods which all formed part of the social wage goods previously means that now the poor have either to pay for it or go without. All in all, the materiality which underlies producers – peasants and pastoralists, proletarians and semi-proletarians, street hawkers selling consumer goods and peddlers selling cooked food, operators and repairers in backyard workshops – in virtually all sectors is the minimizing of their necessary consumption and maximizing their labour. ... This is then the material basis common to all sectors of what I called the working people. (Shivji 2017: 11)

Shivji's elaboration on the notion of 'working people' and Bernstein's 'classes of labour' are important conceptual bases for an idea of 'class-based populism', or 'class struggles with populist consciousness or sensitivities', or indeed *class-conscious left-wing populism.* Each of these formulations can be described as a contradiction in terms, highlighting concrete contradictions that are partly reflective of the current conditions of the peasantry, working class, and precarious section of the middle class. This, in turn, impacts political agency and agrarian politics in various ways, as well as how we understand these (Bernstein 2010; Moyo and Yeros 2005). Shivji's argument is relevant in this context; he claims that 'working people' '*is* a "class" against capital and has great potency in political discourse and mobilisation. … I believe politically, and even conceptually, the term working people has greater potency and validity than worker–peasant alliance'. He elaborates: 'Only real life struggles against the neoliberal phase of capitalist imperialism will help us further to theorise the concept of working people' (Shivji 2017: 12, original emphasis).[35] The notions of 'classes of labour' and 'working people' can partly explain the rise both of left-wing populist groups (Podemos in Spain, Syriza in Greece—before all the internal problems in both parties had erupted), and of right-wing populist groups who are able to conjure a 'community' of such 'classes of labour' or 'working people' using language that is emotive and dramatic: 'the left-behind', 'the pushed aside', 'the forgotten', and so on.

Two further concepts warrant elaboration, namely, socialism and revolution. If the problem is capitalism, it becomes inconceivable to find a positive future within it because capitalism is a system that is fundamentally based on exploitation, oppression, and greed (i.e. profit-making). This brings us back to socialism—defined here in the broadest sense as per Wright (mentioned at the beginning of this text). The immediate and most difficult challenge we face here is similar to the problem of reclaiming 'populism' in the context of agrarian politics: 'socialism' is a term that comes with a lot of baggage. It is a proposition that is passionately contested and debated, where each strand of proposition will surely disappoint many. Aware of this, but needing a reference point for the purpose of this work, I find Wright's proposition for a socialist alternative in the current context helpful. He framed his socialist agenda within a three-pronged strategy of struggles within and against capitalism, namely, ruptural (new emancipatory institutions via a sharp

[35] It is in the context of the current fragmentation of 'classes of labour' (in Bernstein's framing) and of Shivji's 'working people'—amid the rise of various forms of populism—that we also need to take seriously the condition and politics of intersectionality. As Bernstein (2010: 115, original emphasis) reminds us: 'class relations are *universal but not exclusive* "determinations" of social practices in capitalism. They intersect and combine with other social differences and divisions, of which gender is the most widespread and which can also include oppressive and exclusionary relations of race and ethnicity, religion and caste'.

break with existing institutions and social structures; closely associated with revolutionary socialism), interstitial (building new forms of social empowerment in the niches and margins of capitalist society; associated with some strands of anarchism), and symbiotic (using 'the state to extend and deepen the institutional forms of social empowerment in ways which also solve certain practical problems faced by dominant classes and elites', closely associated with social democracy) (Wright 2016: 100–101). This is useful for the argument being developed in this chapter in at least two ways: (a) the agenda is wide enough to accommodate different anti-capitalist progressive social forces, and (b) the institutional, organisational, and political forms and requirements are broad enough to allow for various strands of anti-capitalist and potentially pro-socialist forces (or at least those who are not opposed to socialism, or not opposed to forging alliance with socialists) to converge. As Wright suggests: 'the framework proposed here for a socialism rooted in social empowerment involves a commitment to *institutional pluralism and heterogeneity*' (2016: 104, emphasis added). If we take this framework, and we look at the range of ideologically diverse large transnational movements like La Via Campesina, we can see how such a space can accommodate (to a significant, if not full, extent) the plurality of progressive movements willing and able to identify with a broadly cast socialist agenda, and how it can contribute political resource to such a political project—long-term and short-term goals in the struggles within and/or against capitalism—depending on their capacity and ideological orientation.

The deep social reforms explored earlier around the 5Rs, and the social forces required to struggle for such reforms, can be located within and be made to serve Wright's proposed framework and three-pronged strategy for socialism. Moreover, this is a kind of socialism and strategy around which a broad mass base can be rallied, from the conventional left to anarchist groups, including contemporary antifa groupings and de-growth advocates, from Sanders' typical supporters among the millennials and precarious middle class to the multitude of hard-up classes and sectors that formed the base and supporters of arguably left-wing populist political parties like Podemos in Spain, from radical agrarian movements to urban food justice activists, from radical feminists to climate justice activists, from indigenous peoples' emancipatory movements to progressive race political activists.

The conventional left's notion of a 'vanguard party' or 'corps of cadres' to lead mass movements is unlikely to be effective as a spearhead even though the current situation could benefit from reinvigorated left political parties. Nor is Mao's idea of armed struggle as the principal form of struggle, subordinating all other forms of mass movements, likely to be able to provide an overarching strategy and form of struggle. As Wainwright (2018) reminds us, those working from this tradition will have to seriously rethink how to forge productive conversations with those who subscribe to other traditions, for example, the ideas of anarchism and its stance on anti-authoritarianism and anti-centralism (Bookchin 2015; Springer 2015; see also Roman-Alcalá 2018), the growing advocacy and movement around the idea of de-growth (Boillat et al. 2012; Kallis 2019; Roman-Alcalá 2017), feminist (Harcourt and Nelson 2015) and race politics (White 2017), and so on—keeping in mind that socialist and communist parties' historical track record of engaging with such advocacies and movements has been chequered at best.[36] However, and conversely, without the participation of conventional left-wing communist and socialist parties, the various groupings of non-political party radical movements and initiatives are unlikely to get closer to their goals of transformative social change, as many major waves of such initiatives in the recent past, such as Occupy Wall Street, have shown. The key is to explore and harness potential synergy among these disparate democratic challengers is less about determining which specific political line or idea is superior and thus could be the spearhead, than about determining and building the basis for a *pluralist platform* that could bring such diverse groups to a common political project—however daunting a task that might appear.

The greatest challenge may prove to be how to sustain the initial mild interest in—or at least the non-resistance to—socialism of the types

[36] For those working in the context and tradition of communist and socialist party work, Rosa Luxemburg's earlier ideas of a 'mass party' and 'mass strike' and Antonio Gramsci's notion of 'organic intellectuals' may have renewed relevance and traction, and are worth serious revisiting.

espoused by Sanders' supporters, the agrarian and food sovereignty movements, and the broad environmental and climate justice activist circles, many of which may not have been sensitised to and significantly engaged in socialist politics despite their deep anti-capitalist sentiments. Given that Sanders managed to boldly put forward an openly socialist agenda to be carried out through a 'political revolution' in the context of the United States in the early twenty-first century, and since millions of people across the middle class and working class and across the race, gender, and generational divides came on board, energised and animated, a broader international conversation and political project explicitly framed within socialism become feasible—and so too the vehicle to achieve it, political revolution. The term is used here in a similar way as it is used by Sanders: *a deep commitment against the anti-subversive petty reform incrementalism promoted by social forces behind the 'consensus at the centre'*. This task represents a difficult challenge partly because the justifications, mechanisms, and processes of the 'petty reform incrementalism' have been routinised and normalised to the point that they invisibilise what is wrong with, or disqualify serious interrogation of, this approach and the many issues that matter to people.

Combining class politics and populism is desirable despite the tension and contradictions that such an intersection necessarily internalises. What is promising, along this line, is an enriched and reinvigorated left-wing populism, that is anti-capitalist in character, socialist in orientation, pluralist in ideological orientation, heterogeneous in institutional platforms, and *fundamentally founded* on an alliance between the 'classes of labour' or 'working people' (which necessarily includes the various strata of the peasantry and proletariat) and the precarious strata of the middle class or petty bourgeoisie, *and* their overlapping social groupings based on multiple 'identifiers', including community, gender, race, ethnicity, generation, and religion. This form of populism is very different from an essentially identity-based populism where class is altogether absent or is deliberately negated, a mere footnote, or an afterthought, addressed only if and when it is referenced.

This reclaiming and reformulation constitute a sharp rebuke of certain problematic populisms in the agrarian tradition. Iconic examples include some Indian populist peasant movements (Assadi 1994; Brass 1994, 1995). The problem with these types of movements is not that they are class-blind, because, arguably, they are not. The problem is their consciousness of the antagonistic class interests that exist, and their *deliberate* effort to downplay or cover up the tensions and contradictions within a movement or political agitation to create a single discourse under the dominant class—i.e. rich peasant—banner. When such peasant movements consciously block rural working-class issues such as land reform, landless labourer's work and wage issues, and so on, from getting onto the agenda of populist mobilisations, and instead narrowly focus on 'fairer' (always meaning 'higher') prices for their produce, defence of the private property regime, and so on, they acquire a conservative character (Brass 1994, 1995; Pattenden 2005). This is not a naïve type of agrarian populism; it is a consciously reactionary type of populism. Yet, most contemporary agrarian and food sovereignty movements do not share this type of reactionary agrarian populism.

Finally, a tradition that comes from class-based movements has become more urgent and necessary than ever, namely, an *internationalist perspective*. The emergence of Bernstein's 'classes of labour' or Shivji's 'working people' is a *global* phenomenon. The issues that are addressed by all strands of populism are not confined to a national territory; they are international, and are inserted into the dynamics of global capitalism. Anti-capitalist struggles are to be fought within national territories *and* internationally. Food sovereignty movements have actual and potential contributions to make, albeit unevenly across space and over time, to struggles that are (sub-)national with an international perspective *and* international with (sub-)national roots. *Internationalism* can be a counterweight to the xenophobic nationalism of right-wing populist agitations. An internationalist perspective takes as its starting point that social class and other identity struggles are fought in local communities, but that solidarity struggles are necessary, and necessarily cross-border in character, because of the international interconnection of the causes, conditions, and consequences of the multiple crises caused by capitalism.

I conclude this chapter by going back to Henry Bernstein's appeal for Marxist political economists to take agrarian politics and agrarian populism seriously, dovetailing this appeal with a recent observation by Michael Levien et al. (2018: 853): 'While Marxists have long criticised "populists" for ignoring capitalism and class, populists have charged Marxists with an obsessive concern with accumulation and class, an insensitivity to the contingencies of history and various blindspots regarding gender and identity'. They conclude:

> On the one hand, more 'populist' scholarship – whether focused on land grabs, food sovereignty or land reform – has far more explicitly incorporated Marxian insights about class and the dynamics of capitalism than ever before. On the other hand, much explicitly Marxian scholarship has moved away from its dismissal of peasant political agency; the hyper-structuralism of modes of production debates; and linear or Euro-centric conceptions of history embedded in the transition problematic and 'doomed peasant dogma'. (ibid.: 854)

This intellectual and political reciprocity does not undermine the fundamental standpoint of each camp. Teodor Shanin observed the process between the People's Will intellectuals and Marx, how each treated the other seriously and how each was willing to concede some important elements in their perspective: 'That does not make Marx into a populist or turn members of the People's Will into crypto-marxists. They were political allies, who supported and influenced each other' (Shanin 1983b: 268). Indeed, such a productive encounter is urgently needed in building and reinforcing an agenda for political conversation around the notion of a reconstituted, class-conscious, left-wing populism in order to defeat right-wing populism, and to pave a path towards a possible socialist future.

Acknowledgements This work draws on discussions over many years with comrades in the Philippine left movement, especially Steve Quiambao, Jennifer C. Franco, and Danny Carranza, and with comrades in La Via Campesina, Transnational Institute (TNI), and, in recent years, with colleagues in the ERPI collective and fellow researchers at the ISS, for which I am deeply grateful. I would like to thank Haroon Akram-Lodhi, Alberto Alonso-Fradejas, Henry Bernstein, Daniela Calmon, Ben Cousins, Marc Edelman, Harriet Friedmann, Burak Gurel, Ruth Hall, Francis Isaac, Cris Kay, Michael Levien, Lyda Forero, Jaume Franquesa, Ben Luig, Natalia Mamonova, Phil McMichael,

Sofia Monsalve, Roman Antonio-Alcala, Ian Scoones, Amod Shah, Teodor Shanin, Annie Shattuck, Tony Weis, Ben White, and Wendy Wolford for various comments on earlier drafts, most of which were very critical, some of which were just a few lines of encouragement, but all of which have the overall effect of reaffirming the relevance of pursuing this paper. I also thank the JAC editors for their encouragement and support, and the two anonymous reviewers for critical but very helpful comments and suggestions. Their feedback saved me from embarrassing mistakes, awkward formulations, and ridiculous propositions, and helped improve the level of clarity of its propositions. But still, many of the questions and comments raised by several colleagues regarding various elements of this piece remain unaddressed—not because I decided to ignore these questions and suggestions but because I myself do not know the answers or how to respond at the moment. I would also like to thank the organisers of and participants in the Finnish Society for Development Research (FSDR) conference on 15–16 February 2018 at the University of Helsinki, especially Barry Gills, Bon Juego, and Jesse Ribot, for their comments on the presentation that was based on this text. Various parts of the chapter have been presented at several other international conferences where I greatly benefited from useful feedback and comments from the organisers and participants: University of Bern's Centre for Development and Environment (CDE) in October 2017, University of Indonesia in September 2017, University of North Sumatra in December 2017, Utrecht University in December 2017, and Max Planck Institute in Halle (Germany) in May 2018. I would like to thank Ben Luig of the Rosa Luxemburg Foundation for suggesting that I write a short note on 'authoritarian populism and food sovereignty'. Without Ben's request I would not have thought of embarking on this task, the final outcome of which is this long essay, which is a preliminary outline of an ongoing attempt at making sense of how to understand—and defeat—contemporary right-wing populism from the perspective of the rural world. Finally, I thank Paula Bownas for her excellent copyediting assistance. All remaining errors and shortcomings are mine alone. This text is dedicated to the memory of comrade Gani Serrano.

References

Adaman, F., M. Arsel, and B. Akbulut. 2019. Neoliberal Developmentalism, Authoritarian Populism, and Extractivism in the Countryside: The Soma Mining Disaster in Turkey. *Journal of Peasant Studies* 46 (3): 514–536.

Akram-Lodhi, A.H. 2018. The Promise? Using and Misusing Authoritarian Populism. Paper presented at the ERPI 2018 International Conference Authoritarian Populism and the Rural World, Conference Paper No. 6. March 2018. The Hague: International Institute of Social Studies.

Akram-Lodhi, A.H., and C. Kay. 2010a. Surveying the Agrarian Question (Part 1): Unearthing Foundations, Exploring Diversity. *Journal of Peasant Studies* 37 (1): 177–202.

Akram-Lodhi, A.H., and C. Kay. 2010b. Surveying the Agrarian Question (Part 2): Current Debates and Beyond. *Journal of Peasant Studies* 37 (2): 255–284.

Alinsky, S. 1971. *Rules for Radicals: A Pragmatic Primer for Realistic Radicals*. New York: Vintage.

Anderson, P. 2019. Bolsonaro's Brazil. *London Review of Books* 41 (3): 11–22.

Andrade, D. 2019. Populism from Above and Below: Agriculture and the Political Ambiguities of the Workers' Party in Brazil. ERPI Working Paper Series # 83. The Hague: International Institute of Social Studies; see also Emancipatory Rural Politics Initiative at www.iss.nl/erpi.

Arsel, M., B. Hogenboom, and L. Pellegrini. 2016. The Extractive Imperative in Latin America. *The Extractive Industries and Society* 3 (4): 880–887.

Assadi, M. 1994. 'Khadi Curtain', 'Weak Capitalism' and 'Operation Ryot': Some Ambiguities in Farmers' Discourse, Karnataka and Maharashtra 1980–93. *Journal of Peasant Studies* 21 (3–4): 212–227.

Baud, J.M., and R.A. Rutten, eds. 2004. *Popular Intellectuals and Social Movements: Framing Protest in Asia, Africa, and Latin America*. Cambridge: Cambridge University Press.

Baviskar, A. 1999. *In the Belly of the River: Tribal Conflicts over Development in the Narmada Valley*. Oxford: Oxford University Press.

Bello, W. 2018. Counterrevolution, the Countryside and the Middle Classes: Lessons from Five Countries. *Journal of Peasant Studies* 45 (1): 21–58.

Bernstein, H. 2010. *Class Dynamics of Agrarian Change*. Halifax: Fernwood.

Bernstein, H. 2018. The 'Peasant Problem' in the Russian Revolution(s), 1905–1929. *Journal of Peasant Studies* 45 (5–6): 1127–1150.

Bernstein, H., H. Friedmann, J.D. van der Ploeg, T. Shanin, and B. White. 2018. Fifty Years of Debate on Peasantries, 1966–2016. *Journal of Peasant Studies* 45 (4): 689–714.

Boillat, S., J.F. Gerber, and F.R. Funes-Monzote. 2012. What Economic Democracy for Degrowth? Some Comments on the Contribution of Socialist Models and Cuban Agroecology. *Futures* 44 (6): 600–607.

Bond, B., and Z. Exley. 2016. *Rules for Revolutionaries: How Big Organizing Can Change Everything*. White River Junction, VT: Chelsea Green.

Bookchin, M. 2015. *The Next Revolution: Popular Assemblies and the Promise of Direct Democracy*. New York: Verso.

Borras, S.M. 2007. *Pro-poor Land Reform: A Critique*. Ottawa: University of Ottawa Press.

Borras, S.M., and J.C. Franco. 2009. Transnational Agrarian Movements Struggling for Land and Citizenship Rights. IDS Working Papers, 2009(323). Brighton: University of Sussex, Institute of Development Studies.

Borras, S.M., and J.C. Franco. 2018. The Challenge of Locating Land-based Climate Change Mitigation and Adaptation Politics within a Social Justice Perspective: Towards an Idea of Agrarian Climate Justice. *Third World Quarterly* 39 (7): 1308–1325.

Borras, S.M., T. Moreda, A. Alonso-Fradejas, and Z. Brent. 2018. Converging Social Justice Issues and Movements: Implications for Political Actions and Research. *Third World Quarterly* 39 (7): 1227–1246.

Brass, T. 1994. Post-script: Populism, Peasants and Intellectuals, or what's Left of the Future? *Journal of Peasant Studies* 21 (3–4): 246–286.

Brass, T., ed. 1995. *New Farmers' Movements in India*. London: Frank Cass.

Brass, T. 1997. The Agrarian Myth, the 'New' Populism and the 'New' Right. *Journal of Peasant Studies* 24 (4): 201–245.

Brent, Z.W., C.M. Schiavoni, and A. Alonso-Fradejas. 2015. Contextualising Food Sovereignty: The Politics of Convergence Among Movements in the USA. *Third World Quarterly* 36 (3): 618–635.

Bueza, M. 2018. *Duterte's Satisfaction Rating Rises in Q4 2018—SWS*. Retrieved from https://www.rappler.com/nation/219915-duterte-satisfaction-ratings-sws-december-2018.

Byres, T.J. 1979. Of Neo-populist Pipe-dreams: Daedalus in the Third World and the Myth of Urban Bias. *Journal of Peasant Studies* 6 (2): 210–244.

Byres, T.J. 2004. Neo-classical Neo-populism 25 Years on: Déjà Vu and Déjà Passé. Towards a Critique. *Journal of Agrarian Change* 4 (1–2): 17–44.

Chappell, M.J. 2018. *Beginning to End Hunger: Food and the Environment in Belo Horizonte, Brazil, and Beyond*. Berkeley, CA: University of California Press.

Chayanov, A.V. 1966 [1925]. *The Theory of the Peasant Economy*, ed. D. Thorner et al. Manchester: Manchester University Press.

Chrisafis, A. 2017. The Real Misery is in the Countryside: Support for Le Pen Surges in Rural France. *The Guardian*. Retrieved from https://www.theguardian.com/world/2017/apr/21/counryside-marine-le-pen-forgotten-france-presidential-election-2017.

Chrisman, S. 2016. Want to Understand Trump's Rise? Head to the Farm. *Civil Eats*, 27 October 2016. Retrieved from https://civileats.com/2016/10/27/want-to-understand-trumps-rise-head-to-the-farm.

Claeys, P. 2012. The Creation of new Rights by the Food Sovereignty Movement: The Challenge of Institutionalizing Subversion. *Sociology* 46 (5): 844–860.

Claeys, P., and D. Delgado Pugley. 2017. Peasant and Indigenous Transnational Social Movements Engaging with Climate Justice. *Canadian Journal of Development Studies* 38 (3): 325–340.

Clapp, J., and S.R. Isakson. 2018. *Speculative Harvests: Financialization, Food, and Agriculture*. Halifax: Fernwood; London: Practical Action.

Clark, P. 2017. Neo-developmentalism and a 'Vía Campesina' for Rural Development: Unreconciled Projects in Ecuador's Citizen's Revolution. *Journal of Agrarian Change* 17 (2): 348–364.

Cousins, B., A. Dubb, D. Hornby, and F. Mtero. 2018. Social Reproduction of 'Classes of Labour' in the Rural Areas of South Africa: Contradictions and Contestations. *Journal of Peasant Studies* 45 (5–6): 1060–1085.

Curato, N., ed. 2017. *A Duterte Reader: Critical Essays on Rodrigo Duterte's Early Presidency*. Manila: Ateneo de Manila University Press.

De Schutter, O. 2014. *UN Special Rapporteur on the Right to Food. Report on Agroecology and the Right to Food*. Human Rights Council: Geneva.

Desmarais, A. 2007. *La Via Campesina: Globalization and the Power of Peasants*. Halifax: Fernwood.

Docena, H. 2018. *The Rise of Populist Authoritarianisms in Asia: Challenges for Peoples' Movements*. Bangkok: Focus on the Global South.

Edelman, M. 1999. *Peasants Against Globalization: Rural Social Movements in Costa Rica*. Stanford, CA: Stanford University Press.

Edelman, M. 2018. *Sacrifice Zones in Rural and Non-metro USA: Fertile Soil for Authoritarian Populism*. OpenDemocracy blog entry. Retrieved from https://www.opendemocracy.net/marc-edelman/sacrifice-zones-in-rural-and-non-metro-usa-fertile-soil-for-authoritarian-populism.

Edelman, M., and S.M. Borras. 2016. *Political Dynamics of Transnational Agrarian Movements*. Halifax: Fernwood; Rugby: Practical Action Publishing.

Edelman, M., T. Weis, A. Baviskar, S.M. Borras, E. Holt-Giménez, D. Kandiyoti, and W. Wolford. 2014. Introduction: Critical Perspectives on Food Sovereignty. *Journal of Peasant Studies* 41 (6): 911–931.

Engels, F. 1894. *Peasant Question in France and Germany*. Retrieved from https://www.marxists.org/archive/marx/works/1894/peasant-question/index.htm.

Errejón, I. 2016. Podemos: In the Name of the People. In *Errejón in Conversation with Mouffe*, ed. I. Errejón and C. Mouffe. London: Lawrence and Wishart.

Fairbairn, M. 2014. 'Like Gold with Yield': Evolving Intersections between Farmland and Finance. *Journal of Peasant Studies* 41 (5): 777–795.

Ferguson, J. 2015. *Give a Man a Fish: Reflection on the New Politics of Distribution*. Durham, NC: Duke University Press.

Fox, J. (ed.). 1990. *The Challenge of Rural Democratization*. Lodon: Frank Cass.

Fox, J. 2007. *Accountability Politics: Power and Voice in Rural Mexico*. New York: Oxford University Press.

Franco, J.C. 2011. *Bound by Law: Filipino Rural Poor and the Search for Justice in a Plural-legal Landscape*. Manila: Ateneo de Manila University Press.

Franco, J.C. 2014. *Reclaiming Free Prior and Informed Consent (FPIC) in the Context of Global Land Grabs*. Amsterdam: Transnational Institute (TNI).

Franco, J.C., and S.M. Borras. 2009. Paradigm Shift: The 'September Thesis' and Rebirth of the 'Open' Peasant Mass Movement in the Era of Neoliberal Globalization in the Philippines. In *Agrarian Angst and Rural Resistance in Contemporary Southeast Asia*, ed. D. Caouette and S. Turner, 226–246. London: Routledge.

Franco, J.C., S. Monsalve, and S.M. Borras. 2015. Democratic Land Control and Human Rights. *Current Opinion in Environmental Sustainability* 15: 66–71.

Franquesa, J. 2019. The Vanishing Exception: Republican and Reactionary Specters of Populism in Rural Spain. *Journal of Peasant Studies* 46 (3): 537–560.

Franz, C., M.F. Fratzscher, and A.S. Kritikos. 2018. German Right-wing Party AfD Finds more Support in Rural Areas with Aging Populations. *DIW Weekly Report 7+8* 2018. Berlin: DIW Berlin.

Fraser, N., and A. Honneth. 2003. *Redistribution or Recognition? A Political-philosophical Exchange*. New York: Verso.
Gaventa, J. 1982. *Power and Powerlessness: Quiescence and Rebellion in an Appalachian Valley*. Champaign, IL: University of Illinois Press.
Gaventa, J. 2006. Finding the Spaces for Change: A Power Analysis. *IDS Bulletin* 37 (6): 23–33.
Gerber, J.F. 2014. The Role of Rural Indebtedness in the Evolution of Capitalism. *Journal of Peasant Studies* 41 (5): 729–747.
Giunta, I. 2014. Food Sovereignty in Ecuador: Peasant Struggles and the Challenge of Institutionalization. *Journal of Peasant Studies* 41 (6): 1201–1224.
Griffin, K., A.R. Khan, and A. Ickowitz. 2002. Poverty and the Distribution of Land. *Journal of Agrarian Change* 2 (3): 279–330.
Gudynas, E. 2011. Buen Vivir: Today's Tomorrow. *Development* 54 (4): 441–447.
Gurel, B., B. Kucuk, and S. Tas. 2019. The Rural Roots of the Rise of the Justice and Development Party in Turkey. *Journal of Peasant Studies* 46 (3): 457–479.
Hadiz, V. 2016. *Islamic Populism in Indonesia and the Middle East*. Cambridge: Cambridge University Press.
Hall, S. 1985. Authoritarian Populism: A Reply. *New Left Review* 151: 115–124.
Hall, R. 2011. Land Grabbing in Southern Africa: The Many Faces of the Investor Rush. *Review of African Political Economy* 38 (128): 193–214.
Harcourt, W., and I. Nelson, eds. 2015. *Practising Feminist Political Ecologies: Moving beyond the 'Green Economy.'* London: Zed Books.
Harvey, D. 2003. *The New Imperialism.* New York: Oxford University Press.
Henderson, T.P. 2018. The Class Dynamics of Food Sovereignty in Mexico and Ecuador. *Journal of Agrarian Change* 18 (1): 3–21.
Hobsbawm, E. 1987. *The Age of Empire, 1875–1914*. London: Abacus.
Holt-Giménez, E. 2017. *A Foodie's Guide to Capitalism: Understanding the Political Economy of What We Eat*. New York: Monthly Review Press; Oakland, CA: Food First.
Holt-Giménez, E., and A. Shattuck. 2011. Food Crises, Food Regimes and Food Movements: Rumblings of Reform or Tides of Transformation? *Journal of Peasant Studies* 38 (1): 109–144.
Huizer, G. 1975. How Peasants become Revolutionaries: Some Cases from Latin America and Southeast Asia. *Development and Change* 6 (3): 27–56.
Isakson, S.R. 2014. Maize Diversity and the Political Economy of Agrarian Restructuring in Guatemala. *Journal of Agrarian Change* 14 (3): 347–379.

Jones, S. 2018. Far Right Wins Seats in Spanish Region for First Time Since Franco. *The Guardian*. Retrieved from https://www.theguardian.com/world/2018/dec/03/spain-far-right-vox-party-wins-seats-in-andalucia-for-first-time-since-franco.

Kallis, G. 2019. Socialism without Growth. *Capitalism Nature Socialism* 30 (2): 189–206.

Kautsky, K. 1988 [1899]. *The Agrarian Question, Vol. 1*. London: Zwan.

Kay, C. 2009. Development Strategies and Rural Development: Exploring Synergies, Eradicating Poverty. *Journal of Peasant Studies* 36 (1): 103–137.

Khadse, A., P. Rosset, H. Morales, and B. Ferguson. 2018. Taking Agroecology to Scale: The Zero Budget Natural Farming Peasant Movement in Karnataka India. *Journal of Peasant Studies* 45 (1): 192–219.

Kurtzleben, D. 2016. *Rural Voters Played a Big Part in Helping Trump Defeat Clinton*. National Public Radio. Retrieved from https://www.npr.org/2016/11/14/501737150/rural-voters-played-a-big-part-in-helping-trump-defeat-clinton.

Laclau, E. 2005. *On Populist Reason*. London: Verso.

Lenin, V.I. 2004 [1905]. *Development of Capitalism in Russia*. Honolulu, HI: University Press of the Pacific.

Lerche, J., and A. Shah. 2018. Conjugated Oppression within Contemporary Capitalism: Class, Caste, Tribe and Agrarian Change in India. *Journal of Peasant Studies* 45 (5–6): 927–949.

Levien, M. 2011. Special Economic Zones and Accumulation by Dispossession in India. *Journal of Agrarian Change* 11 (4): 454–483.

Levien, M., M. Watts, and H. Yan. 2018. Agrarian Marxism. *Journal of Peasant Studies* 45 (5–6): 853–883.

Li, T.M. 2007. *The Will to Improve: Governmentality, Development, and the Practice of Politics*. Durham, NC: Duke University Press.

Li, T.M. 2010. To Make Live or let Die? Rural Dispossession and the Protection of Surplus Populations. *Antipode* 41: 66–93.

Li, T.M. 2011. Centering Labor in the Land Grab Debate. *Journal of Peasant Studies* 38 (2): 281–298.

Li, T.M. 2014. *Land's End: Capitalist Relations on an Indigenous Frontier*. Durham, NC: Duke University Press.

Lipton, M. 1977. *Why Poor People Stay Poor: A Study of Urban Bias in World Development*. London: Temple Smith; Canberra: Australian National University Press.

Mamonova, N. 2016. Naive Monarchism and Rural Resistance in Contemporary Russia. *Rural Sociology* 81 (3): 316–342.

Mamonova, N. 2018. Patriotism and Food Sovereignty: Changes in the Social Imaginary of Small-scale Farming in Post-Euromaidan Ukraine. *Sociologia Ruralis* 58 (1): 190–212.
Mamonova, N. 2019. Understanding the Silent Majority in Authoritarian Populism: What Can We Learn from Popular Support for Putin in Rural Russia? *Journal of Peasant Studies* 46 (3): 561–585.
Mao, T.-T. 1975. *Selected Works*, vol. II. Peking: Foreign Language Press.
Martinez-Alier, J. 2014. The Environmentalism of the Poor. *Geoforum* 54: 239–241.
Martinez-Alier, J., L. Temper, D. Del Bene, and A. Scheidel. 2016. Is There a Global Environmental Justice Movement? *Journal of Peasant Studies* 43 (3): 731–755.
Martinez-Torres, M.E., and P.M. Rosset. 2010. La Vía Campesina: The Birth and Evolution of a Transnational Social Movement. *Journal of Peasant Studies* 37 (1): 149–175.
Marx, K. 1983 [1881]. The Reply to Zasulich. In *Late Marx and the Russian Road*, ed. T. Shanin, 123–124. London: Routledge and Kegan Paul.
McKay, B.M. 2017. Agrarian Extractivism in Bolivia. *World Development* 97: 199–211.
McKay, B., R. Nehring, and M. Walsh-Dilley. 2014. The 'State' of Food Sovereignty in Latin America: Political Projects and Alternative Pathways in Venezuela, Ecuador and Bolivia. *Journal of Peasant Studies* 41 (6): 1175–1200.
McKeon, N. 2018. Are Equity and Sustainability a Likely Outcome when Foxes and Chickens Share the Same Coop? Critiquing the Concept of Multistakeholder Governance of Food Security. *Globalizations* 14 (3): 379–398.
McMichael, P. 2008. Peasants Make Their Own History, but Not Just as They Please... *Journal of Agrarian Change* 8 (2–3): 205–228.
Mills, E.N. 2018. Implicating 'Fisheries Justice' Movements in Food and Climate Politics. *Third World Quarterly* 39 (7): 1270–1289.
Monsalve, S. 2013. The Human Rights Framework in Contemporary Agrarian Struggles. *Journal of Peasant Studies* 40 (1): 239–290.
Moore, B. 1967. *Social Origins of Dictatorship and Democracy: Lord and Peasant in the Making of the Modern World*. Boston, MA: Beacon Press.
Moore, J.W. 2017. The Capitalocene, Part I: On the Nature and Origins of Our Ecological Crisis. *Journal of Peasant Studies* 44 (3): 594–630.

Mouffe, C. 2005. The 'End of Politics' and the Challenge of Right-wing Populism. In *Populism and the Mirror of Democracy*, ed. F. Panizza, 50–71. London: Verso.

Mouffe, C. 2016. Podemos: In the Name of the People. In *Errejón in Conversation with Mouffe*, ed. I. Errejón and C. Mouffe. London: Lawrence and Wishart.

Moyo, S., and P. Yeros, eds. 2005. *Reclaiming the Land: The Resurgence of Rural Movements in Africa, Asia and Latin America*. London: Zed Books.

NES. 2014. Statistics: National Election Study 2014. *Economic and Political Weekly* 49 (39): 130–134, Appendix 1.

Newell, P., and J. Wheeler, eds. 2006. *Rights, Resources and the Politics of Accountability*. London: Zed Books.

Nishizaki, Y. 2014. Peasants and the Redshirt Movement in Thailand: Some Dissenting Voices. *Journal of Peasant Studies* 41 (1): 1–28.

Paige, J. 1975. *Agrarian Revolutions*. New York: Free Press.

Panizza, F. 2005. Introduction: Populism and the Mirror of Democracy. In *Populism and the Mirror of Democracy*, ed. F. Panizza, 1–31. London: Verso.

Patel, R. 2009. Food Sovereignty. *Journal of Peasant Studies* 36 (3): 663–706.

Patel, R., and J.W. Moore. 2017. *A History of the World in Seven Cheap Things*. Berkeley, CA: University of California Press.

Pattenden, J. 2005. Trickle-down Solidarity, Globalisation and Dynamics of Social Transformation in a South Indian village. *Economic and Political Weekly* 40 (19): 1975–1985.

Paxton, R. 1997. *French Peasant Fascism: Henry Dorgeres's Greenshirts and the Crises of French Agriculture, 1929–1939*. Oxford: Oxford University Press.

Peluso, N. 1992. *Rich Forests, Poor People: Resource Control and Resistance in Java*. Berkeley and Los Angeles: University of California Press.

Peluso, N.L., and C. Lund. 2011. New Frontiers of Land Control: Introduction. *Journal of Peasant Studies* 38 (4): 667–681.

Popkin, S.L. 1979. *The Rational Peasant: The Political Economy of Rural Society in Vietnam*. Berkeley and Los Angeles: University of California Press.

Putzel, J. 1995. Managing the 'Main Force': The Communist Party and the Peasantry in the Philippines. *Journal of Peasant Studies* 22 (4): 645–671.

Rancière, J. 2016. The Populism That Is Not to Be Found. In *What Is a People*, ed. A. Badiou et al., 100–106. New York: Columbia University Press.

Ribot, J.C. 1999. Decentralisation, Participation and Accountability in Sahelian Forestry: Legal Instruments of Political-administrative Control. *Africa* 69 (1): 23–65.

Ribot, J.C., and N.L. Peluso. 2003. A Theory of Access. *Rural Sociology* 68 (2): 153–181.

Roman-Alcalá, A. 2017. Looking to Food Sovereignty Movements for Post-growth Theory. *Ephemera: Theory and Politics in Organization* 17 (1): 119–145.

Roman-Alcalá, A. 2018. (Relative) Autonomism, Policy Currents and the Politics of Mobilisation for Food Sovereignty in the United States: The Case of Occupy the Farm. *Local Environment* 23 (6): 619–634.

Rosset, P., and M. Altieri. 2018. *Politics and Science of Agroecology*. Fernwood: Halifax; Practical Action Books: London.

Saad-Filho, A. 2018. Brazil: The Collapse of Democracy? *The Bullet*, 27 October. Retrieved from https://socialistproject.ca/2018/10/brazil-the-collapse-of-democracy.

Sanders, B. 2016. *Our Revolution: A Future to Believe in*. New York: Thomas Dunne Books.

Schoenberger, L., A. Beban, and V. Lamb. 2018. Authoritarian Rule Shedding Its Populist Skin: How Loss of Independent Media in the 2017 Crackdown Shapes Rural Politics in Cambodia. Paper presented at ERPI 2018 International Conference Authoritarian Populism and the Rural World, Conference Paper No. 70. The Hague: International Institute of Social Studies.

Scoones, I. 2015. *Sustainable Livelihoods and Rural Development*. Halifax: Fernwood; Rugby: Practical Action Publishing.

Scoones, I., M. Edelman, S.M. Borras, R. Hall, W. Wolford, and B. White. 2018. Emancipatory Rural Politics: Confronting Authoritarian Populism. *Journal of Peasant Studies* 45 (1): 1–20.

Scott, J. 1976. *The Moral Economy of the Peasant*. New Haven, CT: Yale University Press.

Scott, J.C. 1985. *Weapons of the Weak: Everyday Forms of Peasant Resistance*. New Haven, CT: Yale University Press.

Shanin, T. 1972. *The Awkward Class: Political Sociology of Peasantry in a Developing Society Russia 1910–1925*. London and Oxford: Clarendon Press.

Shanin, T. 1983a. Late Marx: Gods and Craftsmen. In *Late Marx and the Russian Road*, ed. T. Shanin, 3–39. London: Routledge and Kegan Paul.

Shanin, T. 1983b. Marxism and the Vernacular Revolutionary Traditions. In *Late Marx and the Russian Road*, ed. T. Shanin, 243–279. London: Routledge and Kegan Paul.

Shattuck, A., C.M. Schiavoni, and Z. VanGelder. 2015. Translating the Politics of Food Sovereignty: Digging into Contradictions Uncovering New Dimensions. *Globalizations* 12 (4): 421–433.

Shivji, I. 2017. The Concept of 'Working People.' *Agrarian South: Journal of Political Economy* 6 (1): 1–13.

Soriano, C.R. 2015. Strategic Activism for Democratization and Social Change. In *The Routledge Handbook of Strategic Communication*, ed. D. Holtzhausen and A. Zerfass, 424–438. London: Routledge.

Springer, S. 2015. Postneoliberalism? *Review of Radical Political Economics* 47 (1): 5–17.

Taggart, P. 2000. *Populism*. Buckingham: Open University Press.

Thompson, E.P. 2013 [1968]. *The Making of the English Working Class*. London: Penguin Modern Classics.

Tolkien, J.R.R. 1980. *Unfinished Tales*. London: George Allen and Unwin.

Tramel, S. 2016. The Road through Paris: Climate Change, Carbon, and the Political Dynamics of Convergence. *Globalizations* 13 (6): 960–969.

Tramel, S. 2018. Convergence as Political Strategy: Social Justice Movements, Natural Resources and Climate Change. *Third World Quarterly* 39 (7): 1290–1307.

Tsikata, D., and J.A. Yaro. 2014. When a Good Business Model Is Not Enough: Land Transactions and Gendered Livelihood Prospects in Rural Ghana. *Feminist Economics* 20 (1): 202–226.

Tsing, A.L. 2000. Inside the Economy of Appearances. *Public Culture* 12 (1): 115–144.

Ulrich-Schad, J.D., and C.M. Duncan. 2018. People and Places Left Behind: Work, Culture and Politics in the Rural United States. *Journal of Peasant Studies* 45 (1): 59–79.

Vanaik, A. 2017a. *The Rise of Hindu Authoritarianism: Secular Claims, Communal Realities*. London: Verso.

Vanaik, A. 2017b. *Hindutva's Forward March*. Retrieved from https://www.jacobinmag.com/2017/09/india-modi-bjp-cow-vigilantism-judiciary-corruption.

Vanaik, A. 2018. *Hindu Authoritarianism and Agrarian Distress*. Retrieved from https://www.opendemocracy.net/openindia/achin-vanaik/hindu-authoritarianism-and-agrarian-distress.

van der Ploeg, J.D. 2013. *Peasants and the Art of Farming: A Chayanovian Manifesto*. Halifax: Fernwood Publishing.

van der Ploeg, J.D. 2018. Differentiation: Old Controversies, New Insights. *Journal of Peasant Studies* 45 (3): 489–524.

Veltmeyer, H., and J.F. Petras. 2014. *The New Extractivism: A Post-neoliberal Development Model or Imperialism of the Twenty-first Century?* London: Zed Books.

Vergara-Camus, L., and C. Kay. 2017. Agribusiness, Peasants, Left-wing Governments, and the State in Latin America: An Overview and Theoretical Reflections. *Journal of Agrarian Change* 17 (2): 239–257.
VTsIOM. 2018. *Vybory Prezidenta Rossii 2018. Otkrytyye Dannyye Issledovaniy Obshchestvennogo Mneniya VTsIOM. Reyting Kandidatov* [Election of the President of Russia 2018. Open Public Opinion Research Data VTsIOM. Rating of Candidates]. Retrieved from https://2018.wciom.ru/index.php?id=1234.
Wainwright, H. 2018. *A New Politics from the Left*. London: Polity Press.
Weis, T. 2007. *The Global Food Economy: The Battle for the Future of Farming*. London: Zed Books.
White, B. 2018. Marx and Chayanov at the Margins: Understanding Agrarian Change in Java. *Journal of Peasant Studies* 45 (5–6): 1108–1126.
White, M.M. 2017. 'A Pig and a Garden': Fannie Lou Hamer and the Freedom Farms Cooperative. *Food and Foodways* 25 (1): 20–39.
Win, K.Z. 2018. *How Populism Directed Against Minorities Is Used to Prop up Myanmar's 'Democratic' Revival*. Retrieved from https://www.opendemocracy.net/khin-zaw-win/how-populism-directed-against-minorities-is-used-to-prop-up-myanmar-s-democratic-reviva.
Wittman, H., Desmarais, A., and Weibe, N., eds. 2010. *Food Sovereignty: Reconnecting Food, Nature and Community*. Halifax: Fernwood.
Wolf, E.R. 1969. *Peasant Wars of the Twentieth Century*. New York: Harper and Row.
Wolford, W. 2005. Agrarian Moral Economies and Neoliberalism in Brazil: Competing Worldviews and the State in the Struggle for Land. *Environment and Planning A* 37 (2): 241–261.
Wolford, W. 2010. *This Land Is Ours Now: Social Mobilization and the Meanings of Land in Brazil*. Durham, NC: Duke University Press.
Wolford, W., S.M. Borras, R. Hall, I. Scoones, and B. White. 2013. Governing Global Land Deals: The Role of the State in the Rush for Land. *Development and Change* 44 (2): 189–210.
Wright, E.O. 2016. Socialism and Real Utopias. In *Alternatives to Capitalism: Proposals for a Democratic Economy*, ed. R. Hahnel and E.O. Wright, 60–91. New York: Verso.
Ye, J., C. Wang, H. Wu, C. He, and J. Liu. 2013. Internal Migration and Left-behind Populations in China. *Journal of Peasant Studies* 40 (6): 1119–1146.
Zasulich, V. 1983 [1881]. A Letter to Marx. In *Late Marx and the Russian Road*, ed. T. Shanin, 98–99. London: Routledge and Kegan Paul.

Index

A. A. R. Ioris and B. Mançano Fernandes (eds.), *Agriculture, Environment and Development*, https://doi.org/10.1007/978-3-031-10264-6

SPRINGER NATURE

GPSR Compliance

The European Union's (EU) General Product Safety Regulation (GPSR) is a set of rules that requires consumer products to be safe and our obligations to ensure this.

If you have any concerns about our products, you can contact us on ProductSafety@springernature.com

In case Publisher is established outside the EU, the EU authorized representative is:

Springer Nature Customer Service Center GmbH
Europaplatz 3
69115 Heidelberg, Germany

FSC
www.fsc.org
MIX
Papier | Fördert
gute Waldnutzung
FSC® C083411

Kolibri 360 GmbH
Ferdinand-Jühlke-Straße 7
99095 Erfurt, Deutschland
produktsicherheit@kolibri360.de